Drone Law 3.0

Evolving the Legal Typology for Commercial Drones in the age of AI

Saheed Babajide Okuboyejo

LL.B. (UNILAG), B.L.(NLS), MBA. (GIBS),. LL.M. (UvA).

Keywords: Commercial Drones; Drone Regulation; Artificial Intelligence; EU Regulations; Drone Privacy; Drone Safety; Drone Surveillance; Drone Liability; Urban Air Mobility; Risk Regulation; Non-Normative Regulation; Airspace Traffic Management; U-Space Airspace, Drone Innovation, Autonomous Drones; Sustainability.

Drone Law 3.0

Drone Law 3.0: *Evolving the Legal Typology for Commercial Drones in the Age of AI* is an in-depth exploration at the crossroads of cutting-edge drone technology and the evolving global regulatory frameworks. As drones redefine industries from agriculture and construction to energy and public services, this book illuminates their transformative potential while curiously addressing safety, privacy, and environmental sustainability threats. Featuring recent case studies from Canada, France, Germany, United Arab Emirates and United Kingdom, it underscores the imperative for adaptive, sophisticated and non-normative regulatory approach. Beyond theory, it offers practical tools such as a step-by-step guide to safe and compliant drone operations, and how to navigate the regulatory environment across key global markets, including the European Union, United States, China and Africa. Drone Law 3.0 dives deep into the detailed analysis of AI-enhanced drone capabilities, and further explores a future where innovation meets adaptive governance. Essential for policymakers, manufacturers, operators, service providers, investors, legal professionals and researchers, this book provides a comprehensive understanding of how technology and law must co-evolve to harness drones' potential while safeguarding public values.

Part of Law 3.0 Series.

Forthcoming titles in the series include:

Autonomous Vehicles Law 3.0

Bioengineering Law 3.0

Blockchain Law 3.0

Clean Tech Law 3.0

Robotics Law 3.0

Singularity Law 3.0

Space Law 3.0

Published in the United States
   by: Staten House
   447 Broadway,
   2nd Floor New York,
   NY 10013

© 2024 Saheed Babajide Okuboyejo

Published in the Netherlands
   by: Babajide B.V.

Publication Date: 12 December 2024

U.S Library of Congress Cataloging-in-publication Data
Names: Okuboyejo, Saheed, author.
Title: Drone Law 3.0 : Evolving the legal typology for commercial drones in the age of ai / Saheed Okuboyejo.
Description: [New York] : Staten House, [2024]. | Includes bibliographical references and index.

ISBN: [979-8-89587-920-7] (hardcover edition)
ISBN: [979-8-89496-482-9] (paperback edition)

Subjects: Commercial Drones; Regulation; Artificial Intelligence; EU Drone Laws; Privacy; Sustainable Mobility; Risk-Based Regulation; Drone Liability; Non-Normative Regulation, Drone Surveillance, U-Space Airspace, Innovation.

Typesetting and cover design by Babajide B.V., The Netherlands.

To Orúnmìlà, the father of evolving intelligence.

Drones are like an expanded consciousness.
They are an extra-eye that allows us to reach places
that otherwise we would never reach.

Architect and Researcher,
Marina Otero Verzier

Preface

Commercial drones, once a projection of science fiction, are swiftly becoming indispensable tools across industries such as agriculture, construction, infrastructure, energy, utilities, and emergency services. They are transforming our use of airspace, reshaping our perception of distance, and expanding our consciousness. At the forefront of this rapid innovation, the drone industry is increasingly integrating artificial intelligence (AI) to enhance capabilities for sustainable development and efficiency. The development, production, and use of this transformative technology are increasingly governed by new legal frameworks in the European Union, North America, Asia, the Middle East and Africa.

Drone Law 3.0 "Evolving the Legal Typology for Commercial Drones in the Age of AI" explores the intersection of drone technology and the rapidly evolving regulatory frameworks that govern them. As drones redefine industries at an unprecedented pace, this book meticulously examines their transformative potential while addressing critical threats to safety, privacy, and environmental sustainability. Through detailed discussions and case studies from Canada, France, Germany, the United Arab Emirates, and the United Kingdom, it highlights significant normative threats posed by drone technologies.

Delving deep into the drone value chain, this book elucidates the roles of various stakeholders—from manufacturers and service providers to regulators and end-users. It explores the impact of artificial intelligence on enhancing drone capabilities, discussing advanced AI technologies like machine learning algorithms for navigation, image recognition for surveillance, and autonomous decision-making processes. The analysis of AI-driven drone incidents, such as those in South Korea, provides a sobering look at the risks associated with these innovations.

Beyond real life discussions, Drone Law 3.0 offers invaluable resources, including a comprehensive step-by-step guide to compliant and safe drone operations while navigating the regulatory environment across key global markets such as the European Union, the United Kingdom, the United States, Canada, China, India, Singapore, South Africa, and Nigeria. Drone Law 3.0 is essential for drone policymakers, manufacturers, service providers, operators, investors, and legal professionals. It offers clear instructions and emphasises the importance of adhering to legal standards to ensure safe and responsible drone use.

Drone Law 3.0 navigates the evolution of EU drone regulations, examining legal principles and common rules, and proposes a shift from normative to non-normative dimensions to keep pace with the rapid innovation of drone technology. The book argues and advocates for adopting agile, dynamic, and sophisticated regulatory frameworks that leverage advanced technologies such as remote identification, geofencing, autonomous flight management systems, obstacle avoidance, fail-safe and return-to-home features, unmanned traffic management systems, and blockchain technologies to effectively navigate the complexities of drone operations in order to make non-compliance practically impossible. Furthermore, through continuous evaluation and international collaboration, the industry with the support of technology lawyers can maintain and even elevate high standards of safety, privacy, and accountability, ensuring that drone technology is harnessed responsibly and beneficially.

For anyone interested in the present and future of the drone industry, within the context of law, regulation and AI, Drone Law 3.0 is an indispensable resource. It offers a nuanced critique of current regulations and a forward-looking perspective on how law, industry, and technology must co-evolve to harness the full potential of transformative technologies while safeguarding public good. This book provides a rare deep-dive into the legal, ethical and technological complexities of the burgeoning drone industry.

I

1. Unlocking the Promise of Drone Technology.

On a crisp morning in Riga, Latvia, in May 2015, the air was charged with anticipation as delegates from across Europe gathered to discuss the future of disruptive technologies poised to reshape society. In the grand conference hall, the European Commission unveiled a vision of skies teeming with drones, performing tasks from mundane parcel deliveries to critical medical supply drops, all with a precision and speed unfathomable with conventional methods. The promise was clear: drones could transform not just commerce and healthcare but also bridge the gaps in accessibility and efficiency that had long challenged remote regions.

Yet, amid the excitement, a crucial caveat was articulated: the integration of drones into daily life must respect the bedrock of safety, privacy and data security. As drones buzzed over test fields outside the venue, demonstrating their capabilities, inside, the conversation turned to the less visible but equally pervasive potential risks. Cybersecurity concerns, privacy breaches, and the environmental impact of widespread drone usage loomed large. The delegates agreed: for drones to soar, robust regulatory frameworks would be essential—not just to nurture innovation but to safeguard the very rights and liberties of citizens across the European continent. This declaration at Riga set the stage for a comprehensive strategy that would steer the evolution of drone technology towards a future where convenience does not come at the cost of security and privacy.

Since the Riga declarations, the European Union (EU) and other continents have embarked on crafting robust regulatory frameworks to manage the rise of drone technology. These regulations not only focus on ensuring compatibility with safety, privacy laws and data protection regulations but also involve integrating drones safely and effectively into national and international airspace systems. The regulatory strategies adopted thus far range from establishing no-fly zones and altitude restrictions to mandating visible identification for drones in flight. These measures aim to prevent unauthorised surveillance and potential misuse, while also mitigating risks related to air traffic conflicts and ground safety.

In recent times, regulatory frameworks have evolved to harmonise drone operations across European Union member states, ensuring a unified approach to safety, privacy, and economic efficiency. This harmonisation is considered pivotal to creating an environment where drones can serve as powerful tools for innovation and growth, all while maintaining public trust and safeguarding individual rights. To achieve these ambitious regulatory goals, the EU has championed pilot projects that delve into advanced traffic management systems for drones, known as U-space. These pioneering systems are designed to enable multitude of drones operating concurrently in urban areas, minimising the risks of collisions and privacy breaches. This coordinating digital infrastructure for advanced drone applications is expected to pave the way for sophisticated services such as drone package deliveries, drone emergency response and urban passenger air taxis, marking a significant leap forward in the integration of drone technology into everyday life.

By championing these advanced pilot projects, regulators in the EU aim to establish a foundation for future technological innovation and to address immediate challenges posed by the integration of drones into civil airspace. As drone applications continue to evolve in the EU and other jurisdictions, addressing questions on legal and ethical frameworks will play a crucial role in guiding their development, ensuring they contribute positively to society and foster an atmosphere of innovation balanced with public welfare and rights protection.

How then do regulators balance the promise of drone technology with the inherent risks it poses to public welfare and fundamental rights? The first step is understanding the technology and providing a robust definition for effective regulation. In 2019, the EU Regulation on

Unmanned Aircraft Systems defined 'Unmanned Aircrafts' or drones as 'aircrafts operating or designed to operate autonomously or to be piloted remotely without a pilot on board.' This definition pinpointed autonomy as the most critical feature distinguishing drone technology. Autonomy introduces new risks compared to standard civil aviation. Without a pilot on board, the potential for accidents and fundamental rights violations increases significantly, presenting unique challenges for regulators and society alike.

In addition to understanding the autonomous nature of the technology, the EU's regulatory framework - along with regulations in United States, United Kingdom, Canada, Singapore and other jurisdictions - have evolved a risk-based approach to managing drone technology by categorising drone operations based on their risk profiles. But how does this nuanced risk-based regulatory approach reconcile the need for transformative technological innovation with the imperative of protecting public interest and safeguarding individual rights? It can be argued that the nuanced risk-based categorisation of drone operations has the potential to effectively bridge the gap between fostering technological innovation and ensuring the protection of public interest and fundamental rights because the categorisation approach tailors regulatory measures to the specific risks each drone operation poses. For low-risk scenarios, lighter regulations encourage innovation by allowing drone operators more flexibility to test and deploy new technologies. Conversely, operations that present higher risks - particularly those that could impact public safety or infringe on fundamental rights - are subjected to stricter regulatory scrutiny.

This system of regulation ensures that drone technologies develop within a framework that prioritises public safety and respects privacy concerns, without stifling innovation. By requiring that higher-risk operations meet more stringent criteria, regulators can prevent substantial harms while supporting advancements in drone technology. Such an approach not only maintains public trust in emerging technologies but also encourages drone operators to prioritise safety and compliance as integral components of innovation. It is clear that this regulatory balance is crucial for the sustainable integration of drones into national and global airspace, aligning technological progress with the fundamental values of society.

With the establishment of the risk-based regulatory approach, the European Union and other technologically advanced jurisdictions are evolving a new way of addressing the risks associated with disruptive technologies such as drone technology and paving the way for more adaptive and dynamic regulatory frameworks. But what does this breakthrough regulatory approach mean for the advancement of drone technology? And how does it prepare society for the increasing integration of these devices? It can be argued that this foresight ensures that as drone technologies evolve to become more advanced and pervasive, they will do so within a framework that recognises, prioritises and mitigates risks, and maintains public trust. The risk-based approach exemplifies a balanced strategy, positioning it at the forefront of drone technology innovation, while attempting to uphold societal values.

At the core of this proactive strategy is the development of U-space - the EU's visionary traffic management system - designed to facilitate the safe operation of numerous drones, especially in densely populated urban areas. But how might U-space address the complexities of drone traffic without compromising on safety or privacy? U-space aims to autonomously manage drone traffic, enhancing the safety of these operations by preventing potential accidents and breaches of fundamental rights which are more likely in urban settings. The intricacies of this system require a nuanced understanding of liability, particularly when autonomous decisions are made without human oversight. In instances of malfunction or failure, determining responsibility involves dissecting the interplay between drone manufacturers, software developers, and operators.

The integration of Artificial Intelligence (AI) into drone operations introduces an additional layer of complexity. AI's capability to learn from vast datasets and make predictive decisions means that drone behaviour can become increasingly difficult to anticipate or control, challenging traditional legal frameworks based on direct human control and clear causation. This raises pivotal questions about accountability and the ethical use of AI-enabled drones in public spaces. Furthermore, the potential for AI-enable drones to be used for surveillance necessitates stringent regulations to protect individuals' privacy and personal freedoms.

For instance, the General Data Protection Regulation (GDPR) in the EU stipulates that drones must operate within strict data handling guidelines to ensure privacy rights are respected. How then are other countries navigating the complexities of drone regulation, ensuring that the rapid AI advancements in drone technology are balanced with the imperatives of public safety and individual privacy rights? Globally, countries like the United States, China, Canada, the United Kingdom,

Australia, South Africa and Nigeria have also adopted risk-based regulatory strategies that include essential components such as pilot certification, pilot oversight, and operational limitations. These measures - designed to safeguard both safety and privacy in the AI era - showcase the intricate interplay between technology, law, and ethics in the realm of drone regulation. The dynamism of disruptive technologies such as drones, highlights the critical need for continuous dialogue and cooperation among lawmakers, technologists, and ethicists. Together, they must work to refine and adapt legal frameworks in the fast paced AI age, ensuring that the integration of drones and other disruptive technologies into society enhances overall well-being, increases abundance and facilitates singularity while respecting individual rights and maintaining public safety.

Transformative Possibilities of Drone Technology

Drones, with their soaring capabilities, are propelling numerous industries into the future, demonstrating unparalleled advantages in efficiency, cost reduction, and innovation. From delivering medical supplies in remote areas to enhancing agricultural practices through precise crop monitoring, drones are unlocking new possibilities. They also play a pivotal role in media and journalism, offering fresh perspectives through aerial photography, and improving safety by conducting inspections in hazardous environments without putting human lives at risk. This surge in drone applications necessitates continued efforts from governments and regulatory bodies to refine policies that maximise their potential while safeguarding public and private interests. As these regulations evolve, they will pave the way for drones to seamlessly integrate into our daily lives, transforming the landscape of modern industry and technological innovations in our age of AI.

Originally developed for military applications, drones have historically played pivotal roles in combat and humanitarian missions. Recently, however, their commercial value has come to the forefront, revolutionising daily life with their efficiency and versatility. Today, a quick search through any consumer technology platform will showcase an array of sophisticated, battery-powered AI drones, available in various shapes, sizes, and colours, often with the promise of 24-hour delivery. This shift from primarily military use to widespread commercial adoption underscores drones' growing importance as a critical mobility technology, spearheading innovations and enhancing productivity in numerous human, economic, and social activities.

Global Drone Market size was valued at US$ 22 Billion in 2022 and is poised to grow from US$ 28 Billion in 2023 to US$ 166 Billion by 2031, growing at a Cumulative Average Growth Rate (CAGR) of 25% during the period between 2024 and 2031. Experts anticipate that by 2040, daily life will be interwoven with drones, as we grow increasingly reliant on rapid services for parcel delivery and other commercial activities. Forecasts suggest that in the European Union alone, the commercial drone sector could generate cumulative benefits exceeding €140 billion by 2050. The European Commission envisions the creation of approximately 150,000 jobs by 2050 through operator services in both commercial and governmental domains. This burgeoning sector promises a transformative impact on society, offering substantial economic advantages and operational efficiencies.

Research indicates that drone delivery services could become the most economically viable method for last-mile delivery of e-commerce purchases in major European markets such as the United Kingdom, Germany, France, and Italy. With an estimated 7 million drones predicted to populate the skies by the mid-2040s, their application will span from urban logistics and air taxi's, to remote healthcare delivery, significantly altering how we interact with goods and services.

Now, one may ask, are these optimistic projections and estimated economic benefits of drones by 2040 truly feasible, or are they just wishful thinking? The likelihood of meeting the ambitious projections for drone technology by 2040 is bolstered by several key factors. Rapid advancements in AI and machine learning continue to enhance the capabilities and efficiency of drones, making them increasingly viable for a wide range of applications. Additionally, the drone industry is experiencing significant investment, indicating strong belief in its potential profitability and growth. Moreover, governments and regulatory bodies are actively working to create frameworks that support safe and efficient drone operations, facilitating broader adoption.

As we project towards 2040, the strategic positioning of existing e-commerce fulfilment centres augments the likelihood of extending drone parcel delivery services to 40 million people in Europe. This feasibility is not just speculative; it's underpinned by current uses of drones that span a variety of sectors. In infrastructure, drones equipped with advanced imaging and sensors are already transforming how we monitor construction progress and inspect power assets for structural defects. In emergency healthcare, drones are delivering life-saving equipment like automated external defibrillators to out-of-hospital cardiac arrest scenes, significantly reducing response times. In agriculture, drones are increasingly utilised for a range of applications, from monitoring crop health and irrigation to the precise application of pesticides and fertilisers, significantly enhancing efficiency and sustainability in farming practices. In urban planning, drones contribute to sustainability by offering new services that reduce the consumption of natural resources and cut costs associated with traditional services. These practical applications not only validate the potential financial forecasts but also highlight the transformative impact of drones across various facets of society.

Agriculture Drones

The rise of drones in agriculture is proving to be a significant paradigm shift, offering a sophisticated blend of technology and traditional farming practices that promises to redefine agricultural efficiency, productivity, and sustainability. The application of drone technology in agriculture, often referred to as precision agriculture, is transforming traditional farming practices by providing precise, real-time data and automated solutions.

One of the primary benefits of drones in agriculture is their ability to perform aerial surveys and create detailed maps of farmland. Drones equipped with high-resolution cameras and multispectral sensors can capture images that help farmers monitor crop health, soil conditions, and irrigation needs. For instance, agriculture specific drones like the DJI Agras T50 offers advanced features for crop spraying and monitoring. The DJI Agras T50 has the capacity to carry up to 50 litres of liquid and cover large areas efficiently, making it ideal for pesticide and fertiliser application. This precision spraying method ensures that crops receive the right amount of nutrients and protection, reducing waste and minimising environmental impact.

The impact of drones in agriculture extends beyond mere observation; they actively contribute to increased agricultural yields and enhanced sustainability. In vineyards across Napa Valley, California drones have been instrumental in increasing grape production by 15%, allowing for early detection of diseases and precise treatment applications. This targeted approach not only conserves resources but also ensures healthier crops, which is vital for both productivity and environmental sustainability.

Drones also play a crucial role in irrigation management. By using thermal imaging cameras, drones can identify areas of a field that are either overwatered or under-watered. This information helps farmers optimise their irrigation systems, ensuring that water is used efficiently and reducing water waste. In arid regions where water is a scarce resource, this capability is particularly valuable. A study conducted in California's Central Valley showed that drone-assisted irrigation management led to a 20% reduction in water usage while maintaining crop health and productivity.

In addition to monitoring and management, drones facilitate planting and seeding operations. Some drones are equipped with seed dispensers that allow for precise planting of crops. This technology is especially beneficial for reforestation efforts and planting in difficult-to-reach areas. For instance, drones used in reforestation projects in British Columbia, Canada have significantly accelerated the planting process, enabling the planting of thousands of trees in a fraction of the time it would take using traditional methods. This rapid reforestation is essential for restoring ecosystems and combating climate change.

Another significant benefit of drones in agriculture is their ability to assess soil health. Drones can collect soil samples and analyse them for nutrient content, pH levels, and moisture. This data helps farmers make informed decisions about soil management practices, such as fertilisation and crop rotation. In one case study, a farm in Iowa used drones to conduct soil health assessments and implemented precision fertilisation based on the data collected. This approach resulted in a 25% increase in corn yield and a reduction in fertiliser use by 30%, demonstrating the economic and environmental benefits of drone technology.

Drones are also instrumental in livestock management. Equipped with thermal imaging and GPS technology, drones can monitor the health and location of livestock, especially in large or remote pastures. This capability allows farmers to detect health issues early, track movements, and ensure the well-being of their animals. For instance, a cattle ranch in Texas utilised drones to monitor the herd, resulting in improved health outcomes and reduced losses due to illness and predation. The drones' ability to cover large areas quickly and efficiently makes them an invaluable tool for livestock management.

Furthermore, drones enhance the safety and efficiency of agricultural operations. Traditional methods of crop monitoring and spraying often involve manual labor, which can be time-consuming and hazardous, especially when dealing with pesticides and fertilisers. Drones eliminate the need for workers to enter potentially dangerous areas, reducing the risk of exposure to harmful chemicals and accidents. This improvement in safety is a significant benefit, particularly for large-scale farming operations.

The use of drones in agriculture is not limited to western countries. In southern regions, drones are being utilised to support smallholder farmers by providing access to advanced agricultural technologies. Organisations and NGOs are deploying drones to assist farmers in optimising their crop production, improving food security, and increasing income. For example, a project in Kenya employed drones to capture high-resolution images of farmland, which are then analysed to assess crop health, monitor growth, and detect areas needing attention. This aerial data helps in identifying issues such as pest infestations, disease outbreaks, and irregular irrigation patterns. With such detailed insights, farmers can apply targeted interventions, significantly reducing resource waste and enhancing crop management efficiency.

The benefits of using drones in Kenyan agriculture are further highlighted by their role in crop spraying. Drones equipped with sprayers were used to cover large areas quickly, applying pesticides and fertilisers more uniformly and precisely than traditional methods. This not only ensures effective crop treatment but also minimises chemical runoff, contributing to environmental sustainability. This approach bolsters the local economy by enhancing crop yields and reducing losses, while fostering a sustainable farming model that other regions with similar agricultural challenges can adopt.

In Nigeria, drones have been harnessed to generate high-resolution aerial images that offer comprehensive overviews of farmlands. This innovative use enables precise assessment of crop health, monitors growth patterns, and identifies areas that require specific interventions such as additional watering or fertilisation. This precise aerial data is particularly crucial in pinpointing issues like pest infestations, disease outbreaks, and uneven irrigation—common challenges in the expansive and diverse Nigerian farmlands. Drones equipped with advanced sprayers are used to cover extensive areas swiftly, applying pesticides and fertilisers in a uniform and precise manner, far surpassing the capabilities of traditional methods. This precision not only ensures that crops receive the exact treatment they require but also significantly reduces the amount of chemicals used, minimising environmental impact and promoting sustainability.

Drones offer a multitude of benefits for agriculture, from precision crop monitoring and irrigation management to soil health assessment and livestock monitoring. Specific agricultural drones like the DJI Agras T50 exemplify the advanced capabilities that these technologies bring to modern farming. Case studies from around the world highlight the positive impact of drones on productivity, sustainability, and efficiency in agriculture. As drone technology continues to evolve, its integration into agricultural practices will undoubtedly play a crucial role in meeting the growing global demand for food while promoting environmental stewardship and economic viability.

Construction Drones

Drones are becoming an integral part of the construction industry, offering numerous benefits that enhance efficiency, safety, and accuracy. By providing aerial perspectives and real-time data, construction drones are transforming traditional practices and setting new standards for project management and execution.

One of the primary benefits of drones in construction is their ability to perform aerial surveys and create detailed maps and 3D models of construction sites. Drones equipped with high-resolution cameras and LiDAR sensors can capture comprehensive visual and geospatial data, which is crucial for site planning and design. For instance, the DJI Phantom 4 RTK, known for its high-precision mapping capabilities, is widely used in the construction industry to produce accurate topographic maps and 3D models. These detailed visualisations help architects, engineers, and project managers understand the site's current conditions, identify potential issues, and make informed decisions about project planning and execution.

A significant advantage of using drones for site surveys is the speed and efficiency with which data can be collected. Traditional ground-based surveys are time-consuming and labor-intensive, often taking days or weeks to complete. In contrast, drones can survey large areas within hours, providing real-time data that accelerates the planning and design process. This rapid data collection capability is particularly beneficial for large-scale construction projects, where timely information is critical for meeting project deadlines.

Using drones for estimates and measurements in construction projects offers numerous advantages, significantly enhancing accuracy and efficiency. Pre-construction site surveys conducted by drones enable project teams to validate costs and ensure all required data is available, transforming a process that traditionally took days into one that can be completed in mere hours. For example, drones can quickly and accurately perform volumetric calculations of stockpiles from the air, determining volume, density, and tonnage with high precision. This ensures that construction sites are stocked with the right amount of materials, reducing overestimates and optimising material supply. Additionally, the use of drones significantly reduces survey times and improves communication among project stakeholders, leading to better-coordinated efforts and more streamlined construction processes.

Drones also play a crucial role in monitoring construction progress. By flying over construction sites regularly, drones can capture images and videos that document the project's advancement. This ongoing monitoring allows project managers to track progress, identify deviations from the schedule, and address issues promptly. For example, the Parrot Anafi FPV, with its thermal imaging and high-zoom capabilities, is used extensively for construction site monitoring. It provides detailed insights into various aspects of the project, including structural integrity, equipment usage, and worker safety.

In addition to progress monitoring, drones enhance safety on construction sites. Construction is inherently hazardous, with risks such as falls, equipment accidents, and structural failures. Drones mitigate these risks by performing tasks that would otherwise require workers to operate in dangerous conditions. For instance, drones can inspect high structures, such as cranes and scaffolding, without putting workers at risk. The DJI Matrice 350 RTK, known for its robust design and advanced imaging capabilities, is often used for such inspections. By reducing the need for manual inspections, drones help prevent accidents and improve overall site safety.

Drones, equipped with advanced navigation systems and high-resolution cameras, offer precise application of coatings, paints, and cleaning solutions, ensuring even coverage and reducing material waste. This method significantly enhances safety by eliminating the need for workers to operate at dangerous heights, and it also cuts down on labor costs and setup time associated with scaffolding. Additionally, drones contribute to more sustainable construction practices by optimising material use and producing less noise and pollution compared to conventional equipment.

Case studies highlight the transformative impact of drones on construction projects. In Route 29 widening project in northern Virginia, Unites States, drones were used to monitor construction progress, manage traffic flow, and improve project delivery by providing real-time data and aerial imagery. This approach facilitated precise earthmoving activities and helped keep the project on schedule and within budget by enabling quicker, more informed decision-making and reducing the risks associated with traditional ground-based survey methods.

Drones are also invaluable for quality control and inspection purposes. They can capture high-resolution images that reveal minute details, allowing for thorough inspections of completed work. This capability is particularly important for projects that require stringent quality standards, such as bridges, tunnels, and high-rise buildings. For example, the Yuneec H520E, with its precision flight and imaging capabilities, is used for detailed inspections of structural elements. These inspections help ensure that construction meets design specifications and safety standards, preventing costly rework and delays.

In the realm of infrastructure maintenance, drones provide a cost-effective and efficient solution for inspecting and maintaining existing structures. Drones can access hard-to-reach areas, such as the undersides of bridges and elevated highways, to perform inspections that would be challenging and risky for human workers. This capability is essential for identifying signs of wear and tear, corrosion, and other structural issues before they become critical problems. By facilitating regular and thorough inspections, drones contribute to the longevity and safety of infrastructure.

Furthermore, drones enable better communication and collaboration among project stakeholders. The real-time data and visualisations provided by drones can be shared with architects, engineers, contractors, and clients, ensuring that everyone has a clear and up-to-date understanding of the project's status. This transparency fosters collaboration and helps resolve issues quickly, reducing the likelihood of misunderstandings and disputes.

In addition to their operational benefits, drones contribute to sustainability in construction. By optimising resource usage and reducing waste, drones help minimise the environmental impact of construction activities. For example, drones can monitor material stockpiles and track resource consumption, ensuring that materials are used efficiently and that excess waste is minimised. This sustainable approach aligns with the growing emphasis on green building practices and environmental stewardship in the construction industry.

One of the most compelling case studies demonstrating the benefits of drones in construction is the Crossrail project in London. Crossrail, one of the largest infrastructure projects in Europe, used drones to monitor the construction of tunnels and stations. The drones provided real-time data on excavation progress, structural integrity, and worker safety. The ability to quickly identify and address issues helped keep the project on track and within budget. The use of drones also improved safety by reducing the need for workers to operate in hazardous underground environments.

Drones offer a multitude of benefits for the construction industry, from aerial surveys and progress monitoring to safety inspections and quality control. Specific construction drones like the DJI Phantom 4 RTK, Parrot Anafi, DJI Matrice 350 RTK, and Yuneec H520E exemplify the advanced capabilities that these technologies bring to modern construction practices. Case studies from around the world highlight the positive impact of drones on efficiency, safety, and sustainability in construction. As drone technology continues to evolve, its integration into construction practices will undoubtedly play a crucial role in shaping the future of the industry, driving innovation and setting new standards for excellence.

Delivery Drones

The global drone package delivery market is experiencing remarkable growth, reflecting a transformative shift in the logistics and supply chain industry. This burgeoning sector, valued at approximately US$ 241.04 million in 2022, is projected to skyrocket to US$ 5,800 million by 2030, growing at a staggering compound annual growth rate (CAGR) of 50% over the forecast period. The increasing adoption of drones for package delivery is not only modernising the logistics landscape but also enhancing efficiency and reducing human effort, marking a significant evolution in the way goods are transported.

Drones offer a swift and efficient means of transporting packages to customers. Unlike traditional delivery methods that rely on human drivers and vehicles, drones deliver packages either autonomously or are controlled remotely by operators on the ground. Drones have the capacity to simplify the delivery process, reduce labor costs, and improve the speed and reliability of deliveries, especially in areas that are difficult to access by conventional means.

The primary driver of this market's growth is the escalating demand for fast and efficient package delivery services worldwide. Consumers' increasingly willingness to pay premium prices for same-day delivery or one-hour orders is accelerating the adoption of drone delivery. Innovations in cargo transportation, coupled with significant investments from logistics and transportation companies, are further propelling the development of high-end drones capable of performing complex delivery tasks. This trend is particularly pronounced in the healthcare sector, where drones are used to transport medical supplies quickly and efficiently, and in the food and beverage industry, where they facilitate rapid delivery from convenience stores and quick-service restaurants. The medical delivery segment holds the largest share of the global drone package delivery market. This dominance is due to the critical need for rapid transportation of medical supplies, including medicines, vaccines, and blood samples, which can be life-saving in emergency situations. The ability of drones to bypass traffic congestion and deliver medical supplies directly to healthcare facilities ensures timely and efficient healthcare services.

In terms of drone types, the hybrid drone segment currently leads the market. Hybrid drones, which combine the capabilities of fixed-wing and rotary-wing drones, offer superior performance for urban deliveries. Their ability to take off and land vertically, like rotary-wing drones, while covering long distances at high speeds, similar to fixed-wing drones, makes them ideal for delivery purposes within city limits.

Geographically, North America and Asia holds the largest share of the global drone package delivery market. The region's advanced logistics infrastructure, coupled with supportive regulatory frameworks, has facilitated the rapid adoption of drone delivery services. Companies in North America are actively investing in drone technology and collaborating with regulatory bodies to develop safe and efficient delivery systems. The continuous innovation of major players in the drone industry, such as DJI, Parrot, Amazon and Google, also contributes to the market's growth in these regions.

The integration of drone infrastructure providers, drone service providers, manufacturers, and drones as service providers is critical to creating lucrative growth opportunities in the global market. Robust research and development activities in drone manufacturing are required to produce innovative equipment capable of carrying heavier loads and operating for longer durations. Improved battery technology is a key focus, enhancing the operational capabilities and range of delivery drones. Ongoing advancements in navigation systems and automation software are pivotal, enabling drones to perform more complex tasks with greater precision and safety. These technological enhancements not only improve the utility of drones in various sectors but also open new markets by increasing the feasibility of drone applications across more challenging environments. This holistic approach to innovation within the drone industry is essential for sustaining growth and expanding the potential.

DJI FlyCart 30

DJI has been making significant strides in the drone delivery sector, notably with their introduction of the FlyCart 30 and the comprehensive DJI DeliveryHub platform. The FlyCart 30, designed for various environmental scenarios, has set a new standard in drone delivery by offering robust features and versatile operational modes.

The DJI FlyCart 30 is capable of carrying up to 30 kg with dual batteries, and up to 40 kg with a single battery, making it suitable for a wide range of delivery tasks. This drone is equipped with advanced safety features, including dual active phased array radar and binocular vision systems for intelligent obstacle sensing, and an integrated parachute for emergency landings. These features ensure that the drone can operate safely in challenging conditions, with the ability to withstand temperatures ranging from -20° to 45° C and winds up to 12 m/s.

One of the most groundbreaking uses of the FlyCart 30 was its mission to Mount Everest, where it successfully delivered supplies to climbers and brought back trash. This mission demonstrated the drone's capabilities in extreme conditions and highlighted its potential for improving safety and efficiency in high-risk environments. The Nepalese government has since contracted DJI for regular supply operations to Everest Camp 1, further validating the drone's effectiveness in critical logistics applications. The DJI DeliveryHub platform complements the FlyCart 30 by providing a one-stop solution for managing aerial delivery operations. This platform offers features like operation planning, real-time status monitoring, and data analysis, integrating seamlessly with external cloud platforms. DJI Pilot 2, a part of this ecosystem, enhances manual flight operations by providing real-time updates on flight and cargo status, ensuring safe and efficient deliveries even in adverse conditions.

DJI's commitment to innovation is evident in the design of the FlyCart 30, which includes multiple safety and operational enhancements. These include a built-in parachute with multiple safeguard functions, intelligent obstacle avoidance technology, and redundant battery systems. The drone also features a flexible delivery system with both cargo and winch modes, allowing it to handle diverse delivery tasks with precision and reliability.

The integration of these technologies into DJI's delivery drones showcases the potential of industry-led standards in shaping the future of drone logistics. By setting high standards for safety, efficiency, and reliability, DJI is not only pushing the boundaries of what drone delivery can achieve but also influencing regulatory frameworks and best practices globally. The company's approach demonstrates how technological innovation, combined with strategic partnerships and regulatory compliance, can drive significant advancements in the logistics industry.

Google Wing

Google's Wing, a subsidiary of Alphabet, is at the forefront of drone delivery innovation. Wing has developed a comprehensive system designed to enhance delivery efficiency and safety through advanced technology. The company operates a fleet of lightweight, autonomous drones that have already completed over 350,000 deliveries across three continents. Wing's drones are designed to operate with high precision and safety. They can carry packages weighing up to 1.5 kg over distances of up to 20 km. Wing's drones are equipped with sophisticated technologies, including the latest advancements in detect and avoid (DAA) systems. These systems allow drones to operate beyond visual line of sight (BVLOS) without the need for visual observers, enhancing operational efficiency and safety. The recent FAA approval for Wing's DAA approach enables these drones to use Automatic Dependent Surveillance-Broadcast (ADS-B) based DAA within complex urban airspaces, such as Dallas-Fort Worth. This approval marks a significant step toward expanding their service capabilities across the United States.

The Wing delivery system integrates seamlessly with existing last-mile logistics services, providing real-time coordination and ensuring safe operations in shared airspace. Their newest drones, capable of carrying double the payload of their predecessors, exemplify the scalability and adaptability of their technology to meet market demands. These drones are designed to support a wide range of delivery needs, from groceries to pharmaceuticals, highlighting the versatility and efficiency of Wing's delivery model. In 2021, Google Wing and Walgreens, a major pharmacy chain, launched a groundbreaking drone delivery service in the Dallas-Fort Worth metropolitan area, marking a significant milestone in the adoption of drone technology for commercial delivery in the United States. This partnership exemplifies how non-normative approaches, such as industry-led initiatives and technological innovations, can reshape traditional logistics and delivery frameworks.

Amazon Prime Air

Amazon Prime Air, the drone delivery arm of Amazon, aims to revolutionise package delivery by offering rapid, reliable services to customers. The latest model, the MK30, is a testament to Amazon's commitment to advancing drone technology. This drone features a range of enhancements over its predecessors, including increased payload capacity, improved flight stability, and enhanced safety features. The MK30 drone is designed to handle diverse weather conditions, ensuring reliable delivery performance. It incorporates advanced sense-and-avoid systems to navigate safely through complex environments, reducing the risk of collisions. This drone is also quieter than previous models, addressing noise concerns and making it more suitable for urban deliveries.

Amazon's Prime Air service focuses on reducing delivery times and improving logistics efficiency. The drones are designed to deliver packages weighing up to five pounds to customers within 30 minutes of ordering. This rapid delivery capability is particularly beneficial for urgent deliveries, such as medical supplies and essential goods.

One of the standout features of the MK30 is its "sense and avoid" also known as "detect and avoid" technology. This system enables the drone to detect and navigate around obstacles such as people, pets, and property, ensuring safe operations even in densely populated areas. The MK30's ability to autonomously make safe decisions in real-time enhances its reliability and operational safety, a crucial factor for gaining public trust in drone delivery systems The MK30 is designed to operate in a wider range of environmental conditions compared to previous models. It can fly in light rain, which significantly expands its usability in various weather conditions. This capability is a considerable improvement, allowing for more consistent delivery services regardless of minor weather disruptions.

Noise reduction is another critical feature of the MK30. The drone's custom-designed propellers reduce noise levels by 25%, making it quieter than many common neighbourhood sounds even during descent. This reduction in noise pollution addresses a significant concern for communities and contributes to the broader acceptance of drone deliveries.

In terms of performance, the MK30 boasts an impressive range, capable of covering twice the distance of its predecessors. It utilises a vertical take-off system before transitioning into horizontal, wing-borne flight. Despite its compact and lightweight design, the MK30 can carry packages weighing up to five pounds, with delivery times typically under an hour. This efficiency is particularly beneficial for time-sensitive deliveries, such as medical supplies. The MK30's integration into Amazon's existing fulfilment network marks a significant step forward in logistics innovation. By deploying from same-day delivery sites, these drones are poised to enhance Amazon's delivery capabilities, offering ultra-fast delivery options to a broader customer base in the U.S., Italy, and the UK.

Amazon has been conducting extensive testing and pilot programs to refine its drone delivery service using the MK30. These tests are demonstrating the feasibility of integrating drones into urban logistics networks, showcasing their potential to streamline supply chains and reduce delivery times significantly. By leveraging advanced AI and machine learning algorithms, Amazon's drones can optimise flight paths and ensure precise deliveries, enhancing the overall customer experience.

In early 2024, Amazon announced that the company is on track to start delivering parcels within an hour in the UK from the end of the year. This drone-based delivery service, successfully tested in California, utilises drones for quicker and more accurate delivery of customer packages. Customers can expect to receive their orders within 20 to 30 minutes for items that previously took about an hour to get from stores and sometimes even longer. Although there are complaints about noise, the benefits are significant, with packages typically weighing no more than 2.27 kg and being monitored from a ground location. Amazon claims that drone delivery is 100 times safer than driving to the store.

Both Google Wing and Amazon Prime Air have been proactive in collaborating with regulatory bodies to ensure their drone operations meet safety and compliance standards. Wing's partnership with the FAA and other aviation authorities has been crucial in securing approvals for BVLOS operations and expanding their service areas.

Similarly, Amazon has worked closely with regulatory agencies to navigate the complex landscape of drone regulations and achieve significant milestones in their delivery service development. These collaborations have not only facilitated the operational expansion of drone delivery services but also contributed to the establishment of industry standards and best practices. The advancements made by Wing and Amazon serve as benchmarks for other companies in the drone delivery space, promoting innovation and setting high safety and performance standards.

The initiatives by Google Wing and Amazon Prime Air highlight the transformative potential of drone technology in the delivery sector. By significantly reducing delivery times, enhancing operational efficiency, and minimising human intervention, drones are poised to become integral to modern logistics networks. The advancements in drone technology also open up new possibilities for delivery in remote and underserved areas, providing essential services to communities that were previously hard to reach.

As these companies continue to innovate and expand their drone delivery capabilities, they set the stage for a future where drones play a pivotal role in the logistics and supply chain industry. The ongoing development and deployment of sophisticated drone systems by Wing and Amazon exemplify the potential of drones to revolutionise the delivery landscape, offering faster, safer, and more efficient services to consumers worldwide. Despite the promising growth, the drone package delivery market faces challenges, including regulatory hurdles and safety concerns. Ensuring compliance with airspace regulations and addressing privacy issues are critical to gaining public trust and widespread acceptance. Additionally, the development of reliable and secure communication systems to control drones remotely is essential to prevent unauthorised access and ensure the safety of delivery operations.

Looking ahead, the market's growth trajectory is expected to continue as advancements in drone technology and supportive regulatory measures pave the way for broader adoption. The integration of artificial intelligence and machine learning in drones is likely to further enhance their capabilities, enabling them to navigate complex environments and make autonomous decisions. These technological advancements will not only improve the efficiency of drone deliveries but also expand their applications across various industries.

Educational Drones

Educational institutions and programs have recognised the potential of drones as engaging learning tools, incorporating them into STEM (Science, Technology, Engineering, and Mathematics) curricula. Drones are used to teach students about aerodynamics, robotics, and programming, providing practical experience with cutting-edge technology. Educational drones, like the Parrot Mambo, come with modular components that allow students to experiment with different functionalities, fostering an interest in engineering and technology from an early age.

One notable example is the "Drones for Schools" program in the United Kingdom. This initiative introduces students to drone technology, offering hands-on experience in building, coding, and flying drones. Students learn about the principles of flight, the mechanics of drone operation, and the applications of drones in various industries. The program not only enhances students' technical skills but also promotes problem-solving and critical thinking abilities.

In the United States, the STEM+C project, funded by the National Science Foundation, integrates drones into middle and high school curricula. This project uses drones to teach coding, mathematics, and engineering principles through interactive lessons and real-world problem-solving scenarios. For instance, students might be tasked with programming a drone to navigate an obstacle course, which requires them to apply their knowledge of geometry, physics, and computer science.

The use of drones in higher education is also gaining traction. Universities are incorporating drones into their research and teaching programs across various disciplines. At the University of Maryland, for example, the Drone Research and Technology Center focuses on developing and applying drone technology for environmental monitoring, agricultural management, and disaster response. Students and researchers use drones to collect data, analyze environmental conditions, and develop innovative solutions to real-world problems.

Case studies highlight the diverse applications of drones in education. In one project, students at Virginia Tech used drones to monitor water quality in local rivers. The drones were equipped with sensors to collect water samples and measure parameters such as temperature, pH, and turbidity. This project provided students with practical experience in environmental science and demonstrated the potential of drones for environmental monitoring and conservation.

Another example is the "AgDrone" project at Purdue University, where students use drones to monitor crop health and optimise agricultural practices. By analysing aerial images captured by drones, students can assess plant health, detect pest infestations, and evaluate the effectiveness of different farming techniques. This hands-on experience with drone technology prepares students for careers in precision agriculture and other emerging fields.

In addition to formal educational settings, drones are used in informal learning environments such as summer camps and after-school programs. These programs often focus on drone racing, where students build and program their drones to compete in timed races. This activity combines elements of engineering, physics, and computer science in a fun and engaging way, encouraging students to explore STEM subjects further.

The integration of drones into augmented reality (AR) and virtual reality (VR) experiences has further expanded their educational uses. AR drone racing games, developed by companies like Edgybees, overlay digital obstacles and race tracks onto real-world environments, offering an immersive mixed-reality experience. VR applications enable users to experience drone flights from a first-person perspective, combining the thrill of flying with the safety of simulation. These interactive experiences enhance gaming and provide new avenues for entertainment, making drones an integral part of the evolving digital landscape.

Beyond the typical classroom applications, drones are also being utilised in special education to support students with diverse learning needs. For example, drones are employed in therapeutic settings to help children with autism improve their social, communication, and motor skills. By engaging students in activities that require them to interact with and control drones, educators can create a motivating and dynamic learning environment that encourages participation and development.

These activities often include guided drone flying exercises that help students practice turn-taking, following instructions, and spatial awareness. Additionally, the use of drones in therapy sessions can be tailored to individual learning plans, providing a personalised and adaptive approach to education that meets the unique needs of each student. This innovative use of drone technology underscores its potential to make education more inclusive and effective for all learners. Drones are proving to be invaluable tools in education, providing students with practical, hands-on experience in STEM subjects and beyond. Through various programs and initiatives, students of all ages are learning to harness the power of drone technology, preparing them for future careers in an increasingly technological world.

Entertainment and Recreational Drones

Drones are significantly impacting the entertainment and recreational industry, offering a multitude of benefits and transforming the way people engage with media and leisure activities. Drones have introduced new possibilities for creativity, audience engagement, and interactive experiences, making them indispensable tools in the entertainment and recreational sectors.

One of the most visible benefits of drones in the entertainment industry is their use in aerial photography, videography and content creation. Drones like DJI Inspire 3 are equipped with high-definition cameras allow filmmakers and photographers to capture stunning aerial shots that were previously only possible with helicopters or cranes. This capability has transformed cinematography by providing unique perspectives and dynamic camera movements, enhancing the visual storytelling of films, television shows, and commercials. Award winning movies like Skyfall, The Greatest Showman, and The Wolf of Wall Street have utilised drone technology to achieve breathtaking aerial sequences, showcasing the versatility and creative potential of drones in film production.

In addition to filmmaking, drones have become essential tools for live event coverage and sports broadcasting. Drones can provide overhead views and close-up shots of sporting events, concerts, and festivals, offering viewers immersive experiences and comprehensive coverage. For instance, during the 2018 Winter Olympics in Pyeongchang, Intel orchestrated a spectacular drone light show that featured over 1,200 drones creating intricate patterns and animations in the sky. In 2024, UVify, a trailblazer in swarm drone technology, redefined the limits of aerial entertainment by setting a new Guinness World Record for the most unmanned aerial vehicles (UAVs) airborne simultaneously. This landmark event took place in Songdo, Korea, where UVify masterfully coordinated a mesmerising display featuring 5,293 IFO drones. These drones lit up the night sky with intricate patterns and captivating visuals, demonstrating the company's advanced programming and control technology. The record-breaking feat not only surpassed previous benchmarks but also highlighted the extensive capabilities of IFO drones in creating impactful entertainment and precise aerial performances. The event, meticulously coordinated with local authorities and air traffic control, ensured safety and regulatory compliance, reflecting UVify's dedication to quality in all technological ventures. This achievement sets a new industry standard and exemplifies the transformative potential of drones in entertainment, pushing the boundaries of synchronized drone performance and offering a glimpse into the future of aerial displays.

Drone light shows have emerged as a popular alternative to traditional fireworks displays, providing environmentally friendly and visually stunning entertainment. These shows involve swarms of drones equipped with LED lights, choreographed to perform synchronised movements and create breathtaking light displays. Companies like Intel have pioneered drone light shows, captivating audiences at major events such as the Super Bowl and the Coachella Valley Music and Arts Festival. Unlike fireworks, drone light shows produce no smoke or loud noises, making them a safer and more sustainable option for public celebrations.

Recreational drone racing has gained traction as a thrilling and competitive sport, drawing enthusiasts and spectators alike. The Drone Racing League (DRL) has popularised this sport by organising high-speed races where pilots navigate drones through complex courses, often at speeds exceeding 100 mph. Equipped with first-person view (FPV) goggles, pilots experience the race from the drone's perspective, adding an adrenaline-pumping element to the competition. DRL races are broadcasted on major sports networks, attracting sponsorships and fostering a growing fan base. The sport's popularity has also spurred advancements in drone technology, with manufacturers developing faster, more agile drones to meet the demands of competitive racing.

Beyond professional applications, drones have become accessible recreational devices for hobbyists and tech enthusiasts. DIY drone kits and open-source software platforms allow individuals to build and customise their own drones, fostering a community of makers and innovators. This hands-on approach to drone technology encourages creativity and experimentation, as enthusiasts modify their drones for activities such as aerial stunts, freestyle flying, and exploring remote areas. The rise of consumer drones from companies like DJI and Parrot has made high-quality drones available to the general public, expanding the recreational drone market and inspiring new uses and applications.

Health Care Drones

Drones are transforming the healthcare services landscape by providing innovative solutions that address critical challenges in medical logistics. One of the most notable benefits of medical delivery drones is their ability to reach remote and underserved areas, ensuring timely access to essential medical supplies, which can be life-saving in emergency situations.

Zipline, a pioneering company in drone-based healthcare delivery, has been at the forefront of this transformation. Operating in Rwanda Japan, Nigeria, Ivory Coast, United States, Kenya and Ghana, Zipline's drones have made over 1 million commercial deliveries, autonomously covering more than 112 million kilometres. These drones are used to transport blood products, vaccines, and essential medications to remote health facilities, significantly reducing the time required for deliveries and improving healthcare outcomes. In Rwanda, drones now carry 35% of the blood supplied for transfusions outside of Kigali, demonstrating their critical role in healthcare logistics.

Another significant example is the Medicine from the Sky initiative by the World Economic Forum in partnership with the State Government of Telangana and Apollo Hospitals in India. This project aimed to create a scalable model for drone-based medical deliveries in South Asia. Since inception, Over 650 drone flights have been executed, delivering more than 8,000 medical products to over 200 patients across a challenging 15,000km ground distance in the region. The project has successfully demonstrated the potential for drones to deliver blood, medical samples, and even organs, bridging the gap in healthcare access in rural and hard-to-reach areas of India.

The technical capabilities of these medical drones are impressive. They can operate in various weather conditions, fly up to 160 kilometres round trip, and deliver supplies with high precision. For instance, Zipline's drones are designed for beyond visual line of sight (BVLOS) operations, allowing them to cover long distances and deliver critical supplies to areas that are inaccessible by road. These drones are equipped with advanced navigation systems and can operate autonomously, ensuring reliable and efficient deliveries.

In the United States, companies like Matternet have partnered with UPS to develop drone delivery networks for medical samples between hospitals. This collaboration aims to reduce the delivery time for critical medical supplies, such as blood and tissue samples, thereby improving diagnostic and treatment outcomes. Matternet's drones are capable of carrying payloads up to 2 kilograms and can travel distances of up to 20 kilometres, making them ideal for urban healthcare logistics.

During the COVID-19 pandemic, the use of drones in healthcare delivery gained significant attention. Drones were deployed to transport COVID-19 test kits, vaccines, and personal protective equipment (PPE) to remote and underserved areas, reducing the risk of virus transmission by minimising human contact during transportation. In Scotland, a project by Skyports demonstrated the effectiveness of drone deliveries by transporting medical supplies and COVID-19 test kits between hospitals in Oban and the Isle of Mull.

The benefits of drone delivery in healthcare extend beyond emergency and pandemic responses. Drones can also support routine healthcare operations, such as delivering medications to patients with chronic conditions in rural areas. This ensures consistent access to necessary treatments and reduces the burden on healthcare facilities. Additionally, drones can assist in telemedicine by delivering diagnostic tools and collecting samples for remote analysis, thereby expanding the reach of healthcare services.

Humanitarian Drones

In the ever-evolving landscape of global humanitarian efforts, drones have emerged as a pivotal tool, transcending traditional boundaries and transforming the reach and impact of aid. Drones, deftly navigating through the most treacherous terrains, have become instrumental in delivering aid where conventional methods falter. Whether it's whisking life-saving supplies to earthquake-stricken regions or providing real-time data to strategise flood relief, drones extend the capabilities of humanitarian missions far beyond previous limitations. Their ability to access the inaccessible not only speeds up response times but also ensures that help reaches the furthest corners of need, reinforcing the efficiency and effectiveness of humanitarian operations worldwide.

In the aftermath of Cyclone Idai in 2019, which devastated parts of Mozambique, Zimbabwe, and Malawi, drones were deployed to assess the extent of the damage. They provided detailed aerial imagery that was crucial for mapping affected areas, identifying the locations of stranded survivors, and prioritising aid delivery. This rapid assessment enabled humanitarian organizations to respond more effectively, ensuring that resources were allocated where they were most needed.

In Europe, drones have significantly enhanced humanitarian efforts, particularly in the aftermath of natural disasters where quick response is crucial. For example, after severe flooding in Ukraine, drones were employed to assess the damage, locate stranded individuals, and deliver essential supplies to inaccessible areas. Drones equipped with high-resolution cameras and thermal imaging created detailed maps of the affected regions, providing rescue teams with real-time data that was critical for effective response and resource allocation. These drones also carried medical supplies and food to isolated communities, demonstrating their utility in reducing response times and improving the efficiency of disaster relief operations.

In Asia, drones have been pivotal in humanitarian operations, particularly in Indonesia, a country prone to natural disasters like earthquakes and tsunamis. Following a devastating earthquake in Sulawesi, drones were rapidly deployed to conduct aerial surveys, helping to assess damage and guide recovery efforts. These UAVs were crucial in mapping devastated areas, identifying safe routes for rescuers, and locating survivors. Additionally, drones were used to deliver emergency supplies to areas cut off by landslides or debris, ensuring timely aid to those in dire need. The agility and ability of drones to operate in challenging environments made them invaluable, significantly enhancing the speed and effectiveness of the humanitarian response. This case not only demonstrates the drones' utility in immediate post-disaster scenarios but also their growing role in ongoing recovery efforts, helping rebuild communities by providing essential data for planning and coordination.

In addition to disaster response and healthcare delivery, drones are used for disaster preparedness and mitigation. For example, in the Philippines, drones have been employed to map areas at risk of natural disasters such as typhoons and floods. By identifying vulnerable regions, these drones help in planning and implementing preventive measures, thereby reducing the potential impact of future disasters.

Recent advancements in drone technology continue to expand their capabilities and applications in humanitarian efforts. For instance, the United Nations World Food Programme (WFP) has been exploring the use of drones to deliver food and medical supplies in disaster-stricken areas. These drones can operate in conditions where traditional vehicles might be hindered by damaged infrastructure, ensuring that aid reaches those in need quickly and efficiently.

Another innovative use of drones is in providing emergency connectivity in disaster-affected areas. The WFP, in collaboration with other organizations, is testing the deployment of drones equipped with communication devices to restore connectivity in regions where communication networks have been disrupted. This capability is crucial for coordinating relief efforts and ensuring that affected communities can communicate with emergency responders.

Drones offer significant benefits for humanitarian efforts, enhancing the speed, efficiency, and effectiveness of aid delivery and disaster response. While there are challenges to overcome, the continued development and deployment of drone technology hold great promise for improving humanitarian operations and ultimately saving lives. By leveraging drones' capabilities, humanitarian organizations can better address the needs of vulnerable populations and respond more effectively to emergencies.

Inspection Drones

Drones are transforming infrastructure inspection, providing unparalleled efficiency, safety, and accuracy in assessing various structures. The ability to capture high-resolution images, videos, and other sensor data from vantage points that are otherwise difficult or dangerous to reach is making drones an invaluable tool in inspecting infrastructure such as transmission lines, solar panels, wind turbines, and hydroelectric power stations.

One of the primary benefits of using drones for infrastructure inspection is their ability to access hard-to-reach areas quickly and safely. Traditional inspection methods often require extensive scaffolding, cranes, or even helicopters, which can be time-consuming, expensive, and hazardous. Drones, on the other hand, can be deployed rapidly and can navigate complex environments with ease, reducing both the time and cost associated with inspections.

Drones are used to inspect power lines, towers, and other components of the electrical grid. Equipped with high-resolution cameras and thermal imaging sensors, drones can detect issues such as damaged insulators, corroded components, and hotspots indicating potential electrical faults. For example, a recent case study by Southern Company, a major utility in the United States, demonstrated the effectiveness of drones in inspecting over 27,000 miles of transmission lines. The drones identified numerous issues that required maintenance, allowing the company to address potential problems before they led to outages or safety hazards.

Drones are also extensively used in the inspection of solar power installations. Solar farms, which can span vast areas, require regular inspections to ensure that the panels are functioning optimally and to identify any defects or damage. Drones equipped with multispectral and thermal cameras can capture detailed images of the solar panels, identifying issues such as cracks, dirt buildup, and electrical faults. A notable example is the use of drones by Duke Energy to inspect their solar farms across the United States. The drones provided high-resolution thermal images that helped identify underperforming panels and areas needing cleaning, thereby improving the overall efficiency of the solar installations.

Drones equipped with specialised cleaning apparatus, like brushes and spray mechanisms, are used in cleaning of solar panels. These drones navigate autonomously over solar installations, performing thorough cleanings to remove accumulations of dirt, dust, bird droppings, and other debris that impede solar efficiency. This automated approach minimises the need for human labor, reducing safety risks and cutting down maintenance costs. Moreover, the precision and efficiency of drones ensure that the solar panels operate at optimal capacity.

Wind turbines, which are often located in remote and challenging environments, also benefit from drone inspections. Inspecting wind turbines manually can be particularly dangerous due to the height of the structures and the harsh conditions in which they operate. Drones can safely fly close to the turbine blades, capturing high-resolution images that reveal cracks, erosion, and other structural issues. Siemens Gamesa, a leading wind turbine manufacturer, has implemented drone inspections across their global wind farms. The use of drones has not only reduced the inspection time significantly but has also enhanced the accuracy of detecting defects, leading to more timely and effective maintenance.

Hydroelectric power stations, which include dams, reservoirs, and spillways, present another critical application for drone inspections. The structural integrity of these facilities is paramount to ensure safety and operational efficiency. Drones can inspect the exterior surfaces of dams, looking for signs of wear, cracks, and other potential issues. In addition to visual inspections, drones equipped with LiDAR sensors can create detailed 3D models of the structures, providing a comprehensive assessment of their condition. A case study from the Hoover Dam highlighted the use of drones to perform a detailed inspection of the dam's face. The drones captured high-resolution images and LiDAR data, enabling engineers to detect and monitor any changes in the dam's structure over time.

Furthermore, drones contribute to the inspection of offshore wind farms and oil rigs, which are particularly challenging due to their location and environmental conditions. Offshore structures require regular inspections to ensure they can withstand the harsh marine environment. Drones can perform these inspections without the need for costly and dangerous manned missions. The use of drones in offshore wind farms, such as those operated by Ørsted, has proven effective in reducing inspection times and improving the accuracy of detecting structural issues, thereby enhancing the reliability and safety of these energy sources.

The integration of advanced technologies, such as artificial intelligence (AI) and machine learning, further enhances the capabilities of inspection drones. AI-powered drones can analyse the captured data in real-time, identifying defects and anomalies with greater accuracy and speed. For instance, Skydio, an AI-driven drone manufacturer, has developed drones that use computer vision to autonomously navigate and inspect infrastructure, providing detailed reports on detected issues. These AI capabilities allow for more efficient inspections and faster decision-making, ultimately improving the maintenance and operation of critical infrastructure.

Drones offer a multitude of benefits for infrastructure inspection, providing a safer, more efficient, and cost-effective solution compared to traditional methods. Case studies from various sectors, including transmission lines, solar power, wind turbines, and hydroelectric power stations, demonstrate the transformative impact of drone technology on inspection practices. As drone technology continues to evolve, its integration into infrastructure inspection will undoubtedly enhance the safety, reliability, and efficiency of these essential structures, ensuring their longevity and optimal performance.

Interstellar Drones

In the quest to explore the vastness of our solar system and beyond, interstellar drones have emerged as indispensable tools, playing a crucial role in the human spacefaring journey. These autonomous and semi-autonomous machines extend our reach, capabilities, and understanding of the universe, allowing us to venture into realms that would be otherwise inaccessible.

The significance of drones in space exploration lies in their ability to operate in environments that are hostile to human life. The harsh conditions of space—extreme temperatures, radiation, vacuum, and the vast distances involved—pose significant challenges for human explorers.

Drones, however, are designed to withstand these conditions, equipped with the necessary technology to perform a wide range of tasks, from scientific research to logistical support.

Voyages to distant celestial bodies such as Mars, Jupiter, Venus, Mercury, and beyond cannot be imagined without the use of drones. These advanced machines are essential for conducting preliminary surveys, gathering critical data, and preparing the groundwork for future human missions. For instance, on Mars, drones can navigate the planet's rugged terrain, collect soil and rock samples, and analyze atmospheric conditions, providing invaluable information that helps scientists understand the planet's geology and climate.

One of the most notable examples of a drone used in space exploration is the Mars Rover series by NASA. The most recent addition, Perseverance, landed on Mars in February 2021, equipped with advanced instruments to search for signs of past life and collect samples for future return to Earth. Accompanying Perseverance is the Ingenuity helicopter, a small, experimental drone designed to demonstrate the feasibility of powered flight in the thin Martian atmosphere. Ingenuity's successful flights have opened new possibilities for aerial exploration on Mars, enabling access to areas that rovers cannot reach.

Similarly, the Juno spacecraft, which has been orbiting Jupiter since 2016, is equipped with a suite of scientific instruments to study the planet's composition, gravity field, magnetic field, and polar magnetosphere. Although not a drone in the traditional sense, Juno operates autonomously, executing pre-programmed instructions to gather data from the gas giant. The insights gained from Juno's mission have deepened our understanding of Jupiter's formation and evolution, shedding light on the broader processes that shape planetary systems.

In the realm of Venus exploration, the concept of drones is being explored through missions like the proposed Venera-D, a collaboration between NASA and Roscosmos. This mission aims to send an orbiter and a lander to Venus, potentially including an aerial platform—a drone—to study the planet's harsh surface and atmospheric conditions. The use of drones in such missions would allow scientists to investigate the dynamic weather patterns and volcanic activity on Venus, providing clues about the planet's past and present.

On Mercury, the European Space Agency's BepiColombo mission, launched in 2018, is en route to study the planet's surface and magnetosphere. While BepiColombo itself is a spacecraft, the data it gathers will inform the design of future drones capable of withstanding

Mercury's extreme temperatures and intense solar radiation. These drones could conduct detailed surface mapping and resource assessments, essential for understanding the planet's history and potential for future exploration. Beyond our solar system, interstellar drones could play a pivotal role in exploring exoplanets and other celestial bodies. Concepts like NASA's Breakthrough Starshot project envision sending swarms of tiny, light-propelled drones to the Alpha Centauri system, our nearest stellar neighbor. These drones, traveling at a significant fraction of the speed of light, could provide humanity's first close-up images and data from another star system, revolutionising our understanding of the universe.

The versatility and resilience of interstellar drones make them indispensable for space exploration. They serve as our eyes and hands in the cosmos, conducting scientific research, scouting for potential landing sites, and performing maintenance on spacecraft and habitats. As technology advances, the capabilities of these drones will continue to expand, enabling more ambitious missions and deeper exploration into the unknown. Interstellar drones are essential for advancing our spacefaring ambitions. They bridge the gap between human limitations and the vast, challenging environment of space, allowing us to explore, understand, and eventually inhabit other worlds. With each new mission, these drones push the boundaries of what is possible, bringing us closer to realising the dream of interstellar travel and exploration.

Maritime Drones

In recent years, the integration of drones into maritime operations is transforming the industry, offering a suite of advancements that enhance both efficiency and safety. These unmanned aerial vehicles are pivotal in transitioning traditional maritime practices by providing rapid data collection, heightened safety measures, cost efficiencies, and superior environmental monitoring. This technological leap not only accelerates operational workflows but also paves the way for a more sustainable interaction with our marine environments.

One notable application of maritime drones is in search and rescue missions. Drones are now deployed quickly and cost-effectively compared to traditional methods like helicopters. For instance, maritime drones equipped with thermal imaging and high-resolution cameras can locate individuals in distress at sea more efficiently. In Australia, a Ripper Lifesaver drone by Flyability successfully deployed a flotation device to two teenagers, allowing them to swim to safety without the need for first responders to risk their lives by going out to sea.

Maritime drones are also used for ship inspections, which traditionally involve significant time and risk. With drones, routine maintenance and inspections can be conducted remotely, providing real-time feedback and significantly reducing the risk to human inspectors. For example, drones like Flyability's Elios can enter confined spaces, such as oil tanks, to perform visual inspections without exposing crew members to dangerous gases or other hazards. This capability not only enhances safety but also speeds up the maintenance process and reduces costs. Drones have proven to be a game-changer for maritime deliveries. Instead of using launch boats, which can be costly and time-consuming, drones can deliver essential items such as documents, medicine, or spare parts directly to ships at sea. This method reduces delivery costs by up to 90% and saves significant time, thereby ensuring that ships do not have to return to port for supplies.

The first drone delivery at sea occurred in March 2016, conducted by Maersk Tankers near Kalundborg, Denmark. This groundbreaking delivery was part of a test to evaluate the feasibility of using drones in the maritime supply chain. The drone, supplied by the French company Xamen Technologies, was launched from a tugboat due to adverse weather conditions and successfully delivered a parcel by dropping it from five meters above the deck of the vessel. This test demonstrated the potential for drones to significantly reduce the costs of delivering small parcels such as urgent spare parts, mail, or medicine to ships at sea, compared to traditional methods using barges.

Moreover, drones are playing a critical role in environmental monitoring and protection. They are used to detect illegal fishing activities, monitor pollution levels, and gather data on marine wildlife. For instance, the European Maritime Safety Agency (EMSA) has deployed drones to assist with border control, search and rescue, and pollution monitoring. This initiative is part of a broader strategy to enhance the surveillance and safety of European waters, demonstrating the effectiveness of drones in maintaining maritime security.

Moreover, ports like Antwerp and Rotterdam have been at the forefront of integrating drone technology into their operations. These ports use drones to monitor security, inspect infrastructure, and respond to emergencies. For example, Port of Antwerp uses drones for real-time surveillance during emergencies, such as container fires and rescues at sea. Similarly, Port of Rotterdam has converted sections of its quay into drone ports to test long-range drones, demonstrating how drones can enhance port safety and efficiency. These long-range drones are being tested for a variety of applications, including the inspection of infrastructure, delivery of parts and crucial documents, and possibly even for environmental monitoring. The ability to quickly dispatch drones across the port's extensive area allows for rapid responses to incidents, regular maintenance checks, and efficient transportation of small cargo items, potentially reducing the need for manned vehicles and boats.

The technical capabilities of maritime drones continue to evolve, with models like the JOUAV CW-15 and CW-25 offering advanced features for surveillance and marine biology research. These drones come equipped with long flight times, high-speed data transmission, and sophisticated imaging technology, making them ideal for a wide range of maritime applications (JOUAV). Maritime drones will continue to transform the industry by enhancing safety, reducing costs, improving efficiency, and supporting environmental protection efforts.

Urban Passenger Air Drones

Urban Air Taxis are emerging as a revolutionary mode of transportation, integrating cutting-edge technologies like electric vertical take-off and landing (eVTOL) aircraft, which are designed to alleviate urban congestion and provide faster, more efficient travel within and between cities. Companies such as Joby Aviation, Lilium, and Volocopter are leading the charge in this new frontier, with extensive research, testing, and initial operations underway.

Joby Aviation, for instance, has positioned itself as a major player in the urban air mobility (UAM) space, with a focus on on-demand passenger flights. Their electric aircraft are designed for low-noise operations, making them suitable for densely populated areas. Joby aims to make aerial ride-sharing as affordable as ground-based services over time, aiming for mass adoption. The company has already conducted numerous test flights and is working closely with regulatory bodies like the FAA to secure certification for commercial operations by 2024.

Lilium, another prominent player, is developing eVTOL aircraft with a range of up to 300 km, focusing on both urban air taxis and regional air mobility. Their innovative approach combines electric propulsion with vertical and horizontal flight capabilities, allowing for flexible and efficient transport between cities. Lilium's ambition includes creating regional flight networks, connecting urban centers and reducing travel times dramatically. They are in active collaboration with regulators in

Europe and North America to bring their vision to life.

Volocopter, a pioneer in UAM, has made significant strides in public demonstrations and regulatory approvals. Their collaboration with Skyports to develop prototype "vertiports" in Singapore exemplifies their forward-thinking approach to infrastructure. Volocopter's vision involves building a network of landing pads across major cities to facilitate air taxi operations, with trials in cities like Dubai and Singapore showcasing the potential for this technology to transform urban travel.

Despite these advancements, challenges remain, particularly around public acceptance, safety concerns, and the regulatory environment. The European Union Aviation Safety Agency (EASA) and the FAA are working to establish robust frameworks that can ensure the safe integration of air taxis into existing airspace systems. These agencies are adapting regulations, initially designed for traditional aviation, to accommodate the specific needs of eVTOLs.

Public perception studies suggest that while there is growing acceptance, concerns about safety and data privacy need to be addressed through stringent regulation and transparent communication. Several cities are conducting pilot programs to test the feasibility of urban air taxis. For example, in Dubai, Volocopter successfully demonstrated autonomous air taxi flights as early as 2017, showcasing the potential for large-scale adoption. Similarly, in the U.S., Joby Aviation is working toward FAA certification and has conducted thousands of test flights as part of its preparation for commercial operations.

The future of urban air mobility is promising, but it requires ongoing collaboration between technology companies, regulators, and infrastructure developers. By addressing the regulatory, safety, and operational challenges head-on, urban air taxis are set to become an integral part of the transportation ecosystem, offering a sustainable and efficient alternative to traditional travel. As companies like Joby, Lilium, and Volocopter continue to innovate, the vision of air taxis providing rapid, congestion-free travel across cities is becoming increasingly tangible.

Urban Planning Drones

As urban populations continue to expand, the complexity and demand for smarter, more efficient urban planning intensifies. Drones, specifically used for urban planning, are emerging as invaluable tools due to their ability to quickly and accurately gather high-resolution data from extensive areas. This technological integration is transforming how urban planners and developers approach city development and management.

One of the fundamental advantages of using drones in urban planning is their capability to conduct large-scale aerial surveys. These surveys capture detailed images and generate rich data sets that can significantly enhance the accuracy of urban development plans. For example, drones can efficiently monitor traffic patterns, land use, and infrastructure, providing a clear, comprehensive overview that supports effective decision-making and strategic planning.

Moreover, drones facilitate the assessment of urban spaces without disrupting daily activities, which is a critical advantage in densely populated areas. This minimises the inconvenience typically associated with traditional survey methods that often require partial or complete shutdowns of active sites. Additionally, drones can access hard-to-reach areas, making them perfect for inspecting high-rise buildings and other complex structures, thus ensuring that all parts of the urban fabric are considered in development projects.

This integration of drone technology in urban planning not only supports the efficiency and sustainability of urban development but also propels cities towards becoming smarter and more responsive to the needs of their growing populations. Drones in urban planning has the ability to collect high-resolution data quickly and efficiently. Equipped with advanced sensors, cameras, and LiDAR drones can capture detailed aerial imagery and create accurate 2D and 3D maps of urban areas. This capability allows urban planners to gather comprehensive data on topography, infrastructure, and land use without the need for extensive field surveys. For instance, in complex urban environments, drones can easily navigate and map out intricate cityscapes, providing planners with the necessary information to make informed decisions about land use and development.

Drones also play a crucial role in the visualisation of urban planning projects. By generating detailed 3D models and simulations, drones enable planners to visualise how new developments will integrate with existing structures and landscapes. This visualisation helps identify potential design issues and assess the impact of new buildings or infrastructure on the urban environment. It also facilitates public engagement by providing realistic visual representations of proposed projects, allowing citizens to better understand and provide feedback on development plans. For example, augmented reality (AR) and virtual reality (VR) technologies, when combined with drone imagery, can create immersive experiences for stakeholders, enhancing their understanding of planning proposals.

In addition to data collection and visualisation, drones contribute significantly to monitoring and managing urban infrastructure. They can be used to inspect bridges, roads, and buildings, capturing high-resolution images and sensor data that reveal structural issues and maintenance needs. This proactive approach to infrastructure management helps identify problems early, reducing the risk of costly repairs and ensuring the safety and durability of urban assets. For example, in San Francisco, drones are used for regular inspections of critical infrastructure, allowing for timely interventions and maintenance.

Drones also support environmental monitoring and sustainability efforts in urban planning. They can be equipped with various sensors to monitor air quality, temperature, and other environmental parameters. This data is essential for assessing the environmental impact of urban development and for implementing measures to mitigate negative effects. In Amsterdam, a pioneering initiative known as the "Amsterdam Smart City" project illustrates the transformative role of drones in urban environmental management. A key component of this initiative involves the deployment of drones equipped with sensors to monitor air quality across different parts of the city. These drones collect real-time data on various pollutants, which is then analysed to assess air quality trends and identify pollution hotspots. The project is a collaboration between the City of Amsterdam, environmental agencies, and technology companies specializing in drone and sensor technology. This collaborative effort not only enhances the city's ability to manage air quality but also supports the development of policies aimed at reducing urban pollution effectively. Moreover, drones enhance the efficiency of urban planning processes by reducing the time and cost associated with traditional surveying and data collection methods. Automated drone flights can cover large areas in a fraction of the time it would take a human surveyor, significantly speeding up the planning process. This efficiency is particularly beneficial in rapidly urbanising areas where timely data collection is crucial for effective planning and development.

The integration of drones with Internet of Things (IoT) technologies further amplifies their benefits in urban planning. Drones can act as mobile data collection platforms, transmitting real-time information to central databases for analysis. This integration enables continuous monitoring and dynamic management of urban environments. For example, in smart city initiatives like those in Singapore, drones and IoT sensors are used to manage traffic flow, monitor infrastructure health, and optimise waste collection routes, resulting in more efficient and responsive urban management.

Drones offer a multitude of benefits for urban planning, including efficient data collection, enhanced visualisation, proactive infrastructure management, environmental monitoring, and increased process efficiency. By leveraging these capabilities, urban planners can create more sustainable, livable, and resilient cities, addressing the challenges of rapid urbanisation and ensuring a better quality of life for urban residents. The continued advancement of drone technology and its integration with other smart city solutions will undoubtedly play a critical role in shaping the future of urban planning.

II

2. Navigating the Cons of Drone Technology

Normative Threats and Emerging Risks

Dawn breaks over the city of Amsterdam, a network of drones begins their daily routines, buzzing through the skyline with the precision of clockwork. Beneath their harmonious hum lies a web of complexities and risks that stretch from the crowded airspace above to the dense urban landscapes below. In this world, where the sky teems with unmanned flyers, the line between innovative utility and unforeseen hazard is razor-thin. The labyrinth of safety and security challenges exemplifies the risk of the very technology designed to elevate our capabilities could also press us to confront new vulnerabilities.

As drone technology rapidly becomes a staple in our daily lives, its ascent is accompanied by considerable challenges that necessitate careful regulation. The widespread use of drones introduces significant normative threats—risks that go beyond mere accidents and technical failures to raise profound concerns about safety, security, privacy, and environmental impact. These issues are particularly acute in Europe, where dense populations and busy airspaces magnify the potential for harm. The most pressing safety concerns include the risk of collisions between drones and manned aircraft, particularly near airports and aerodromes, as well as the dangers drones pose to individuals and property on the ground. Operating at low altitudes, drones not only risk crashing into residential areas but also bring up complex questions about privacy intrusion and the inadvertent collection of data, all of which must be navigated with a balance of innovation and caution.

The diversity and versatility of drones creates peculiar communication, navigation, and surveillance problems for the air traffic management (ATM) systems. The risk of cyber attacks on data generated by drones is becoming more prominent as drones are increasingly digitally connected with cameras and sensors onboard. Drones equipped with advanced technologies can read Internet Protocol addresses, infiltrate networks, track Radio Frequency Identified devices, collect unencrypted data, and even set up fake access points to access vulnerable devices.

Threat to privacy and data infringement are also significant risk associated with the use of drones. Most modern drones are now equipped with a high-resolution camera and visual sensors making it easy to infringe on the privacy of location, space, body and, association. In addition, The potential invisibility and undetectability of drones reduce the chances of knowledge when privacy is being violated or when one is being surveyed. This lack of knowledge has been considered to result in a chilling effect, which occurs when individuals perform a form of self-preservation or self-censorship by restricting their behaviour when they are, or believe that they are being watched.

Nowadays, drones, when combined with advanced technologies and applications, can generate an enormous amount of data in the form of image data, location data, sound data and, biometric data. Drones can, without consent, collect personal data, make video recordings in high resolution, as well as store and, if required, transfer such data to a ground station in real-time. Drones change the nature of surveillance, magnifying it when combined with other technologies like satellites, aircrafts, helicopters, and CCTVs.

The absence of human pilots and lightweight miniatured forms of drones allow flight operations into wildlife and natural environments that are otherwise difficult, dangerous, or inaccessible. Notwithstanding these benefits, drones have been shown to cause wildlife disturbance, a situation in which the behavioural and physiological response of animals to the presence of stimuli, such as a predator or anthropogenic causes such as humans and vehicles. Studies show that in the context of bird species, the adverse impact that may arise includes changes in distribution (e.g. short-term movement or displacement), behaviour (e.g. flight response or increased vigilance), demography (e.g. reduced survival), and changes to population size.

Safety and Security Threats

The skies are now getting more crowded. Critical infrastructures, such as airports, power grids and communication networks, face potential vulnerabilities from unauthorized drone access. Public safety is also at stake, as incidents of drones interfering with emergency services would likely become the norm. Addressing these issues is paramount for the seamless integration of drones into our daily lives, ensuring that their benefits are not overshadowed by the risks they introduce.

One of the most pressing safety threats posed by drones is the risk of collision with manned aircraft. This risk is especially acute near airports and aerodromes, where the presence of drones can interfere with the flight paths of commercial and private planes. For instance, In October 2017, a significant drone collision incident took place in Quebec City. A small commercial plane operated by SkyJet, a charter service, struck a drone while approaching Jean Lesage International Airport. The collision occurred at an altitude of 1,500 to 1,700 feet, involving a Beech King Air 100 aircraft and a yellow drone approximately 16 inches by 4 inches in size.

Beyond airports and high altitudes, drones flying at low altitudes pose risks to buildings and individuals on the ground. There has been numerous reports of drones crashing into buildings, vehicles, and even people, causing property damage and personal injuries. For example, in 2015, a drone crashed into the stands during the U.S. Open tennis tournament, narrowly missing spectators. Such incidents highlight the need for stringent operational guidelines and fail-safe mechanisms to prevent drones from causing harm in populated areas.

The diversity and versatility of drones introduce unique challenges for Air Traffic Management (ATM) systems. Unlike traditional aircraft, drones vary widely in size, flight capabilities, and operational patterns. This diversity complicates the task of monitoring and controlling air traffic, especially in urban environments where drones are increasingly used for delivery and surveillance. The integration of drones into ATM systems requires sophisticated digital technologies that can track multiple drones simultaneously and manage their flight paths to avoid collisions. The complexity of this task is further heightened by the need to coordinate with existing manned aircraft operations, necessitating robust communication and navigation systems.

Another significant safety and security threat associated with drones is the potential for cyber-attacks. As drones become more digitally connected, they are increasingly vulnerable to hacking and other forms of cyber intrusion. Cybercriminals can exploit weaknesses in drone systems to gain unauthorized access, take control of the drone, or intercept the data it collects. This threat is particularly concerning for drones equipped with cameras and sensors, which can capture sensitive information. For instance, drones used in critical infrastructure inspections or surveillance operations can be targeted to gather intelligence or disrupt services. The potential for cyber-attacks necessitates robust cybersecurity measures, including encryption, secure communication protocols, and regular software updates to protect drone systems from malicious activities.

The use of drones for malicious purposes is another growing concern. Drones can be used to carry out espionage, smuggle contraband, or even launch attacks. For instance, in 2019, drones were used to drop explosives on an oil facility in Saudi Arabia, causing significant damage and highlighting the potential for drones to be used as weapons. Such incidents have raised alarm among security agencies, prompting the development of counter-drone technologies and strategies to detect and neutralise rogue drones. These measures include radar systems, radio frequency jammers, and anti-drone lasers designed to disable or destroy unauthorized drones.

In addition to intentional misuse, the inadvertent misuse of drones can also lead to significant safety and security risks. Hobbyist drone operators, often unaware of the regulations governing drone flights, may inadvertently fly their drones into restricted airspace or engage in unsafe flying practices. This lack of awareness can lead to dangerous situations, such as drones interfering with firefighting operations or rescue missions. To address this issue, there is a need for comprehensive education and training programs for hobbyist drone content creators, ensuring they understand the rules and responsibilities associated with drone use.

The proliferation of drones has also led to concerns about surveillance. Drones equipped with high-resolution cameras and other sensors can capture detailed images and videos, raising concerns about unwarranted intrusion into private spaces. This capability makes drones powerful surveillance tools, exacerbating fears of a surveillance society where individuals are constantly monitored. The potential for drones to be used for mass surveillance necessitates stringent privacy regulations and oversight mechanisms to protect individuals' rights and prevent abuse.

While drones offer numerous benefits, their increasing utility raises significant safety and security threats that need to be addressed through adaptive regulatory frameworks and technological solutions. Ensuring the safe integration of drones into the airspace requires robust measures to prevent collisions, protect against cyber-attacks, and mitigate the risks associated with malicious and inadvertent misuse. As drone technology continues to evolve, ongoing assessment and adaptation of safety and security measures will be essential to manage emerging risks and ensure the responsible use of drones.

Privacy and Data Protection Threats

The whir of drones in the sky signals a new era of technological advancement, yet it also stirs concerns over privacy and data protection. Picture a scene where drones, equipped with high-resolution cameras and sophisticated sensors, glide silently above, capturing more than just picturesque landscapes. They collect data — sometimes intimate, sometimes sensitive — raising alarms about the potential for privacy invasion and data misuse. As these aerial devices become integral to various sectors, the challenge lies in balancing their benefits with the imperative to safeguard individual privacy and data security. How do we harness the potential of drone technology without compromising our privacy? The answer lies in a meticulous approach to understanding and mitigating these emerging risks, ensuring that the sky remains not just the domain of innovation but also a safe space for all.

One of the most significant privacy threats posed by drones is their ability to conduct extensive surveillance. Modern drones are often equipped with high-resolution cameras and various sensors that can capture detailed images and videos from a distance. This capability allows drones to monitor private properties and individuals without their knowledge or consent. For instance, a drone hovering over a residential area can capture footage of backyards, windows, and other private spaces, leading to a significant invasion of privacy. This surveillance capability is not limited to visible light; drones can also be equipped with thermal imaging cameras, night vision, and other advanced sensors that can capture data in various conditions, further enhancing their surveillance capabilities.

The undetectability and invisibility of drones exacerbate privacy concerns. Drones can operate at high altitudes or from concealed positions, making it difficult for individuals to detect when they are being monitored. This lack of awareness creates a chilling effect, where people alter their behaviour out of fear of being watched, even if they are not aware of any specific surveillance activity. This chilling effect can lead to self-censorship and a reduction in the freedom to express oneself openly, significantly impacting personal liberties and freedoms.

Data protection is another critical issue associated with drone technology. Drones can collect vast amounts of data, including images, videos, location data, and even biometric data such as facial recognition. This data is often transmitted in real-time to ground stations or stored on the drone for later retrieval. The collection, storage, and transmission of such sensitive information raise significant concerns about data security. Unauthorized access to this data can lead to privacy violations, identity theft, and other forms of misuse. For example, if a drone collecting data for urban planning inadvertently captures images of individuals, those images could be accessed and misused by unauthorized parties if not properly secured.

The risk of data breaches is heightened by the increasing connectivity of drones. Many drones are connected to other drones, and to the internet, and other digital networks, making them vulnerable to hacking and cyber-attacks. Cybercriminals can exploit vulnerabilities in drone systems to gain access to the collected data, take control of the drone, or disrupt its operations. In a notable incident in 2013, security researcher Samy Kamkar demonstrated the vulnerability of drones to cyber-attacks by hijacking a Parrot AR Drone using a custom-built device called "SkyJack." Kamkar's device exploited weaknesses in the drone's communication protocol, allowing him to take control of any drone within its range. By sending malicious commands to the drone's Wi-Fi network, Kamkar could intercept and override the drone's communication with its original controller, effectively hijacking it.

This demonstration underscored the significant risks associated with drone technology, particularly regarding cyber-attacks and data security. It highlighted how easily drones could be compromised, posing threats to both their operational integrity and the data they collect. Such vulnerabilities have implications not just for individual drone operators but also for industries and services that increasingly rely on drones for critical operations. Kamkar's experiment brought to light the urgent need for robust security measures in drone technology to prevent unauthorized access and ensure safe, reliable drone operations. The potential for drones to be used for mass surveillance is another significant concern. Government agencies and private entities can deploy fleets of drones to monitor large areas, collecting extensive data on the movements and activities of individuals. This capability raises concerns about the erosion of privacy rights and the potential for abuse by authorities. For instance, law enforcement agencies in some countries have used drones for crowd control and monitoring during public protests, raising concerns about the impact on civil liberties and the right to peaceful assembly.

Regulatory frameworks and legal protections are essential to address these privacy, data protection and human rights threats. Many countries have implemented regulations that govern the use of drones, particularly concerning data collection and privacy. For example, the European Union's General Data Protection Regulation (GDPR) imposes strict requirements on the processing of personal data. These regulations require that data be collected lawfully, transparently, and for specific purposes, and that individuals' privacy rights be respected.

In addition to regulatory measures, technological solutions can help mitigate privacy and data protection risks. Privacy-enhancing technologies, such as anonymization and encryption, can protect the data collected by drones from unauthorized access and misuse. For instance, drones used for urban planning can be equipped with software that automatically blurs faces and other identifiable features in the captured images, ensuring that personal data is not inadvertently collected. Encryption can protect data during transmission and storage, preventing unauthorized parties from accessing sensitive information.

Education and awareness are also crucial in addressing privacy and data protection threats. Drone operators must be educated about the legal and ethical implications of data collection and the importance of respecting individuals' privacy rights. Public awareness campaigns can inform individuals about their rights and the measures they can take to protect their privacy in the face of increasing drone surveillance.

While drones offer numerous benefits, their increasing use raises significant privacy and data protection threats. Addressing these threats requires sophisticated approaches that includes adaptive regulatory frameworks, technological solutions, and accessible education and awareness initiatives. By taking these measures, we can ensure that the promise of drone technology is realised without compromising individual privacy and data security. As drone technology continues to evolve, ongoing sandbox assessment and adaptation of privacy and data protection measures, and dynamic regulatory approaches will be essential to manage emerging risks and safeguard personal freedoms.

Environmental Sustainability Threats

Drones, in their silent flight over serene landscapes, often become uninvited guests in the habitats of wildlife, causing subtle yet significant disturbances. The disruption of natural environments and the intrusion into the tranquil realms of bird species introduce an unsettling dichotomy: the very technology designed to advance our lives may inadvertently unsettle the delicate ecological balance we strive to protect. Moreover, the incessant buzz of drones contributes to a growing concern about noise pollution, a modern malaise that stretches its tendrils into both urban and rural sanctuaries, affecting not just wildlife, but human tranquility as well.

One of the most pressing environmental concerns is the disturbance drones cause to wildlife. Drones can provoke behavioural and physiological stress responses in animals, especially bird species. Birds, highly sensitive to disturbances, may exhibit altered behaviours such as increased vigilance, disrupted foraging patterns, and even flight initiation, where they leave their nests or feeding grounds prematurely. These disturbances can lead to broader ecological consequences, including changes in distribution, where birds may abandon their traditional habitats, and impacts on demography, potentially reducing survival rates and affecting population sizes. Studies have documented instances where birds exhibit significant stress responses to drone presence, leading to long-term impacts on their health and reproductive success.

For example, a study conducted on waterbirds demonstrated that drone flights caused noticeable distress, resulting in increased energy expenditure and reduced feeding efficiency. Such disruptions can be particularly detrimental during breeding seasons when energy demands are highest. Similarly, raptor species, which are top predators and crucial for ecosystem balance, may abandon their nests if disturbed by drones, leading to decreased breeding success and potential long-term population declines.

Drones also pose a threat to natural environments by facilitating access to otherwise inaccessible areas. While this capability is beneficial for environmental monitoring and research, it also opens up these sensitive areas to potential ecological disturbances. The presence of drones can lead to habitat degradation, particularly in fragile ecosystems such as wetlands, forests, and marine environments. For instance, drones flying over wetland areas can disturb the delicate balance of these ecosystems, affecting the flora and fauna that depend on them.

The ability of drones to penetrate dense forests or remote wildlife reserves can inadvertently introduce human presence and activity into these environments, causing disruptions that can alter natural behaviours and ecological interactions. This disruption is particularly concerning in protected areas where the goal is to minimise human impact to preserve biodiversity. The intrusion of drones can lead to unintended consequences, such as the introduction of invasive species or pollutants, further stressing already vulnerable ecosystems.

Another significant environmental threat posed by drones is noise pollution. Drones, especially larger models with powerful rotors, generate substantial noise, which can be disruptive to both wildlife and human communities. Noise pollution from drones can interfere with animal communication, predator-prey interactions, and mating behaviours. For example, many bird species rely on vocalisations to communicate with each other, establish territories, and attract mates. The noise generated by drones can mask these sounds, leading to communication breakdowns and reduced reproductive success.

In marine environments, noise pollution from drones can affect aquatic species, particularly those that rely on sound for navigation, foraging, and communication. Marine mammals, such as dolphins and whales, are especially vulnerable to noise pollution. The introduction of drone noise into their habitats can disrupt their echolocation abilities, leading to disorientation, stress, and even standings. Human communities are also impacted by noise pollution from drones. In urban areas, the constant buzzing of drones can contribute to noise levels, affecting the quality of life for residents. This is particularly problematic in areas with strict noise regulations or in communities where peace and quiet are valued. Prolonged exposure to drone noise can lead to increased stress, sleep disturbances, and other health issues among residents.

The environmental sustainability threats posed by drones are multifaceted and complex. Addressing these threats requires a sophisticated approach that includes adaptive regulations, technological innovations, and public awareness. Regulatory frameworks need to be developed and enforced to limit drone operations in sensitive areas, protect wildlife from disturbances, and mitigate noise pollution. Technological advancements, such as quieter drone models and automated flight paths that avoid sensitive habitats, can help reduce the environmental impact of drones. Public awareness campaigns can educate drone operators about the potential environmental impacts of their activities and promote responsible drone use.

Although, drones offer significant benefits for various applications, their environmental sustainability threats cannot be overlooked. Wildlife disturbance, natural environment disruption, and noise pollution are critical issues that need to be addressed to ensure that the use of drones does not come at the expense of ecological balance and biodiversity. By implementing comprehensive strategies to mitigate these threats, we can harness the potential of drone technology while safeguarding the environment for future generations.

Pivotal Incidents

Canada: Quebec City Drone Incident

In October 2017, a landmark event underscored the latent dangers posed by drones operating within commercial airspace. This incident marked the first confirmed collision between a commercial aircraft and a drone in North America, drawing significant attention to the potential hazards and the necessity for stringent regulatory measures.

The incident occurred on October 12, 2017, near Jean Lesage International Airport in Quebec City. A Beech King Air 100 aircraft operated by SkyJet was on its final approach when it collided with a drone at an altitude of approximately 1,500 to 1,700 feet. The drone, identified as a yellow quadcopter about 16 inches by 4 inches in size, struck the aircraft's wing, causing damage but fortunately not resulting in any injuries to the six passengers and two crew members onboard.

As the aircraft descended towards the runway, the pilots noticed a small object rapidly approaching. Before they could react, the drone struck the wing of the plane. Despite the impact, the pilots managed to land the aircraft safely, avoiding what could have been a catastrophic event. The incident immediately triggered an investigation by Transport Canada, the regulatory authority responsible for civil aviation safety.

Transport Canada's investigation revealed that the drone was operating well above the permitted altitude for recreational drones and within the restricted airspace around the airport. Canadian Aviation Regulations explicitly prohibit drone flights within 5.6 kilometres of an airport without special authorisation, and drones are generally not allowed to exceed an altitude of 90 meters (approximately 300 feet). The collision occurred at an altitude more than five times the legal limit for drone operations, highlighting a severe breach of these regulations.

In response to the incident, Canada's Minister of Transport, reiterated the critical need for drone operators to adhere to existing regulations. He emphasised that incidents involving drones and manned aircraft could have devastating consequences, and regulatory compliance was essential for maintaining aviation safety.

The Quebec City drone incident had significant implications for both regulatory frameworks and public awareness regarding drone operations. It highlighted the urgent need for robust enforcement of existing regulations and potentially stricter measures to prevent similar occurrences in the future. In the wake of the incident, Transport Canada intensified its efforts to educate drone operators about the rules governing drone usage, including information on legal flying altitudes, restricted zones, and the severe penalties for non-compliance. The regulatory body considered introducing harsher penalties for violations, including substantial fines and potential jail time for egregious breaches that endanger public safety.

The incident accelerated discussions about implementing technological solutions such as geo-fencing, which would automatically prevent drones from entering restricted airspace. Additionally, mandatory identification systems for drones were considered to aid in tracking and accountability. The aviation and drone industries, as well as the general public, responded to the incident with a mix of concern and calls for action. Airline pilots and industry experts underscored the potential for severe accidents, advocating for immediate and effective regulatory measures. Drone enthusiasts and professional operators also expressed their concerns, recognising that irresponsible drone usage could lead to tighter restrictions that might stifle innovation and responsible drone use.

The Quebec City drone incident served as a pivotal moment in the realm of aviation safety, demonstrating the real and present dangers that drones pose when operated irresponsibly. It underscored the necessity for stringent regulations, effective enforcement, and continuous public education to ensure the safe integration of drones into shared airspace. This event has become a case study in aviation safety courses and regulatory discussions, exemplifying the challenges and responsibilities that come with the rapid adoption of drone technology.

France: Paris Drone Incident

In January 2015, Paris witnessed a significant drone incident that underscored the potential disruptions and security risks posed by unauthorized drone activities in urban areas. This event involved multiple drone sightings over key landmarks, including the Eiffel Tower and the U.S. Embassy, which led to heightened security concerns and operational challenges for law enforcement.

Over the course of several nights, unidentified drones were observed flying over various sensitive sites in Paris. These sightings occurred in rapid succession and involved drones operating in restricted airspace without authorisation. The French authorities were particularly alarmed by the potential implications of these unauthorized flights, especially given the heightened security context following the Charlie Hebdo attacks earlier that month.

The drones flew over some of the most iconic and heavily guarded locations in Paris, including the Eiffel Tower, the U.S. Embassy, and the Place de la Concorde. The sightings primarily took place during nighttime, making it difficult for authorities to track and intercept the devices. Despite deploying police helicopters and ground units, the operators of the drones remained elusive, leading to a prolonged state of heightened alert in the city.

The Paris drone incident had several far-reaching impacts, highlighting the multifaceted challenges associated with unauthorized drone operations. The presence of drones over critical sites raised immediate concerns about espionage, terrorism, and the vulnerability of important landmarks to aerial threats. The inability to identify and apprehend the drone operators exacerbated these fears, prompting a reassessment of existing security protocols.

The incident necessitated a significant deployment of police and military resources to monitor and attempt to neutralise the drones. This diversion of resources strained the capacity of law enforcement to address other security needs and underscored the operational challenges posed by rogue drones. The visibility of drones over well-known landmarks created a palpable sense of unease among residents and visitors, contributing to public anxiety and a sense of insecurity, particularly given its timing shortly after the Charlie Hebdo attacks.

In response to the Paris drone incident, French authorities implemented several measures to enhance drone regulation and improve security. These included stricter enforcement of existing no-fly zones, increased penalties for unauthorized drone operations, and the deployment of advanced drone detection and mitigation technologies. The incident also spurred broader discussions on the need for comprehensive drone regulations at the national and EU levels.

From a regulatory perspective, the Paris incident highlighted several key lessons for the development of effective drone policies. There is a critical need for advanced technologies capable of detecting and mitigating unauthorized drones in real-time. This includes the deployment of radar systems, radio frequency jammers, and other counter-drone measures to safeguard sensitive sites. Establishing clear and enforceable legal frameworks is essential to deter unauthorized drone activities. This involves defining no-fly zones, mandating drone registration, and imposing stringent penalties for violations.

Raising public awareness about the legal and safety implications of drone operations is crucial. Educational campaigns can help inform drone users about regulatory requirements and encourage responsible behaviour. Effective drone regulation requires collaboration between national authorities, international bodies, and industry stakeholders. Sharing best practices and developing standardised approaches can enhance the overall effectiveness of regulatory frameworks.

The Paris drone incident serves as a stark reminder of the potential risks and challenges associated with unauthorized drone operations in urban areas. By examining this case, policymakers and industry stakeholders can draw valuable insights to inform the development of robust and adaptive drone regulations that ensure public safety while fostering innovation in the drone industry.

Germany: Frankfurt Peregrine Falcon Case

In In June 2020, during the peregrine flacon breeding season, a drone pilot flew a commercial drone to the height of 300 meters. The drone reached the roof of the Commerzbank Tower in Frankfurt and killed a peregrine falcon. Investigations by the police led to a 30-year old resident of Berlin. Criminal proceedings were brought against the defendant before the Frankfurt District Court.

The public prosecutor argued that it is common knowledge that peregrine falcons nest on the roof of high-rise buildings in Frankfurt and prayed for the court to order a penalty and a conditional resolution. The defence attorney argued that because the defendant is a Berliner, he could not have known that peregrine falcons nest on the roof of the

Commerzbank Tower in Frankfurt.

The District Court held that because the drone pilot did not yet have a criminal record, the killing of the peregrine falcon would be considered a 'minor violation' of the Reichstierschutzgesetz (Reich Animal Protection Act of 1933). The judge, the public prosecutor and the defence attorney agreed to discontinue the proceedings in exchange for a monetary payment of €1.500 to the Naturschutzbund Deutschland conservation.

The peregrine falcon case highlights the complexities surrounding the use of commercial drones and the need for effective regulation to protect wildlife and prevent harm. The case illustrates the potential dangers of drones to wildlife and the need for drone pilots to be aware of the risks and take necessary precautions to avoid harm. One of the issues raised by this case is the lack of awareness and education of drone pilots. The defence attorney argued that the defendant, being a Berliner, could not have known that peregrine falcons nest on the roof of the Commerzbank Tower in Frankfurt. This argument raises questions about the extent to which drone pilots are educated and informed about the regulations and the potential risks of flying drones in populated areas. Drone pilots need to be aware of the locations of wildlife habitats and sensitive areas where flying drones can cause harm. Education and awareness-raising campaigns are necessary to ensure that drone pilots are well-informed and equipped with the knowledge to fly their drones safely and responsibly.

Another issue highlighted by this case is the need for effective regulation of commercial drones. The judge, the public prosecutor and the defence attorney agreed to discontinue the proceedings in exchange for a monetary payment of €1.500 to the Naturschutzbund Deutschland conservation. This penalty highlights the inadequacy of the current regulations in addressing the harm caused by drones to wildlife. The Reichstierschutzgesetz (Reich Animal Protection Act of 1933) is outdated and does not specifically address the use of drones. This highlights the need for modern and effective regulations that take into account the potential risks and harm caused by drones to wildlife and the environment.

Moreover, the case raises questions about the liability of drone manufacturers and operators for harm caused by drones. In this case, the drone pilot was held liable for the killing of the peregrine falcon. However, the manufacturer of the drone, as well as the operator of the Commerzbank Tower, could also be held liable for harm caused by the drone. There is a need for clear regulations on the liability of drone manufacturers and operators for harm caused by drones. Manufacturers should be required to ensure that their drones are safe and do not pose a risk to wildlife and the environment.

The peregrine falcon case also highlights the need for effective enforcement of regulations. Despite the criminal proceedings brought against the defendant, the penalty was only a monetary payment to a conservation organization. This penalty is unlikely to deter other drone pilots from flying drones in areas where they can cause harm to wildlife. There is a need for effective enforcement of regulations and penalties that are sufficient to deter drone pilots from engaging in reckless and harmful behaviour.

The peregrine falcon case illustrates the complexity of commercial drones and the need for effective regulation to protect wildlife and prevent harm. The case highlights the lack of awareness and education of drone pilots, the need for modern and effective regulations, the liability of drone manufacturers and operators, and the need for effective enforcement of regulations. The case serves as a wake-up call for policymakers and stakeholders to take action to ensure that the use of drones is safe and responsible, and does not pose a risk to wildlife and the environment.

United Arab Emirates: Dubai Airport Drone Incident

On June 11, 2016, Dubai International Airport, one of the busiest aviation hubs in the world, experienced a significant disruption due to unauthorized drone activity. The incident unfolded late in the morning, prompting immediate safety protocols and resulting in substantial operational and economic impacts.

Around 11:36 am local time, an unauthorized drone was detected within the airspace of Dubai International Airport, which led to the immediate closure of the airport's airspace. This preventive measure was essential to ensure the safety of incoming and outgoing flights, given the significant risks drones pose to aircraft. The closure lasted until 12:45 pm, during which time all flight operations were halted.

The implications of this disruption were far-reaching. Incoming flights destined for Dubai were rerouted to nearby airports, including Al-Maktoum International Airport in the Jebel Ali area, Sharjah, and Al-Ain. Outbound flights were delayed, creating a cascade of logistical challenges. The economic impact of this disruption was immense, with estimates suggesting a loss of AED 3.7 million ($1 million) per minute for the emirate's economy, as more than a dozen flights were affected.

In the aftermath of the incident, it was revealed that the drone had encroached upon a drone no-fly zone, which extends 5 kilometres around critical infrastructures such as airports. This incursion highlighted the need for stricter enforcement of existing regulations and the implementation of advanced monitoring technologies to prevent similar incidents in the future. The General Civil Aviation Authority (GCAA) responded by reinforcing the prohibition of flying drones within 5 kilometres of airports and other sensitive areas. This regulation included not only Dubai International Airport but also Al-Maktoum International Airport, the Al-Minhad air base, and Palm Jumeirah around Skydive Dubai. The incident underscored the importance of these measures and the need for public awareness regarding drone regulations to prevent such costly and dangerous disruptions.

In response to this incident, Dubai has since invested in enhanced drone detection and mitigation systems, including radar and radio frequency jammers, to protect its airspace. These technologies are designed to detect unauthorized drones early and neutralise them before they can pose a threat to aircraft.

The June 11, 2016, drone incident at Dubai International Airport serves as a critical case study for the aviation industry worldwide. It highlights the severe risks drones can pose to airspace security and the importance of robust regulatory frameworks and technological solutions to mitigate these risks. This incident has paved the way for more stringent drone regulations and increased collaboration between aviation authorities and drone operators to ensure the safe integration of drones into the airspace.

United Kingdom: Gatwick Airport Drone Incident

In December 2018, Gatwick Airport, one of the busiest in the United Kingdom, faced an unprecedented disruption caused by unauthorised drone activity. Over three days, from December 19 to December 21, multiple drone sightings near the airport's runways led to significant operational chaos, grounding hundreds of flights and affecting thousands of passengers. This case study delves into the Gatwick Airport drone incident, highlighting the severe disruptions that drones can inflict on critical infrastructure.

The initial sighting occurred on the evening of December 19, prompting an immediate suspension of all flights. Despite attempts to reopen the runways, further drone sightings continued to disrupt operations, necessitating repeated closures. Authorities responded by deploying substantial resources, including police units, helicopters, and even military assistance equipped with counter-drone technology. Despite these efforts, the operators of the drones remained elusive, and the airport struggled to maintain normal operations amid the ongoing threat.

The Gatwick incident had extensive consequences, underscoring several key areas of disruption caused by unauthorised drone activity. The operational disruption was immense, with the closure of Gatwick's runways leading to the cancellation of over 1,000 flights and affecting approximately 140,000 passengers. The operational chaos extended beyond Gatwick, impacting air traffic across Europe and causing a ripple effect of delays and cancellations. Economically, the disruption was substantial. Airlines incurred significant losses due to cancelled flights, passenger compensation, and rerouting expenses. Gatwick Airport faced operational costs related to heightened security measures and crisis management, while the broader economic impact included lost business for the surrounding areas dependent on the airport's operations.

Security concerns were starkly highlighted by the incident. It exposed vulnerabilities in airport security regarding drone threats. The potential for drones to carry out surveillance, interfere with aircraft operations, or even carry dangerous payloads underscored the urgent need for robust counter-drone measures. The inability to quickly neutralise the threat highlighted significant gaps in existing security protocols. Public safety and perception were also profoundly affected. Although no physical harm occurred, the possibility of a drone colliding with an aircraft during takeoff or landing presented a serious safety risk. The incident heightened public awareness about the potential dangers of unregulated drone use, affecting the perception of drone technology. The regulatory and technological responses to the Gatwick incident were immediate and far-reaching. In the short term, Gatwick Airport implemented enhanced security measures, including the deployment of military counter-drone technology to detect and neutralise drones within the airport's vicinity. In response to the incident, the UK government introduced stricter drone regulations, extending the no-fly zone around airports from 1 km to 5 km and mandating registration and competency tests for drone operators. These measures aimed to enhance accountability and deter unauthorised drone activities near sensitive areas.

Airports across the UK and globally began investing in advanced drone detection and mitigation systems. Technologies such as radar, radio frequency jammers, and anti-drone lasers were explored to enhance the capability to detect and respond to drone threats. These systems are designed to provide real-time alerts and enable swift action to mitigate risks. Public awareness campaigns were also launched to educate drone users about safe practices and regulatory requirements, aiming to reduce the risk of accidental violations and promote responsible drone use.

The incident underscored the importance of collaboration between regulatory bodies, law enforcement, and industry stakeholders. Coordinated efforts were necessary to develop effective strategies for monitoring, detecting, and responding to unauthorised drone activities. This collaboration also involved sharing best practices and technological innovations.

The Gatwick Airport drone incident serves as a stark reminder of the potential disruptions that drones can cause to critical infrastructure. The extensive operational, economic, and security impacts of the incident highlight the urgent need for robust regulatory frameworks and technological solutions to manage the risks associated with drone operations. The incident prompted significant regulatory and technological advancements, including stricter legislation, enhanced security measures, and increased public awareness. By understanding the lessons learned from Gatwick, regulators and industry stakeholders can better prepare for and mitigate similar threats in the future, ensuring the safe and responsible integration of drones into the airspace.

III

3. Drone Value chains and Key stakeholders

The global commercial drone value chain is a complex ecosystem involving multiple stakeholders. It starts with OEM manufacturers who design and produce drones, followed by OEM suppliers providing essential components. Software developers and data analysts enhance the functionality and data processing capabilities of these drones. Pilots and operators manage flight operations, while service providers offer various operational support services. Maintenance centres ensure the drones remain in optimal condition, and training and certification centres equip personnel with necessary skills and credentials. Testing centres validate drone performance, while research, development, and innovation centres drive technological advancements. End-users across various sectors utilise these drones for diverse applications. Investors provide the necessary capital to fuel research, development, and expansion efforts within the commercial drone industry, thereby accelerating technological advancements and market growth.

The drone landscape is highly competitive and features global vertically integrated hardware companies such as DJI, Parrot, Amazon, Skydio, AeroVironment, SenseFly, Yuneec, Autel Robotics and many others. These companies have introduced new models of drones with enhanced features, such as longer battery life, improved camera systems, powerful sensors, AI-integration, and increased stability and agility.

Operators are critical actors in the commercial drone value chain. These are individuals or companies that use drones for various purposes, including delivery, surveying, mapping, and inspection. Operators must obtain the necessary certifications and authorisations from relevant regulatory authorities to fly drones in specific areas. Drone operators must comply with the regulatory framework in the relevant jurisdiction. For example in the EU operators must comply with the rules outlined in the EU Common Rules in the Field of Civil Aviation, Unmanned Aircraft Systems Regulation, and the Rules for the Operation of Unmanned Aircraft including specific rules outlined in member states.

As drone hardware becomes increasingly affordable, the future growth of the industry will pivot away from hardware manufacturing towards service provision. Companies that operate and manage drones, handle drone data, and perform maintenance will generate the most value. End-user companies are likely to outsource these functions to third parties. Telecommunications companies, for example, might sell drone data communication services to guide drones and relay collected data. This shift highlights the crucial role of service providers within the commercial drone value chain, as they offer critical drone services in real time.

Maintenance centres play a critical role in the commercial drone value chain by ensuring that drones remain in optimal condition, conducting routine inspections, repairs, and upgrades to prevent any operational failures. Training and certification centres are equally vital, as they equip personnel with the necessary skills and credentials to safely and effectively operate drones, adhering to stringent regulatory standards. Testing centres are responsible for validating the performance and safety of drones, conducting rigorous assessments under various conditions to ensure reliability and compliance. Meanwhile, research, development, and innovation centres drive technological advancements in drone technology. They focus on developing new capabilities, enhancing existing functionalities, and pushing the boundaries of what drones can achieve in various applications. These centres often collaborate with academic institutions, industry partners, and government agencies to foster innovation and bring cutting-edge drone technologies to market. Collectively, these entities ensure the growth and sustainability of the commercial drone industry, supporting its expansion into new sectors and applications.

End-users in the commercial drone value chain use drones for specific purposes, such as delivery, agriculture, construction, inspection, mapping amongst others. End-users can purchase or rent drones from manufacturers or operators and use them for various applications. The adoption of drones by end-users has been significant in the EU, particularly in the agriculture and construction industries.

For example, the agricultural sector has witnessed significant growth in the adoption of drones in recent years. Drones have been used for crop monitoring, soil analysis, and spraying pesticides and fertilisers. The use of drones in agriculture has improved productivity, reduced costs, and minimised the environmental impact of conventional farming practices. In the construction industry, drones have been used for surveying, mapping, and inspection of buildings and structures. Drones have also been used for search and rescue operations and environmental monitoring.

In the EU, The commercial drone value chain is dynamic and constantly evolving. The growth of the industry has led to the emergence of new business models and services, such as 3D Mapping for construction and urban planning, and inspection of power infrastructure. The market for drone services is expected to grow significantly in the coming years, with several industries adopting drones for various purposes and new services becoming commercially viable. The adoption of drones is being facilitated by the regulatory framework that governs drone operations, which provides guidelines and standards for safe and efficient drone use.

The global commercial drone value chain is a dynamic and rapidly evolving ecosystem comprising various actors, including manufacturers, operators, service providers, investors, and end-users. The growth of the industry will be driven by the emergence of new business models and services, with several industries adopting drones for various purposes. The regulatory framework governing drone operations will continue to evolve in response to the innovative business models to provide guidelines and standards for safe and efficient drone use, to facilitate the smart adoption of drones into modern cities and industries.

Drones and the EU Internal Market

Commercial drone application is projected to become a significant source of value for the EU internal market. Substantial value is currently being created in hardware, software and services value chains. The hardware chain is reliant on specialised components and system partners spread across global value chains. Supply and services agreements are a central building block in the EU internal market.

The software and services chain require specialised digital technologies to support a wide range of commercial drone applications. For instance, drone delivery is emerging as a niche sector in the service chain due to the growth of multinational e-commerce companies. Recent test cases show that drone delivery has the potential to reduce greenhouse gas emissions and energy use in the commercial package delivery sector. In addition to delivery drones, there is a growing demand for commercial drone applications in the field of agriculture, infrastructure inspection, and emergency services. In agriculture, drones provide precise monitoring of crops, allowing for targeted spraying and early detection of plant diseases. Infrastructure inspection is another area where drones are increasingly being used, particularly in the inspection of power lines, bridges, and wind turbines. In emergency services, drones have been used to search for missing persons, monitor forest fires, and assist in disaster relief efforts.

The use of drones in these various industries has led to the development of specialised software and services to support their operations. For example, software that allows for the creation of 3D maps and models of crop fields or infrastructure can provide valuable insights to farmers and engineers. Additionally, specialised software can assist in the planning and coordination of emergency response efforts, allowing for more efficient and effective use of resources.

The services value chain for commercial drones also includes a variety of activities such as training, repair and maintenance, and consulting services. Training services are essential for the safe and effective operation of commercial drones, particularly for pilots who are new to the technology. Repair and maintenance services are also important, as drones are subject to wear and tear and can be damaged during operation. Consulting services are also in demand, particularly for companies that are new to the use of drones and require guidance on regulatory compliance and best practices.

One of the key technology challenges in the commercial drone value chain in the EU is the development of a reliable and secure communications infrastructure to support drone operations. Drones require high-speed data transmission and low-latency connectivity to operate effectively, particularly in applications that require real-time data collection and processing. Additionally, the use of drones raises concerns about privacy and security, particularly in cases where drones are used to collect data or monitor individuals.

To address these challenges, the European Union has established regulations for the operation of commercial drones, including requirements for certification and licensing of operators, as well as restrictions on where and when drones can be flown. The EU has also established guidelines for the safe and secure operation of drones, including requirements for the use of geofencing technology to prevent drones from flying into restricted areas.

The commercial drone value chain in the EU is evolving rapidly, driven by advancements in technology and increasing demand for drone-based services. While the hardware value chain remains a critical component of the commercial drone ecosystem, the software and services chain is also essential for the safe and effective operation of drones. As the technology continues to evolve, it is likely that new applications and services will emerge, creating new opportunities for businesses and entrepreneurs in this rapidly growing industry.

The potential to reduce greenhouse gas emissions brings drones closer to the the toolbox for addressing climate change. The European Climate Law establishes the framework for the irreversible and gradual reduction of greenhouse gas emissions by 2050. The climate law creates legal effect for the growth strategy set out in the European Green Deal, which lays down Union goals for greenhouse gas emission reduction and recognises the sustainability benefits of digitalisation.

The potential for drones to reduce greenhouse gas emissions aligns with the EU's climate goals and highlights the importance of sustainable technologies in achieving these goals. The European Green Deal sets a target of reducing greenhouse gas emissions by at least 55% by 2030, compared to 1990 levels. To achieve this, the EU has adopted a multi-dimensional approach that includes, among others, decarbonisation of energy, transport, and industrial sectors. The use of drones in commercial operations presents a significant opportunity for reducing carbon emissions in the transport sector.

Furthermore, the EU has recognised the sustainability benefits of digitalisation and is promoting the integration of digital technologies to achieve sustainable development goals. The European Commission's Digital Single Market strategy seeks to ensure that Europe embraces digitalisation in a way that enhances sustainability and supports climate objectives. In this context, the use of drones in commercial applications can be seen as a sustainable digital solution that can help achieve these objectives.

Drone delivery has the potential to transform the package delivery industry by reducing carbon emissions and energy use associated with traditional delivery methods. A recent study by the World Economic Forum found that using drones for last-mile deliveries could reduce carbon emissions by up to 22% per delivery, compared to traditional delivery methods. This reduction in emissions is primarily due to the reduced need for delivery vehicles, which are typically responsible for the majority of carbon emissions in the package delivery sector. In addition, drone delivery can also reduce traffic congestion and improve delivery times, which can further reduce emissions and energy use.

However, as earlier discussed, the widespread adoption of drone delivery in the EU also poses some challenges that need to be addressed. One of the main concerns is the potential impact on privacy and data protection. As drones collect data during their flights, there is a risk that sensitive information could be captured and used for nefarious purposes. To mitigate this risk, the EU has implemented strict data protection laws, such as the General Data Protection Regulation (GDPR), which requires companies to obtain explicit consent from individuals before collecting their personal data.

Another concern is the potential impact on traditional delivery jobs. As drone delivery becomes more widespread, there is a risk that traditional delivery jobs could be replaced by autonomous drones. To address this, the EU has emphasised the importance of re-skilling and up-skilling the workforce to adapt to the changing nature of work.

In addition, the use of drones in commercial operations also poses a risk to public safety. Drones flying in populated areas could collide with buildings or people, causing injury or damage. To mitigate this risk, the EU has implemented strict regulations for the operation of drones, including mandatory training and registration requirements for operators, as well as restrictions on where and when drones can be flown.

The EU has also established a harmonised regulatory framework for drones, which aims to ensure a high level of safety and security for drone operations across the EU. The framework also includes provisions for the use of geo-fencing technology, which restricts drones from flying in designated no-fly zones, such as airports and military installations.

In this context, drones are placed within the transport system as a whole and are considered to be part of smart and sustainable transport infrastructure including electrical and hydrogen-powered aircraft. The Communication on the European Green Deal announced, therefore, a strategy for sustainable and smart mobility, which was adopted by the Commission in December 2020. In the Sustainable and Smart Mobility Strategy, EC has set the objective to adopt a Drone Strategy 2.0 in 2022 to reap the full potential offered by drones to contribute to the safeguarding of a well-functioning single market. Drone Strategy 2.0 aims to enable drones to contribute, through digitalisation and automation, to a new offer of sustainable services and transport, while accounting for technological synergies.

The Sustainable and Smart Mobility Strategy aims to ensure that the transport system is more efficient and environmentally friendly while remaining competitive in the global market. The strategy highlights the need to explore innovative technologies such as electric and hydrogen-powered aircraft, which could potentially replace fossil fuel-based aircraft. Drones are an integral part of this strategy due to their potential to contribute to sustainable services and transport.

The Drone Strategy 2.0 builds on the achievements of the initial Drone Strategy, which was launched in 2018, and aims to establish a regulatory framework that enables the development and use of drones in the EU. The strategy focuses on four key areas: safety, privacy and data protection, security, and the environment. The objective is to ensure that drones can be used safely, securely and in a manner that respects the privacy of individuals and the protection of personal data, while also being environmentally sustainable.

The European Commission has recognised the potential of drones to contribute to sustainable transport and has included them as part of its Clean Aviation Initiative. Clean Aviation aims to reduce the environmental impact of aviation by promoting the use of cleaner and more efficient aircraft, including drones. The initiative will support the development and deployment of drones that are environmentally sustainable, while also promoting the use of drones for the delivery of goods and services.

Drones have the potential to reduce the carbon footprint of the transport industry by reducing the number of delivery vehicles on the road. The use of drones for last-mile delivery has the potential to significantly reduce the number of delivery vehicles required, which in turn reduces traffic congestion and emissions. Additionally, drones can be used to inspect and monitor infrastructure, reducing the need for traditional inspection methods that require the use of vehicles.

The use of drones in the transport industry is not without its challenges, however. The proliferation of drones has raised concerns about safety and security, as well as privacy and data protection. The Drone Strategy 2.0 addresses these concerns by establishing a regulatory framework that ensures the safe and secure operation of drones while protecting the privacy of individuals and the protection of personal data. The use of drones in the transport industry is also subject to national regulations. National regulations may vary in terms of scope and requirements, which can create challenges for companies operating across borders. The European Commission is working to establish a harmonised regulatory framework for drones in the EU, which will make it easier for companies to operate across borders while ensuring that drones are operated safely and securely.

The use of drones in the transport industry has the potential to transform the way goods and services are delivered. Drones offer a sustainable and efficient alternative to traditional delivery methods, and their use is being actively promoted by the European Commission as part of its Clean Aviation initiative. However, the proliferation of drones has raised concerns about safety, security, privacy and data protection. The regulatory framework for drones in the EU is designed to ensure the safe and secure operation of drones while also allowing for innovation and development in the drone industry. The European Commission's Drone Strategy 2.0 aims to establish a regulatory framework that enables the development and use of drones in the EU while ensuring that drones are operated safely and securely.

Drone Value Chain Legal Agreements

Contracts and agreements form the backbone of the drone industry, encompassing a wide array of stakeholders including OEM manufacturers, software developers, service providers, and end-users. These legal frameworks are essential for defining responsibilities, managing risks, and ensuring compliance with regulatory standards.

The first layer of contracts typically involves OEM manufacturers and suppliers. These agreements cover the design, production, and supply of drone hardware, ensuring that components meet specified standards and timelines. Companies like DJI, Parrot, and Yuneec often engage in such contracts to secure high-quality materials and components that form the basis of their advanced drones. These contracts are pivotal in maintaining the competitive edge and reliability of the final products.

Beyond hardware, software developers play a crucial role in the drone ecosystem. Contracts with these developers focus on creating, testing, and deploying software that enhances drone functionality. These agreements often include clauses related to intellectual property, data security, and performance benchmarks. For instance, software that enables autonomous navigation or real-time data analysis must comply with stringent performance and security standards, often detailed in these contracts.

Service providers, who operate drones for various commercial applications such as delivery, inspection, and emergency response, enter into contracts that define the scope of services, compliance with regulatory requirements, and liability issues. These agreements ensure that service providers adhere to safety protocols and operational guidelines, minimising risks associated with drone operations. For example, in the agriculture sector, service providers might use drones for precision spraying and crop monitoring under strict contractual obligations to ensure safety and efficiency.

Maintenance and training centers also form part of the contractual landscape. Contracts with these entities ensure that drones are kept in optimal condition and that operators are adequately trained and certified. These agreements often include service level agreements (SLAs) that stipulate response times for repairs and maintenance, as well as training modules that must be covered to ensure compliance with regulatory standards.

Investors are another critical component of the value chain, often enter into investment agreements to participate in various segments of the drone industry. Investment agreements detail the terms of funding, expected returns, and strategic goals. In countries like the U.S., China, and Israel, significant investments are directed towards innovative drone technologies and applications. For instance, venture capital funds and corporate ventures have heavily invested in startups focusing on autonomous solutions and business intelligence software, driving growth and innovation in the industry.

How do these contracts and agreements balance innovation with regulation? The first step is understanding the core of the technology. Contracts often include compliance clauses that ensure all parties adhere to local and international regulations. For example, data collected by drones must comply with GDPR in the EU, necessitating explicit consent and secure data handling practices, which are explicitly detailed in contractual agreements.

These contracts and agreements not only facilitate the smooth functioning of the drone value chain but also address potential risks and liabilities. For instance, in the case of drone malfunctions or accidents, liability clauses in the contracts determine who is responsible—be it the manufacturer, software developer, or operator. This clarity is essential for mitigating risks and ensuring accountability.

Moreover, contracts in the drone industry are evolving to include provisions for emerging technologies such as AI and machine learning. These agreements specify the use of AI in enhancing drone capabilities, such as obstacle detection and autonomous decision-making, while ensuring compliance with ethical and legal standards.

In essence, the intricate network of contracts and agreements in the drone industry plays a vital role in balancing innovation with safety and compliance. These legal frameworks ensure that all stakeholders, from manufacturers to service providers and investors, operate within a structured and regulated environment, fostering growth and technological advancement in the rapidly evolving drone sector.

Value chain structure and private actors

The European commercial drone market is connected to global hardware, software and services value chains. The hardware chain consists of the design and commercialisation of drone platforms, drone components and systems, passenger drones and drone components and systems. The hardware chain is specifically characterised by the manufacturing and assembly of components parts which typically occur in different local and global locations. The European drone value chain structure chart below was developed by the author to provide a snapshot of the private sector actors.

The software chain consists of the combination of digital solutions for piloting flights, fleet and operation management solutions, navigation, artificial intelligence, drone transport management systems, and data analytics services. The services chain is expected to generate the majority of the value from commercial demand. The service chain consists of drone service providers and business internal services; education, simulation, training and certification; market research and consulting; system integration, engineering, research and development and advisory; coalition, organisations and initiatives; maintenance, insurance, user groups, suppliers and retailers, test sites, universities and research centers, shows and conferences.

Commercial drone hardware is designed to have multiyear useful lives, although the exact replacement cycles will differ by industry domain based on the intensity of use and capital availability. Passenger drones are expected to have much longer useful lives and may only be replaced every 20 years with ongoing maintenance and updated sensory equipment as needed. Related pricing for drones components and systems is expected to evolve with, decreases over time anticipated to be around 4% per annum. Price decrease for drone component and systems is particularly significant since a significant portion of the component cost is spent on sensors. In general, drone platforms are still in the tens of thousands of euros, including sensory, and in most cases, passenger drones rates are in the millions of euros.

Drones-as-a-service and data assets produced by leveraging drone technologies is the real intention of commercial users. Reports and data tools for climate change mitigation, agriculture, energy infrastructure, parcel delivery represents some of the service components to the drone industry. Financial services are inherently directly connected to these sources of value and therefore dependent on the specific application and domain versus the type of drone used.

Drone service providers may choose to operate their fleet of drones or enter into service contracts with certified pilots to deliver internal business services. Companies may also be drone providers in their internal business operations to maximise the benefit of vertical integration. Economic value for drone services is generated – e.g., in agriculture where farms pay over 10 Euros per hectare on each flyover and by energy and construction companies that pay service contracts for site visits. For most drone services, the economic value is still to be defined, and assumptions are generally made around potential savings the drone service could bring to companies. This includes civil aviation, where the application of artificial intelligence will dictate whether drone technologies for cargo and passenger transport will be offered as hardware products or as a service.

The scalable nature of data processing and analytics means leaders in this area will have export opportunities as well. Overall, the benefits presented by these value-added services and their geographic scalability should be prioritised to unlock growth. The impact of services is also displayed by drone piloting and operations, which are suggested to represent half the job opportunities. These operations, especially for near-term missions mostly completed within visual line of sight, highlights another area for Europe to prioritise as for the drone market to reach its full potential, many pilots still need to be trained to operate drones safely. Most of these trainees will not represent new job creation as drone responsibilities will simply be an extension of an existing role (e.g. land surveyors).

Finally, insurance and maintenance represent additional extensions to the commercial drones market. Similar to other vehicles and aircraft, drones will need to be maintained and repaired and must be insured for liability purposes against potential damages. The aspects related to the maintenance and insurance of drones create a marketplace over EUR 1 billion. The maintenance portion of the valuation is varied by industry based on the intensity of use with a range of 25 percent to 75 percent of annual depreciation estimated for smaller specific drones and between 30 percent and 40 percent for the more expensive certified drones. Insurance plans covering limited liability for specific units are projected to be around EUR 21.30, whereas insurance premiums for certified operations are forecasted to range in the multiple thousands euros if benchmarked on the current spend by airlines.

The drone ecosystem is complex and constantly evolving encompassing, each segment plays crucial role in the development, deployment, and operation of drones. This analysis dives deeper into the different segments of the drone value chain, including manufacturing, software development, service providers, and end-users. We will also profile major industry players and startups, their contributions to the market, and their roles within the value chain. Furthermore, we will discuss the collaborations and partnerships between companies and their impact on the industry.

Design, Assembly, and Manufacturing

The first segment of the drone value chain involves the design, assembly, and manufacturing of hardware components. This segment is foundational to the drone industry, providing the essential building blocks required for drone operation.

Components and Systems: Manufacturers like FLIR Systems and Velodyne Lidar are at the forefront, providing specialised equipment essential for drone operations. FLIR Systems, renowned for its thermal imaging cameras, equips drones with the capability to detect heat signatures. This is particularly useful in search and rescue missions, surveillance, and agricultural monitoring. Thermal cameras can identify temperature variations in crops, detect livestock in large pastures, and even locate individuals in disaster-stricken areas, showcasing the broad applicability of FLIR's technology.

Velodyne Lidar, on the other hand, specialises in Lidar sensors, which are crucial for navigation and environmental sensing. Lidar technology uses laser pulses to create detailed 3D maps of the environment. This capability is indispensable for autonomous navigation, as it allows drones to detect and avoid obstacles in real-time. Lidar sensors are also employed in topographical mapping, urban planning, and infrastructure inspection, providing high-resolution data that is vital for accurate analysis and decision-making.

Motors and batteries are the lifeblood of drones, determining their flight capabilities and operational range. Companies focusing on high-performance motors ensure that drones can achieve the necessary thrust and manoeuvrability, even under challenging conditions. Advanced battery technology, meanwhile, extends flight times and enhances reliability. Innovations in lithium-polymer and lithium-sulfur batteries have led to lighter, more efficient power sources, enabling drones to carry heavier payloads and operate for longer periods.

The integration of these components is pivotal for the advanced functionalities of drones. High-resolution cameras facilitate detailed imaging and video capture, which are essential for applications ranging from filmmaking to security surveillance. Sensors for environmental monitoring collect data on air quality, weather conditions, and pollution levels, contributing to environmental research and public health initiatives.

Moreover, the synergy between these components fosters the development of sophisticated drone systems capable of performing complex tasks autonomously. For instance, combining thermal cameras, Lidar sensors, and powerful processors enables drones to conduct detailed infrastructure inspections, identifying structural weaknesses that are invisible to the naked eye. This capability not only enhances safety but also reduces the need for human intervention in hazardous environments. The competitive landscape of component manufacturers drives continuous innovation. Companies are investing heavily in research and development to improve the performance, efficiency, and integration of their products. This ongoing advancement ensures that drones remain at the cutting edge of technology, capable of meeting the evolving demands of various industries.

Drone Platforms: Drone platforms are pivotal in integrating various components and systems, enabling drones to perform multifaceted functions across diverse industries. Major players in this space, including DJI, Parrot, Amazon, Skydio, Yuneec, AeroVironment, SenseFly, and Autel Robotics, have developed specialised platforms that cater to different applications, ranging from delivery services to agricultural monitoring.

DJI, a global leader in the drone industry, offers platforms like the DJI Phantom and Mavic series, which are renowned for their versatility and advanced features. DJI drones are equipped with high-resolution cameras and sophisticated navigation systems, making them ideal for aerial photography, videography, and industrial inspections. DJI's Agras series, specifically designed for agricultural use, features precision spraying capabilities and multispectral imaging, significantly enhancing crop management and monitoring.

Parrot, another key player, has developed platforms such as the Parrot Anafi series, known for their high-resolution imaging and thermal cameras. These drones are extensively used in agriculture for crop health monitoring, infrastructure inspections, and environmental research. Parrot's focus on integrating AI and sensor technologies allows for advanced data collection and analysis, providing valuable insights for various industries.

Amazon's Prime Air delivery service aims to revolutionise package delivery by utilizing drones. These drones are designed for rapid, reliable delivery within 30 minutes of an order. They incorporate advanced navigation systems, robust safety mechanisms, and efficient battery management. Amazon's efforts highlight the potential of drones to streamline logistics and enhance customer satisfaction.

Skydio has garnered attention for its autonomous drones, which utilise advanced AI for navigation and obstacle avoidance. The Skydio 2 drone, with its array of visual sensors, can navigate complex environments autonomously, making it ideal for infrastructure inspections and emergency response scenarios. This autonomy reduces the need for manual control and enhances operational safety.

Yuneec offers a range of drones suited for both consumer and commercial use. The Yuneec Typhoon series, for instance, features drones equipped with 4K cameras and sophisticated flight controls. These platforms are employed in media production, agriculture, and public safety. Yuneec's user-friendly designs and robust performance make them accessible to a wide audience.

AeroVironment specialises in small unmanned aircraft systems for military and commercial applications. Their RQ-20 Puma drone is used for reconnaissance and surveillance, equipped with advanced sensors and imaging systems that provide real-time data essential for tactical operations and environmental monitoring.

SenseFly, a subsidiary of Parrot, focuses on professional fixed-wing drones like the eBee series, which are widely used in mapping, surveying, and agriculture. These drones are lightweight and can cover large areas efficiently, providing high-resolution aerial maps that are crucial for precision agriculture and land management.

Autel Robotics is known for its EVO series, combining portability with advanced features like 8K cameras and enhanced imaging capabilities. These drones are used in sectors such as real estate, construction, and environmental monitoring. Autel's platforms offer high-definition aerial photography and video, crucial for various industrial applications.

The integration of high-resolution cameras, Lidar sensors, and powerful processors allows these drone platforms to perform complex tasks autonomously. For instance, combining thermal cameras with Lidar technology enables drones to conduct detailed infrastructure inspections, identifying structural weaknesses that are invisible to the naked eye. This capability not only enhances safety but also reduces the need for human intervention in hazardous environments.

The drone industry is characterised by rapid innovation and intense competition. Companies are investing heavily in research and development to improve the performance, efficiency, and integration of their products. This ongoing advancement ensures that drones remain at the cutting edge of technology, capable of meeting the evolving demands of various industries. The synergy between different components and systems fosters the development of sophisticated drone platforms that are essential for modern applications.

Drone platforms developed by companies like DJI, Parrot, Amazon, Skydio, Yuneec, AeroVironment, SenseFly, and Autel Robotics are at the forefront of the industry, driving innovation and expanding the capabilities of drones. These platforms integrate advanced technologies that enable drones to perform a wide array of functions, from delivery and photography to environmental monitoring and infrastructure inspection. As the industry continues to evolve, these companies will play a crucial role in shaping the future of drone technology.

Passenger Drones and Air Taxis: This emerging segment includes companies like Volocopter and EHang, which are developing drones capable of carrying passengers. These air taxis represent a significant leap in urban mobility, aiming to reduce traffic congestion and provide an alternative mode of transportation.

Volocopter, a German aviation company, has been pioneering the development of electric vertical takeoff and landing (eVTOL) aircraft. Their flagship model, the Volocopter 2X, is designed to carry two passengers and is powered by 18 rotors. This design provides stability, safety, and efficiency, making it an ideal candidate for urban air mobility. The company envisions a future where these air taxis operate on predetermined routes, similar to how public transportation functions today. They have conducted several successful test flights in cities like Dubai, Singapore, and Paris, showcasing the potential of their technology. Volocopter aims to launch commercial services by the mid-2020s, positioning themselves as a leader in the emerging market of urban air mobility.

EHang, a Chinese company, has developed the EHang 216, an autonomous aerial vehicle (AAV) designed to carry passengers. This drone can accommodate two passengers and is equipped with 16 rotors for vertical takeoff and landing. EHang has been aggressively testing its air taxis in various countries, including China, the United States, and Austria. Their approach to urban air mobility is cantered around automation, with the EHang 216 being fully autonomous, reducing the need for pilot intervention. This focus on autonomy aims to enhance safety, efficiency, and scalability of urban air transportation. EHang's vision includes establishing a network of "vertiports" across cities, where passengers can board and disembark from these air taxis, creating a seamless urban mobility experience.

The introduction of passenger-carrying drones represents a leap forward in urban mobility. These air taxis offer a potential solution to the ever-growing problem of traffic congestion in metropolitan areas. By taking to the skies, they can bypass the gridlock on the ground, providing faster and more efficient transportation. This innovation not only promises to reduce travel times but also contributes to environmental sustainability by utilizing electric propulsion systems, thereby reducing the carbon footprint associated with urban transportation.

Moreover, the development of passenger-carrying drones opens up new economic opportunities. The infrastructure required for urban air mobility, including vertiports and maintenance facilities, will create jobs and stimulate economic growth. Additionally, the operation of these air taxis will generate revenue through ticket sales, creating a new stream of income for cities and transportation companies.

Navigation and Positioning Systems: This includes the development of GPS modules, inertial measurement units (IMUs), and other systems that enable precise navigation. Companies like Trimble Navigation and u-blox are key providers of navigation technologies essential for drone operation. Navigation and positioning systems are the backbone of precise drone operations, crucial for both autonomous flight and remote piloting. This segment includes the development of GPS modules, inertial measurement units (IMUs), and other advanced positioning systems. Companies like Trimble Navigation and u-blox lead the way in providing these essential technologies.

Trimble Navigation is renowned for its high-precision GPS modules that are indispensable for accurate drone navigation. These modules enable drones to maintain stable flight paths, perform precise landings, and execute complex flight manoeuvres. Trimble's GPS solutions are particularly valuable in applications requiring exact positioning, such as surveying, mapping, and agricultural monitoring.

U-blox, another key player, specialises in developing compact and efficient GPS and GNSS modules. Their products are designed to deliver high accuracy and reliability, even in challenging environments. u-blox modules are integrated into drones to ensure seamless navigation and positioning, supporting various applications from delivery to environmental monitoring.

In addition to GPS, inertial measurement units (IMUs) are critical components in drone navigation systems. IMUs provide data on the drone's orientation, acceleration, and angular velocity, allowing for precise control and stability. This data is crucial for maintaining balance during flight, especially in adverse weather conditions or when performing intricate tasks.

These navigation and positioning technologies collectively enable drones to operate with a high degree of autonomy and precision. The integration of GPS and IMUs ensures that drones can follow pre-programmed routes accurately, avoid obstacles, and perform real-time adjustments based on environmental conditions.

The continuous advancements in navigation technology enhance the overall capabilities of drones. Improved accuracy and reliability of GPS modules and IMUs contribute to the safe and efficient operation of drones in various sectors, including delivery, agriculture, and infrastructure inspection. These technologies are pivotal in expanding the potential applications of drones, making them more versatile and effective tools in the modern landscape. The field of drone navigation and positioning systems requires continuous research and development efforts focus on increasing the precision, reducing the size, and improving the energy efficiency of these components. This ongoing innovation ensures that drones remain at the cutting edge of technology, capable of meeting the evolving demands of various industries.

Power Systems: Innovations in power systems, such as advanced batteries and power management systems, are crucial for extending the flight time and efficiency of drones. Companies like Tesla are exploring the potential of their battery technologies for drone applications.

The advancement of drone technology hinges significantly on innovations in power systems. Extending flight time and enhancing operational efficiency are critical for expanding the functional range and applications of drones. Companies like Tesla are exploring their advanced battery technologies to meet the growing demands of the drone industry. One of the primary challenges in drone operations is limited flight time, often constrained by the capacity and efficiency of batteries. Innovations in battery technology, such as those developed by Tesla, aim to address this limitation. Tesla's expertise in lithium-ion batteries, known for their high energy density and longevity, is being leveraged to create more efficient power solutions for drones.

These advanced batteries are designed to provide longer flight durations, enabling drones to perform extended missions without the need for frequent recharges. This is particularly beneficial in applications such as delivery, surveillance, and agricultural monitoring, where operational efficiency is paramount. Improved battery technology also enhances the payload capacity of drones, allowing them to carry heavier equipment and supplies. In addition to battery advancements, power management systems play a crucial role in optimising the energy consumption of drones. These systems ensure that the power generated by the batteries is used efficiently, reducing waste and maximising flight time. Power management involves regulating the power supply to various components of the drone, such as motors, sensors, and communication systems, ensuring that each component receives the necessary energy without overconsumption.

Companies are developing sophisticated power management algorithms that dynamically adjust power distribution based on real-time operational demands. This adaptive approach not only extends the operational life of the batteries but also enhances the overall performance and reliability of the drones. Moreover, the improved efficiency and reliability of drones open up new possibilities for their use in emergency response, environmental monitoring, and infrastructure maintenance. Drones equipped with advanced power systems and positioning capabilities can quickly respond to disasters, monitor environmental changes, and inspect critical infrastructure, providing valuable data and support in real-time.

Testing, and Deployment

The second segment involves the software development and testing required to bring drones from the prototype stage to deployment and full operations.

Flight, Fleet, and Operations Management: An integral part of the drone value chain is the sophisticated management of flight, fleet, and real-time operations. This segment is vital for ensuring that drone missions are planned, executed, and monitored efficiently, with a focus on safety and compliance with regulatory standards.Leading companies like AirMap and Skyward provide advanced platforms designed to facilitate various aspects of drone operations. These platforms offer robust tools for flight planning, fleet management, and real-time operations, ensuring that drone operators can conduct missions seamlessly while adhering to airspace regulations.

Flight planning software is crucial for preparing and executing drone missions. These tools allow operators to create detailed flight plans, taking into account factors such as airspace restrictions, weather conditions, and mission objectives. By using these platforms, operators can ensure that their drones follow predefined routes, minimising the risk of entering restricted areas or encountering adverse conditions.

Fleet management systems are essential for operators managing multiple drones. These systems provide comprehensive oversight of the entire drone fleet, including maintenance schedules, battery status, and operational readiness. By leveraging fleet management software, operators can streamline their operations, ensuring that each drone is in optimal condition and ready for deployment when needed. Real-time operations management is a critical component of drone operations, particularly for missions that require constant monitoring and adjustments. Software platforms like those provided by AirMap and Skyward offer real-time tracking and telemetry data, enabling operators to monitor their drones' positions, speeds, and altitudes. This real-time data is invaluable for ensuring that drones remain on course and can adapt to changing conditions or unexpected obstacles.

One of the significant challenges in drone operations is ensuring compliance with airspace regulations. Platforms like AirMap and Skyward integrate real-time airspace information, providing operators with up-to-date data on temporary flight restrictions (TFRs), no-fly zones, and other regulatory requirements. This integration helps operators avoid violations and maintain compliance with local, national, and international airspace rules. The integration of flight planning, fleet management, and real-time operations into a unified platform enhances the safety and efficiency of drone missions. Operators can manage their fleets more effectively, reduce the risk of accidents, and ensure that missions are completed successfully. These platforms also facilitate better communication and coordination among team members, improving overall operational efficiency.

For instance, in the construction industry, companies use these platforms to manage fleets of drones that conduct site surveys, monitor progress, and inspect structures. By using flight planning software, construction firms can schedule regular flights over project sites, ensuring consistent and accurate data collection. Fleet management systems help keep track of drone maintenance and readiness, while real-time operations tools allow for immediate adjustments if weather conditions or site changes occur. In agriculture, farmers and agronomists use these platforms to plan and execute flights for crop monitoring, spraying, and mapping. The ability to manage multiple drones and access real-time data helps optimise agricultural operations, leading to better crop yields and resource management.

As drone technology continues to advance, the importance of sophisticated flight, fleet, and operations management will only grow. Future developments may include more automated and AI-driven systems, enhancing the ability of drones to operate autonomously while maintaining safety and compliance. The integration of these advanced management systems will be pivotal in expanding the applications and efficiency of commercial drones across various industries.

Navigation and Artificial Intelligence: Navigation and Artificial Intelligence (AI) are pivotal in enhancing the autonomy and intelligence of modern drones. Companies like Skydio are leading the charge by integrating sophisticated AI and machine learning algorithms into their drone systems. This integration allows drones to perform complex tasks such as obstacle avoidance, autonomous flight, and advanced navigation, making them significantly more capable and versatile.

AI and machine learning are at the core of these advancements. By using vast amounts of data to train algorithms, drones can learn to interpret their surroundings, make decisions, and execute tasks without human intervention. This capability is particularly crucial for applications that require high levels of precision and reliability, such as in disaster response, surveillance, and infrastructure inspection. One of the standout features enabled by AI is obstacle avoidance. Skydio, for instance, uses AI-driven vision systems that allow drones to navigate complex environments autonomously. These drones are equipped with multiple cameras and sensors that feed data into the AI algorithms, enabling real-time analysis and decision-making. This technology ensures that drones can avoid obstacles such as trees, buildings, and other structures, significantly reducing the risk of collisions.

Autonomous flight is another significant advancement brought about by AI. Drones equipped with AI can plan and execute flights with minimal human input, following pre-determined routes or dynamically adjusting their paths based on real-time data. This capability is essential for applications like agriculture, where drones can autonomously monitor crop health, spray pesticides, and collect data across large fields. In logistics, autonomous drones can optimise delivery routes, ensuring timely and efficient parcel delivery. Enhanced navigation capabilities are critical for the precise operation of drones in various environments. AI algorithms process data from GPS modules, inertial measurement units (IMUs), and other navigation systems to ensure accurate positioning. Companies like u-blox and Trimble Navigation provide advanced GPS and IMU systems that, when combined with AI, allow drones to maintain stable and precise flight paths even in challenging conditions.

In agriculture, AI-driven drones can autonomously monitor crop health, identify areas needing attention, and even apply treatments without direct human control. This not only increases efficiency but also reduces labor costs and improves yield outcomes. Similarly, in infrastructure inspection, AI-powered drones can autonomously scan structures for defects, providing detailed analysis and reports that facilitate timely maintenance and repairs.

AI enables drones to continuously improve their operational capabilities. Through machine learning, drones can analyze past missions, learn from mistakes, and optimise future performance. This iterative learning process ensures that drones become more adept at handling various tasks, from navigation to data collection and processing.

The integration of AI and ML in drone technology is set to advance further, with future drones expected to exhibit higher levels of autonomy and intelligence. As these technologies evolve, drones will be able to undertake more complex and diverse missions, ranging from urban logistics to environmental monitoring, with greater efficiency and safety.

Drone Transport Management: Drone transport management is a critical component of the commercial drone ecosystem, focusing on the logistics and transportation of goods. Companies like Matternet and Zipline are at the forefront of developing sophisticated systems for drone delivery networks, particularly targeting healthcare and emergency supplies.

Matternet, a Silicon Valley-based company, has developed a comprehensive logistics platform that integrates drones, ground stations, and cloud-based management systems. Their platform is designed to facilitate the efficient and reliable transport of medical supplies, such as blood samples, vaccines, and medications, between hospitals, clinics, and laboratories. Matternet's drones are equipped with advanced navigation systems and can carry payloads up to 2 kilograms over distances of 20 kilometers. The company's technology has been implemented in several countries, including Switzerland and the United States, where it has significantly improved the speed and efficiency of medical deliveries.

Zipline, another pioneer in drone logistics, has revolutionised the delivery of blood products and medical supplies in remote areas. Founded in 2014, Zipline has established a robust delivery network in Rwanda and Ghana, where their drones have flown thousands of missions, delivering critical supplies to remote health facilities. Zipline's drones can carry up to 1.75 kilograms of cargo and travel distances of up to 80 kilometers, making them ideal for reaching isolated regions with poor road infrastructure. The company's automated distribution centers manage the entire delivery process, from order processing to drone dispatch and delivery, ensuring that medical supplies reach their destinations quickly and efficiently.

The impact of these drone delivery networks extends beyond healthcare. During emergencies, such as natural disasters, drones can rapidly transport essential supplies to affected areas, bypassing damaged infrastructure and reducing the response time. This capability was demonstrated during the COVID-19 pandemic when Zipline's drones were used to deliver personal protective equipment (PPE) and medical supplies to healthcare facilities in need. The integration of drone transport management systems into the broader logistics network involves several key elements. First, it requires the development of reliable and scalable drone technology that can operate in various weather conditions and terrains. Second, it necessitates robust software platforms for managing the logistics operations, including route planning, real-time tracking, and inventory management. Third, regulatory compliance is crucial to ensure safe and legal drone operations, particularly in densely populated areas and near critical infrastructure.

Furthermore, the success of drone transport management depends on collaboration between various stakeholders, including drone manufacturers, logistics companies, healthcare providers, and regulatory bodies. These partnerships are essential for developing and implementing standardised protocols and procedures that ensure the safety, efficiency, and reliability of drone delivery networks.

Data Analytics and AI: The integration of data analytics and AI into the drone value chain is pivotal for transforming raw aerial data into actionable insights. Companies like PrecisionHawk are leading this charge by offering advanced platforms that utilise AI to analyze data collected by drones, addressing a variety of applications such as agriculture, infrastructure inspection, and environmental monitoring. PrecisionHawk's platform is equipped with machine learning algorithms that process high-resolution images and sensor data to deliver precise analytics. In agriculture, this technology helps farmers monitor crop health, optimise irrigation, and predict yields by analysing multispectral and thermal imagery. By detecting early signs of disease or pest infestation, these analytics enable timely interventions, which can significantly enhance crop productivity and reduce resource wastage.

For infrastructure inspection, AI-driven data analytics allow for the identification of structural anomalies and potential failures in bridges, roads, and power lines. PrecisionHawk's platform can analyze vast amounts of data to pinpoint areas that require maintenance, thereby preventing costly repairs and enhancing safety. This capability is particularly valuable for utilities and construction companies that manage extensive networks of infrastructure. Environmental monitoring is another critical application of drone data analytics. Drones equipped with various sensors can gather data on air quality, vegetation health, and water resources. AI algorithms process this data to monitor environmental changes, assess the impact of human activities, and support conservation efforts. For example, in forest management, AI can help track deforestation and assess the health of ecosystems, guiding reforestation and conservation strategies.

Moreover, the integration of AI and data analytics in drones extends to disaster management and humanitarian efforts. In the aftermath of natural disasters, drones can rapidly survey affected areas and collect data that AI platforms analyze to assess damage, prioritise rescue operations, and distribute resources efficiently. This approach enhances the effectiveness of disaster response and recovery efforts. The ability to process and analyze data in real-time is crucial for the success of these applications. Companies like PrecisionHawk leverage cloud computing and edge computing technologies to ensure that data analytics are performed swiftly and accurately. This real-time analysis is essential for applications where timely decisions are critical, such as emergency response and precision agriculture.

The evolution of AI and data analytics in the drone industry is driven by continuous advancements in machine learning, computer vision, and big data technologies. These innovations are enabling more sophisticated and reliable analytics, enhancing the capabilities of drones across various sectors. As AI algorithms become more refined and datasets grow larger, the precision and scope of drone analytics are expected to expand, offering even more valuable insights and solutions.

The integration of data analytics and AI into the drone value chain is transforming how industries leverage aerial data. Companies like PrecisionHawk are at the forefront of this transformation, providing platforms that turn raw drone data into actionable intelligence. This capability not only enhances efficiency and productivity across various sectors but also supports critical applications in agriculture, infrastructure, environmental monitoring, and disaster management. As technology continues to advance, the role of AI and data analytics in the drone industry will only become more integral and impactful.

Commercial Operations and Research

The third segment involves the services provided by various stakeholders to support the commercial operation and research of drones.

Drone Services Providers: Companies like Zipline, Zwoop Aero, Wing, and Manna are spearheading a transformative shift. These companies are not just deploying drones; they are pioneering a future where aerial delivery becomes an integral part of our daily lives, showcasing the immense potential and tangible benefits of drone technology.

Zwoop Aero stands out with its focus on last-mile delivery. Envision a future where packages—whether life-saving medical supplies or everyday consumer goods—arrive at your doorstep with unprecedented speed and precision. Zwoop Aero is making this vision a reality. Their drones, engineered for a wide range of payloads and capable of navigating intricate urban environments, epitomise efficiency and reliability. By drastically reducing delivery times and enhancing logistical accuracy, Zwoop Aero is setting new benchmarks for the industry.

Wing, a subsidiary of Alphabet, pushes the boundaries of drone delivery. Their approach is not merely about replacing traditional methods but seamlessly integrating drones into existing delivery networks. Wing's drones have been deployed to deliver everything from consumer products to critical medical supplies, showcasing their versatility and adaptability. Through Wing's innovative operations, the potential for drones to become a seamless part of our logistical infrastructure is increasingly evident.

Manna, with its commitment to suburban and semi-rural deliveries, tackles a unique set of challenges. Traditional delivery services often falter in these regions, but Manna's fleet of drones is designed to overcome these obstacles. By providing fast and reliable deliveries to even the most remote locations, Manna is extending the reach of drone technology and demonstrating its capability to serve diverse communities effectively.

These companies are more than service providers; they are the vanguard demonstrating the practical applications of drone technology. Their efforts show that drones can significantly reduce delivery times, lower operational costs, and minimise environmental impact. As Zwoop Aero, Wing, and Manna continue to innovate and refine their technologies, they are not just shaping the future of delivery; they are redefining the possibilities of logistics and transportation in the modern world.

Distributors and Suppliers: These are entities responsible for ensuring that drones and their components reach consumers and businesses alike. These companies form the backbone of the industry, facilitating the availability and accessibility of cutting-edge drone technology.

Among the prominent players in this space are Adorama and B&H Photo, renowned for their extensive range of drone products. Adorama has built a reputation for providing a diverse selection of drones, from consumer-friendly models to high-end professional equipment. Their comprehensive inventory caters to hobbyists, commercial users, and industry professionals, making them a go-to source for all things drone-related. Similarly, B&H Photo is a well-established retailer known for its vast assortment of drones and drone components. Whether it's the latest model from leading brands or specialised accessories and parts, B&H Photo offers a one-stop-shop experience for consumers and businesses seeking reliable and high-quality drone products. In addition to these giants, other notable distributors and suppliers have emerged, enriching the market with a variety of offerings. DroneNerds is a key player known for its extensive catalog and excellent customer service, serving both the retail and enterprise sectors. DJI Store, the official retailer of DJI products, provides direct access to one of the most trusted names in the drone industry, ensuring customers get authentic and up-to-date products. GetFPV is another significant retailer, catering specifically to the FPV (First Person View) drone community. Their focus on providing specialised FPV drones and components has made them a favorite among enthusiasts looking to customise and optimise their flying experience.

These distributors and suppliers are essential to the drone ecosystem. They not only supply the market with the latest technology but also offer critical support, from product education to after-sales service. Their role in stocking a wide range of drones and accessories, coupled with their ability to meet the demands of both casual users and professionals, underscores their importance in driving the industry's growth and innovation. By bridging the gap between manufacturers and end-users, they ensure that the benefits of drone technology are readily available to a broad audience, fostering the continued expansion and evolution of the drone landscape.

Education and Training: Comprehensive training programs and certification courses are crucial for ensuring safe and competent drone operations. Numerous companies around the world are leading the charge in providing these essential services, each catering to the unique regulatory and practical needs of their regions.

Drone Pilot Ground School in the United States offers a range of certification courses that cover both the theoretical and practical aspects of drone operation. Their programs are designed to meet FAA standards, preparing students for the Part 107 certification, essential for commercial drone pilots in the US.

In Europe, European Flyers based in Madrid, Spain, provides extensive training programs not just for airplane and helicopter pilots but also for drone pilots. Their curriculum is comprehensive, covering industry standards and practical flight experience, ensuring graduates are well-prepared for a career in aviation and drone operations.

COPTRZ in the UK offers various courses, including the General Visual Line of Sight Certificate (GVC) and the A2 Certificate of Competence (A2 CofC). These certifications are aligned with the new UK drone regulations, ensuring that operators are up-to-date with the latest legal and safety standards. COPTRZ also provides specialised training for complex UAV projects, making it a significant player in the European drone training market.

Global Drone Solutions in Australia delivers CASA certified training courses, which are essential for obtaining a drone pilot license in Australia. They offer a mix of classroom and practical flight training, complemented by a suite of resources to support new drone businesses. This comprehensive approach ensures that students are not only skilled pilots but also knowledgeable about the operational and business aspects of drone services.

In India, several DGCA-approved organizations such as DroneAcharya Aerial Innovations and Garuda Aerospace provide extensive training and certification programs. These institutions focus on various applications, including agricultural drones and aerial surveys, reflecting the diverse needs of the Indian market. Their programs are designed to enhance technical skills and ensure compliance with national regulations . Droneversity offers a unique global perspective by providing training programs across several countries including Nigeria, Zambia, and the UAE. Their focus on integrating drone technology into STEM education and promoting entrepreneurship makes them a key player in advancing drone education worldwide.

DARTdrones, also in the US, offers a range of courses from beginner levels to advanced certifications like the Trusted Operator Program. Their real-world flight planning and execution programs are designed to prepare pilots for practical challenges they may face in the field, ensuring comprehensive operational readiness.

These education and training providers play a pivotal role in the drone industry by ensuring that operators are well-equipped with the knowledge and skills needed for safe and efficient drone operations. Their efforts not only enhance individual capabilities but also contribute to the overall growth and professionalisation of the drone sector globally.

Market Research and Consulting: These firms provide the data and analysis that businesses need to make informed decisions and stay competitive in a rapidly evolving landscape. Drone Industry Insights (DRONEII) is a prominent player in this field. DRONEII offers detailed market reports and strategic insights that help businesses understand the complexities of the drone market. Their reports cover various segments, including hardware, software, and services, providing a comprehensive overview of the industry. DRONEII's expertise is particularly valuable for companies looking to make short and long-term strategic decisions, as their data is often used to inform product development and market entry strategies.

DroneAnalyst is another key firm that provides in-depth research and analysis on the drone market. Their annual market sector reports are highly regarded for their detailed coverage of drone buyers, service providers, and software services. DroneAnalyst helps businesses navigate the hype versus reality in the drone industry, offering actionable insights that are essential for strategic planning and operational improvement.

Frost & Sullivan and Gartner are well-known global market research firms that also cover the drone industry extensively. These companies offer reports that analyze market dynamics, competitive landscapes, and technological advancements. Their insights help businesses identify growth opportunities and understand regulatory changes that could impact their operations. Frost & Sullivan's detailed market analysis and

Gartner's strategic advice are invaluable for companies looking to stay ahead of industry trends .

Grand View Research provides comprehensive market research reports that forecast market growth and analyze key trends within the drone industry. Their reports highlight the increasing application of drones in various sectors, such as agriculture, logistics, and military, and provide detailed segment analysis based on payload capacity, power source, and regional trends. This granular approach helps businesses understand specific market segments and tailor their strategies accordingly.

StartUs Insights offers a data-driven perspective on the drone industry, focusing on investment trends, startup ecosystems, and innovation hubs. Their reports emphasise the importance of supporting the startup ecosystem and investing in research and development to drive innovation in drone technologies. StartUs Insights also identifies key geographical hubs for drone innovation, such as the United States, United Kingdom, India, Canada, and Australia, which are crucial for understanding regional market dynamics.

Together, these market research and consulting firms provide the knowledge and expertise that drive the drone industry's growth and innovation. By leveraging their insights, businesses can navigate the complexities of the market, anticipate regulatory changes, and capitalise on emerging opportunities to maintain a competitive edge.

Organisations and Public Interest: Non-profit organizations and advocacy groups play a crucial role in shaping public policy and promoting the responsible use of drones. These entities work tirelessly to ensure that the integration of drone technology into various sectors is safe, efficient, and beneficial to society. Here are some of the key players globally.

AUVSI is the world's largest non-profit organization dedicated to advancing the use of un-crewed systems and robotics. Representing members from over 60 countries, AUVSI works across defence, civil, and commercial markets to promote the safe and responsible use of drone technology. The organization provides a platform for industry professionals to collaborate, share knowledge, and influence regulatory frameworks to support innovation and growth in the drone sector.

Founded in 2016, the CDA is an industry-led 501(c)(6) non-profit association in the Unites States that advocates for the safe integration of commercial drones into the National Airspace System. The CDA works closely with the federal government, including agencies like the FAA and NASA, to reduce regulatory barriers and promote the benefits of drone technology in terms of safety, security, and economic impact. By providing a unified voice for the commercial drone industry, CDA helps shape policies that support technological advancement and industry growth.

The DSPA represents the interests of small and medium-sized drone service providers. Established by industry veterans, this non-profit organization focuses on advocating for reasonable regulations and providing resources to help drone businesses thrive. DSPA's efforts include engaging with regulatory bodies, educating government officials about the needs of drone service providers, and promoting best practices within the industry.

The Association of Remotely Piloted Aircraft Systems UK (ARPAS-UK) is the trade association for the drone industry in the United Kingdom. It represents commercial drone operators, manufacturers, and stakeholders, working to ensure that the industry grows in a safe and sustainable manner. ARPAS-UK engages with regulatory authorities, including the Civil Aviation Authority (CAA), to influence policy and regulatory developments. It also provides members with access to resources, training, and networking opportunities.

Based in India, NADDSO is a leading organization dedicated to advancing the use of drones and aerial systems. The association focuses on industry development, policy advocacy, global trade development, and talent development. NADDSO works to align business strategies with government policies and fosters innovation in drone technology through research and development initiatives.

These organizations play vital roles in advocating for the drone industry, ensuring that the regulatory environment is conducive to growth while promoting the safe and responsible use of drone technology. By collaborating with governments, providing education and resources, and representing the interests of their members, these non-profits help drive the industry forward and unlock the full potential of drones across various sectors.

Maintenance and Insurance: As the utilization of drones expands across various industries, the necessity for robust maintenance and insurance services becomes increasingly critical. These services are essential for managing the operational risks associated with drone flights and ensuring the longevity and reliability of the equipment.

Global Aerospace is a leading provider of drone insurance, offering comprehensive coverage tailored specifically for unmanned aircraft systems (UAS). Their policies cover third-party liability, physical damage to the drone (hull coverage), and associated equipment. They cater to both single drones and entire fleets, providing flexibility for operators of different scales. Global Aerospace's extensive experience in aviation insurance ensures that drone operators can rely on them for managing a wide range of risks effectively.

SkyWatch.AI stands out for its flexible insurance options, allowing operators to purchase coverage on an annual, monthly, or even hourly basis. This flexibility is particularly beneficial for both frequent flyers and occasional hobbyists. SkyWatch.AI provides liability coverage up to $10 million, with policies underwritten by Starr Insurance, a company known for its financial strength and reliability.

Coverdrone, based in Europe, offers specialised insurance for both commercial and recreational drone operators. Their policies include coverage for data protection, invasion of privacy, and physical damage to the drone and its equipment. Coverdrone also provides a companion app, Flysafe, which delivers real-time safety data and integrates with global unmanned traffic management systems to ensure safe and compliant flights.

DJI Care provides an alternative to traditional insurance by offering repair and replacement services for DJI drones. Their plans cover accidental damage during normal use, ensuring that operators can maintain their equipment in top condition. DJI Care's services are available globally, reflecting the widespread use of DJI drones.

In addition to these, Allianz offers comprehensive drone insurance policies that cover a broad spectrum of risks for both private and commercial operators. Allianz's global presence and extensive range of insurance products make it a reliable choice for drone operators looking to secure their investments and manage liabilities.

Assured Partners provides drone insurance with a focus on commercial applications, including coverage for manufacturers, distributors, and flight schools. Their policies are designed to meet the diverse needs of drone operators, offering both liability and hull coverage to protect against physical damage and third-party claims.

In the maintenance sector, companies like DJI provide factory maintenance programs that ensure drones remain in optimal condition throughout their lifecycle. These programs are essential for maintaining the performance and safety of drones, especially for operators who rely on them for critical applications.

Overall, the integration of maintenance and insurance services into the drone industry is crucial for its sustainable growth. Providers like Global Aerospace, SkyWatch.AI, Coverdrone, DJI Care, Allianz, and AssuredPartners play a pivotal role in supporting drone operators worldwide by mitigating risks and ensuring the reliability of their operations. These services not only enhance the safety and efficiency of drone flights but also instils confidence in operators and stakeholders, driving further adoption of drone technology across various sectors.

System Integration and Engineering: The field of system integration and engineering is vital for the seamless operation of drones, as it involves the integration of various components and systems, ensuring they function harmoniously. This is particularly critical for applications in both defence and commercial sectors, where reliability and precision are paramount.

Boeing and Lockheed Martin are renowned for their large-scale drone system integration, primarily focusing on defence applications. These companies leverage their extensive experience in aerospace to develop sophisticated drone systems that meet stringent military standards. Boeing's investments in advanced drone technologies and Lockheed Martin's development of autonomous systems highlight their commitment to leading the drone integration field.

OpenWorks Engineering, based in the UK, specialises in autonomous technology for surveillance and air defence applications. Their SkyWall products provide physical capture solutions for counter-UAS (Unmanned Aerial Systems) missions, showcasing their expertise in integrating autonomous systems for critical national infrastructure protection and defence applications.

SPH Engineering from Latvia offers comprehensive integration services that include UAV software solutions and custom developments. Their UgCS (Universal Ground Control Software) is a leading flight planning solution used globally, and their integration services span various sensors, including Ground Penetrating Radars (GPR), Methane detectors, and Magnetometers. This flexibility allows for tailored solutions across multiple industries such as geophysics, environmental monitoring, and industrial inspections.

Draganfly, a Canadian company with over two decades of innovation in UAV hardware and software, excels in integrating custom drone solutions. Their products cater to diverse applications from emergency response to industrial inspections, emphasising their role in enhancing drone capabilities through robust engineering and system integration.

Terra Drone, a Japanese company, has made significant strides in integrating UAS Traffic Management (UTM) systems. Their collaboration with Aloft Technologies aims to enhance the efficiency and scalability of global drone operations by harmonising operational standards and leveraging advanced UTM technology.

Equinox Innovative Systems, recently acquired by TCOM L.P. in the USA, specialises in advanced multi-rotor drone technology. Their focus on tethered and heavy-lift drone systems supports applications in military operations, emergency response, and infrastructure inspection. This acquisition allows TCOM to expand its range of drone solutions, integrating various capabilities to meet diverse operational needs.

SPH Engineering also supports complex integration projects involving multiple sensors and UAV platforms, facilitating advanced drone applications for surveying, data collection, and environmental monitoring. Their UgCS Integrated Systems are designed for sensor interchangeability, making it easier to adapt to different project requirements.

Overall, the landscape of drone system integration and engineering is marked by significant contributions from global players who are driving advancements in drone technology through innovative integration solutions. Companies like Boeing, Lockheed Martin, OpenWorks Engineering, SPH Engineering, Draganfly, Terra Drone, and Equinox Innovative Systems are at the forefront, ensuring that drones operate seamlessly and efficiently in various complex environments.

Shows, Conferences, and Events: Industry events such as InterDrone and the Commercial UAV Expo provide essential platforms for stakeholders to showcase innovations, share knowledge, and network. These events are crucial for fostering collaboration and driving growth within the drone industry.

InterDrone is a prominent event in the United States, known for its comprehensive program that includes workshops, keynote speeches, and an extensive exhibition hall. Held annually, it draws professionals from various sectors of the drone industry, including commercial, government, and military sectors. InterDrone offers an excellent opportunity for participants to learn about the latest technologies, regulatory updates, and market trends.

UMEX and SimTEX 2024 in Abu Dhabi, UAE, is the Middle East's premier event for drones, robotics, and unmanned systems. From January 22-25, this event will showcase the latest technologies and innovations, offering a platform for industry leaders to explore new business opportunities and partnerships.

AUVSI XPONENTIAL is one of the largest gatherings dedicated to unmanned systems and robotics. Organised by the Association for Uncrewed Vehicle Systems International (AUVSI), this event features keynote speeches, workshops, and outdoor demonstrations. The 2024 edition was held from April 22-25 in San Diego, California. XPONENTIAL attracts a global audience, providing a platform for discussing the latest advancements in drone technology and their applications in various sectors.

International Drone Show in Odense, Denmark, was held on May 29, 2024, features live flight demonstrations, panel discussions on industry themes, and B2B matchmaking events. This show emphasises legislation, operational safety, and new technologies, making it a vital event for those looking to stay ahead in the drone industry.

Energy Drone & Robotics Summit, held in Houston, Texas, from June 10-12, 2024, is the largest event in the world for uncrewed and autonomous systems in energy and industrial operations. This summit offers forums and workshops focusing on drones and robotics in energy applications, attracting professionals looking to leverage technology for operational efficiency.

Commercial UAV Expo is another major event, typically held in Las Vegas, Nevada. This international trade show focuses on the commercial use of drones across industries such as construction, agriculture, energy, and public safety. The 2024 edition is scheduled for September 3-5, and it promises to feature an array of new technologies and solutions, extensive networking opportunities, and insights from leading industry experts.

DroneX Tradeshow & Conference, held in London, UK, is Europe's largest business event dedicated to the UAV industry. Scheduled for September 24-25, 2024, DroneX brings together key innovators, suppliers, and buyers from the commercial, military, and future flight sectors. The event features live demos, keynote sessions, and extensive networking opportunities.

Global Drone Conference & Exhibition, co-located with the iNNOVATE Tech Show in Kuala Lumpur, Malaysia, is scheduled for October 23-24, 2024. This event focuses on future possibilities, applications, and commercialisation in the drone market, providing a comprehensive overview of the latest developments in the industry.

These events and shows are crucial for the growth and development of the drone industry, providing platforms for stakeholders to collaborate, innovate, and stay updated on the latest trends and technologies. Whether you are a manufacturer, service provider, or end-user, participating in these events can significantly enhance your understanding and engagement with the evolving drone landscape.

Investors and Venture Capitalists

Investment in the drone industry is concentrated primarily in China, Israel, and the United States, each with a distinct focus within the market. China, led by DJI, dominates the consumer market and hardware solutions, accounting for a significant share of North American consumer drone sales. The U.S. companies excel in developing specific commercial hardware solutions and end-to-end software for commercial applications.

Since 2012, investments in the drone industry have totalled close to $1.5 billion, driven by decreasing drone component prices, the massive commercial market potential, and advancements in AI and analytics. Investments continue to grow, with significant funding directed towards autonomous solutions and business intelligence software. In 2017 alone, 52 deals worth $216 million were recorded, with projections indicating a record high by the end of the year.

Investment trends highlight a concentration on early-stage ventures, reflecting the developmental phase of various drone industry sectors. Notable deals include $53 million for 3D Robotics, $34 million for Swift Navigation, $32 million for Airobotics, and $29 million for Echodyne. Despite the nascent stage of the industry, there have been 34 exits and IPOs since 2012, with more anticipated in the next five to ten years.

Venture capital funds and corporate ventures are major players in drone industry investments. Lux Capital, Andreessen Horowitz, and other "smart money" VCs have made substantial contributions, viewing the drone industry as highly lucrative. Corporate ventures, including Qualcomm Ventures, Google Ventures, and Intel Capital, have also been active, particularly in early-stage companies, focusing on technologies like mapping, pipe inspection, delivery, and autonomous solutions.

These investments underscore the dynamic growth potential of the drone industry, highlighting the crucial role of investors in fuelling research, development, and market expansion. As the industry matures, the strategic involvement of venture capitalists and corporate ventures will continue to shape the landscape, driving innovation and commercial viability in drone technology.

Collaborations and Partnerships

Collaborations and partnerships between companies and public sector across the value chain are essential for driving innovation and addressing complex challenges in the drone industry.

Cross-Sector Collaborations: Cross-sector collaborations are essential for the rapid development and integration of drone technologies. These partnerships between hardware manufacturers, software developers, and service providers lead to the creation of comprehensive and innovative solutions tailored to various industries. By pooling their expertise, these collaborators can address complex challenges and accelerate the deployment of advanced drone applications globally.

DJI and DroneDeploy exemplify a successful collaboration where DJI, a leader in drone manufacturing, partners with DroneDeploy, a leading provider of drone software. This collaboration results in integrated solutions for agriculture, construction, and inspection, allowing users to leverage DJI's robust hardware with DroneDeploy's sophisticated mapping and analysis software. Such synergies enable precision agriculture practices, detailed construction site surveys, and thorough infrastructure inspections, enhancing efficiency and productivity in these sectors.

HevenDrones, known for their hydrogen-powered drones, collaborates with Swarmer, a company specializing in multi-drone mission control, to enhance the operational capabilities of their drones for complex missions. This partnership focuses on developing advanced multi-drone systems that can execute coordinated tasks autonomously, beneficial for large-scale agricultural operations and search-and-rescue missions.

In the maritime sector, F-drones partners with various shipping companies to deploy their drones for ship-to-shore deliveries. These drones are designed to transport essential supplies and documents quickly and efficiently, reducing the reliance on traditional delivery methods that are often slower and more costly. This collaboration highlights the potential for drones to revolutionise logistics in maritime industries.

In India, the Digital Sky program has been instrumental in fostering collaborations between governmental bodies, industry players, and stakeholders. This initiative aims to streamline regulations and promote safe and efficient drone operations across various sectors, including agriculture, infrastructure, and logistics. The program's success demonstrates the importance of regulatory support in nurturing an ecosystem conducive to innovation and growth.

Globally, collaborations between tech companies and governmental agencies drive significant advancements in drone technology. For example, the partnership between India's Directorate General of Civil Aviation (DGCA) and the United States Federal Aviation Administration (FAA) focuses on developing regulatory frameworks and standards for safe drone operations. Such international collaborations ensure that drone technologies are implemented safely and effectively, facilitating their integration into critical sectors like healthcare, disaster management, and environmental conservation.

Zipline and Walmart have teamed up to explore drone delivery solutions for healthcare and retail. Zipline's drones are used to deliver medical supplies in remote areas, while Walmart leverages these drones for last-mile delivery of retail goods. This partnership illustrates how cross-sector collaborations can address logistical challenges and improve access to essential services.

In the energy sector, companies like GeoNadir collaborate with oil and gas firms to provide drone-based data mapping and inspection services. These collaborations help monitor and maintain infrastructure, ensuring safety and operational efficiency. By integrating drones with advanced data analytics, these partnerships can detect issues early and prevent costly disruptions.

These examples of cross-sector collaborations highlight the transformative potential of drones across various industries. By combining their strengths, hardware manufacturers, software developers, and service providers can develop integrated solutions that drive innovation, improve operational efficiency, and unlock new market opportunities. As the drone industry continues to evolve, such partnerships will be pivotal in shaping its future.

Public - Private Partnerships: Public-private partnerships (PPPs) play a crucial role in advancing drone technology and developing regulatory frameworks. These collaborations between government agencies and private companies are essential for exploring innovative drone applications, ensuring safety, and integrating drones into national airspaces.

United States: FAA's UAS Integration Pilot Program: In the United States, the Federal Aviation Administration (FAA) collaborates with industry players through initiatives like the UAS Integration Pilot Program. This program aims to explore advanced drone operations and integrate drones safely into the national airspace. By working closely with private companies, the FAA can test new technologies, refine regulations, and gather data on various use cases. These partnerships help bridge the gap between innovation and regulation, fostering a conducive environment for drone technology to thrive.

India: Digital Sky Program: India has made significant strides in fostering public-private partnerships through its Digital Sky program. This initiative streamlines regulations and encourages collaboration between government bodies, industry players, and stakeholders. The program aims to create a supportive environment for drone innovation, facilitating the integration of drones into sectors such as agriculture, logistics, and infrastructure. Partnerships with international organizations like the Directorate General of Civil Aviation (DGCA) and the United States Federal Aviation Administration (FAA) have further enhanced India's regulatory framework, promoting global interoperability and safe drone operations.

Canada: P3 Model for Infrastructure Projects: Canada's mature P3 market provides valuable lessons in best practices for public-private partnerships. The country has successfully implemented numerous P3 projects, which have demonstrated better schedule and cost performance compared to traditional methods. These projects, including those in the drone industry, benefit from clear governance, shared risk allocation, and integrated resources. For example, Canada has established agencies to oversee the growth and accountability of P3 opportunities, driving transparency and consistent approaches in project delivery.

Europe: Collaborative Efforts for Regulatory Harmonisation: In Europe, public-private partnerships have been instrumental in developing harmonized regulatory frameworks for drones. Collaborative efforts between government agencies, private companies, and international organizations like the International Civil Aviation Organization (ICAO) have led to the creation of standardised protocols and ethical practices for drone integration. These partnerships have been crucial in advancing drone technology and ensuring safe and efficient operations across various European countries.

Global Collaborations: Enhancing Connectivity and Disaster Management: Globally, collaborations between tech companies, government bodies, and stakeholders are driving transformative advancements in drone technology. Initiatives such as joint research programs, regulatory dialogues, and knowledge-sharing platforms are pivotal catalysts for innovation. For instance, the ITU's partnerships with organizations like UNICEF aim to improve connectivity in rural areas using satellite technology, which complements drone operations in remote regions. These collaborations enhance early warning systems and disaster management, showcasing the potential of public-private partnerships in addressing global challenges.

Public-private partnerships are essential for the growth and integration of drone technology. By leveraging the strengths of both sectors, these collaborations foster innovation, ensure regulatory compliance, and enable the safe and efficient use of drones across various industries. As the drone industry continues to evolve, the strategic involvement of government agencies and private companies will remain crucial in shaping its future and maximising its potential.

Industry Consortia: Industry consortia play a vital role in the development of global standards for Unmanned Traffic Management (UTM) systems, ensuring the interoperability and scalability of drone technologies. By bringing together stakeholders from various sectors, these groups facilitate collaboration, standardise practices, and drive innovation across the drone industry.

One of the most prominent consortia is the Global UTM Association (GUTMA). This organization includes members from aerospace companies, software developers, and regulatory bodies who work together to create and implement global UTM standards. GUTMA's efforts are crucial for harmonising drone traffic management worldwide, enabling seamless and safe drone operations across borders. Their work includes developing frameworks for data exchange, communication protocols, and operational guidelines that are essential for integrating drones into national airspaces.

In Europe, the SESAR (Single European Sky ATM Research) Joint Undertaking is another key initiative that involves public-private partnerships to enhance UTM systems. SESAR focuses on integrating drones into the European airspace by developing and validating new technologies and operational procedures. This program involves collaboration with various stakeholders, including air navigation service providers, drone operators, and technology companies, to ensure that the European airspace is equipped to handle the increasing number of drones safely and efficiently.

In Asia, the Japan UAS Industrial Development Association (JUIDA) works towards similar goals within Japan. JUIDA collaborates with government agencies, academic institutions, and private companies to advance drone technology and establish regulatory frameworks. This association plays a significant role in promoting the safe and effective use of drones across different sectors, including agriculture, logistics, and disaster management.

Australia's RPAS (Remotely Piloted Aircraft Systems) Consortium is another example of an industry group that brings together diverse stakeholders to develop and standardise drone operations. The consortium includes representatives from government agencies, industry leaders, and research institutions. Their focus is on creating a robust framework for drone integration, ensuring that Australian airspace can accommodate growing drone usage while maintaining safety and efficiency.

The Energy Drone & Robotics Coalition focuses on the energy sector, bringing together companies involved in drone and robotics technology to address industry-specific challenges. This coalition works on developing solutions for using drones in energy operations, including inspections, monitoring, and data collection. By fostering collaboration between energy companies and drone technology providers, the coalition helps to advance the application of drones in the energy sector, driving efficiency and safety improvements.

Red Cat Holdings recently launched the Robotics and Autonomous Systems Consortium to foster collaboration in developing advanced drone technologies for military and commercial applications. This consortium aims to integrate robotic hardware and software, enhancing the capabilities of autonomous systems. Such collaborations are pivotal in driving innovation and setting industry standards for emerging drone technologies.

Industry consortia like GUTMA, SESAR, JUIDA, the RPAS Consortium, the Energy Drone & Robotics Coalition, and Red Cat Holdings' consortium are essential for advancing the drone industry. These groups bring together diverse stakeholders to develop global standards, promote regulatory harmonization, and drive technological innovation. By fostering collaboration and standardising practices, these consortia ensure that the drone industry can grow sustainably and safely, benefiting various sectors worldwide.

The drone value chain is a dynamic and multifaceted ecosystem involving various segments, from hardware manufacturing to software development and service provision. Major industry players and startups each contribute to the growth and innovation of the market through their specialised roles. Collaborations and partnerships across the value chain are essential for addressing regulatory challenges, driving technological advancements, and ensuring the safe and effective integration of drones into society. As the industry continues to evolve, these collaborative efforts will play a pivotal role in shaping the future of drone technology and its applications.

Public actors in the value chain

European Union

The European Union Aviation Safety Agency (EASA) is a key executive agency of the European Commission (EC), tasked with regulatory and executive responsibilities in civil aviation. EASA is empowered to propose technical regulations and frameworks to govern drones of all sizes within Europe. Its primary mission is to ensure the safety, privacy, and environmental protection of the European airspace, leveraging the expertise of international players in the drone domain. EASA promotes innovative regulatory approaches for disruptive technologies within its purview, ensuring that the regulatory environment keeps pace with technological advancements.

National Aviation Authorities (NAAs) are essential in implementing EU laws and safety requirements at the national level. They are responsible for enforcing the Unmanned Aircraft Systems Regulation and rules for operating unmanned aircraft within their respective countries. NAAs collaborate closely with EASA and National Qualified Entities (NQEs), organizations with the technical skills and experience required to implement European drone regulations. NQEs assist member states in enforcing drone rules and often operate under the oversight of NAAs.

National Market Surveillance Authorities play a critical role in ensuring that mass-produced drones comply with harmonization requirements across the EU. They verify that drones placed on the market bear the 'CE marking' and are accompanied by necessary safety information. This surveillance ensures that drones meet the safety standards required for consumer protection and regulatory compliance.

EUROCONTROL is an intergovernmental organization tasked with the coordination and planning of Air Traffic Control (ATC) across Europe. It supports member states in maintaining safe and efficient air traffic management and plays a pivotal role in developing new technologies for integrating drones into the airspace. EUROCONTROL's efforts are crucial for ensuring that drones operate safely alongside manned aircraft, thereby enhancing overall airspace management.

SESAR (Single European Sky ATM Research), established by the EC and EUROCONTROL, focuses on the technical and operational aspects of the Single European Sky initiative. SESAR Joint Undertaking (SESAR JU) is responsible for developing a modern air traffic management system and has been mandated to lead the development of the Drone Traffic Management System (U-Space). U-Space aims to enable the use of fully automated drones in low-level airspace, ensuring safe and efficient integration into the broader airspace system.

JARUS is a collective of experts from NAAs tasked with recommending a unified set of technical, safety, and operational requirements for the certification and safe integration of Unmanned Aircraft Systems (UAS) into the European airspace. JARUS provides guidance materials, including mandatory risk assessment templates, which are crucial for maintaining consistent safety standards across Europe.

The European Innovation Council and Small and Medium-sized Enterprises Executive Agency (EISMEA), which evolved from the European Agency for Small and Medium Enterprises (EASME), informs SMEs and entrepreneurs about applicable drone regulations, including safety, insurance, privacy, and data protection. EISMEA manages research and innovation projects and programs to support SMEs, fostering innovation and ensuring that smaller enterprises can navigate the regulatory landscape effectively.

EASA, in collaboration with NAAs, EUROCONTROL, SESAR, JARUS, and EISMEA, forms a robust regulatory and operational framework for drones within the European Union. These entities work together to ensure that drone operations are safe, innovative, and aligned with international standards. Their collective efforts support the sustainable integration of drones into European airspace, benefiting various sectors and driving technological advancement.

United States

The Federal Aviation Administration (FAA) is the central authority responsible for regulating all aspects of civil aviation in the United States, including drone operations. The FAA's responsibilities include developing and enforcing regulations, providing oversight, and ensuring the safety of the national airspace. This encompasses setting standards for drone operations, certifying drone pilots, and facilitating the integration of drones into the national airspace system.

The FAA actively collaborates with a range of stakeholders, including industry leaders, academic institutions, and other government agencies, to develop a comprehensive regulatory framework. One significant initiative in this regard is the UAS Integration Pilot Program (IPP). This program enables state, local, and tribal governments to partner with private sector entities to accelerate the safe integration of drones. The IPP is designed to test and evaluate various operational concepts such as beyond-visual-line-of-sight (BVLOS) operations, night operations, and flights over people, providing valuable data to inform regulatory decisions.

Moreover, the FAA Reauthorisation Act of 2024 introduces several key provisions that further support the integration and growth of the drone industry. The Act mandates the establishment of a performance-based regulatory pathway for BVLOS operations and continues the BEYOND program, which focuses on using private industry testing and data to refine drone regulations. The Act also allocates significant funding towards infrastructure inspections and drone education, emphasising the role of drones in enhancing public safety and operational efficiency .

The FAA also engages with international bodies like the International Civil Aviation Organization (ICAO) and the Joint Authorities for Rulemaking on Unmanned Systems (JARUS) to harmonise global drone regulations. This international cooperation is crucial for facilitating cross-border drone operations and ensuring a consistent regulatory environment worldwide.

The Drone Advisory Committee (DAC) is another vital element of the FAA's approach, providing a venue for stakeholders to discuss key issues and develop recommendations on matters related to drone integration. The DAC includes representatives from various sectors such as aviation, technology, labor, and state and local governments, ensuring that diverse perspectives are considered in the regulatory process.

In addition to these initiatives, the FAA hosts events like the FAA Drone and Advanced Air Mobility (AAM) Symposium, in collaboration with the Association for Unmanned Vehicle Systems International (AUVSI). This symposium brings together stakeholders to discuss safety, emerging technologies, and the integration of drones and advanced air mobility into the national airspace .

Through these efforts, the FAA ensures that its regulatory framework not only promotes innovation and growth in the drone industry but also maintains the highest standards of safety and efficiency. This proactive and collaborative approach positions the United States as a leader in the global drone industry, fostering a robust ecosystem that supports both commercial and recreational drone operations.

Canada

In Canada, the regulation of drones is managed by Transport Canada the federal institution responsible for overseeing transportation policies and programs. This agency plays a critical role in developing and enforcing regulations that ensure the safe integration of drones into Canadian airspace, aiming to balance technological innovation with safety.

Transport Canada's regulatory framework is comprehensive, focusing on both recreational and commercial drone operations. The Canadian Aviation Regulations (CARs) Part IX outlines specific requirements for operating Remotely Piloted Aircraft Systems (RPAS). This includes pilot certification through Basic and Advanced Operations

Certificates, depending on the complexity of the drone activities. To obtain these certifications, pilots must pass a knowledge test, and for advanced operations, a practical flight review. All drones weighing between 250 grams and 25 kilograms must be registered with Transport Canada, with the registration number clearly marked on the drone. Additionally, operational restrictions are in place to ensure safety, such as maximum flight altitudes, minimum distances from people and properties, and prohibitions on flying near airports and critical infrastructure.

Transport Canada actively engages with industry stakeholders, government agencies, and the public through consultations to gather feedback on proposed regulations. This collaborative approach ensures that the regulatory framework evolves in response to the needs and concerns of the drone community. By participating in international forums and working with organizations like the International Civil Aviation Organization (ICAO), Transport Canada aligns its regulations with global standards, facilitating the integration of Canadian drones into international airspace systems.

Canada's dedication to fostering a safe and innovative drone industry is evident through several key initiatives and collaborations. The establishment of a dedicated RPAS Task Force within Transport Canada ensures ongoing development and implementation of drone regulations. This task force collaborates closely with industry partners to address emerging challenges and promote technological advancements.

The DRONE Centre of Excellence in Alma, Quebec, serves as a hub for research, testing, and certification of drone technologies. This center works alongside Transport Canada and industry leaders to drive innovation and support the development of safe drone applications. Additionally, the National Research Council (NRC) collaborates with Transport Canada on research projects, providing technical expertise on drone integration and safety, which informs regulatory decisions.

Transport Canada's proactive approach to drone regulation, characterised by robust stakeholder engagement and international cooperation, positions Canada as a leader in the global drone industry. The agency's efforts to balance innovation with safety, coupled with initiatives like the RPAS Task Force and the DRONE Centre of Excellence, demonstrate a comprehensive strategy for integrating drones into national and international airspaces. By engaging with stakeholders and staying informed about global developments, Transport Canada ensures its regulatory framework supports the responsible and sustainable growth of the drone sector.

Australia

In Australia, the Civil Aviation Safety Authority (CASA) is responsible for regulating civil aviation, including the operation of drones. CASA's mandate includes developing and enforcing regulations to ensure the safety of national airspace and promoting the responsible use of drones. This is achieved through a collaborative approach involving industry stakeholders, government agencies, and the public.

CASA's regulatory framework is outlined in Part 101 of the Civil Aviation Safety Regulations (CASR), which consolidates rules for various unmanned aeronautical activities, including drones, model aircraft, and other remotely piloted aircraft (RPA). This framework ensures that drone operations are conducted safely and in compliance with established standards.

One of the cornerstone initiatives by CASA is the development of the Remote Pilot Licence (RePL) and the Remote Operator Certificate (ReOC). These certifications are mandatory for commercial drone operators, ensuring they possess the necessary skills and knowledge to operate drones safely. The RePL requires passing an examination and, for more complex operations like beyond-visual-line-of-sight (BVLOS), additional qualifications are needed. This structured certification process underlines CASA's commitment to maintaining high safety standards.

CASA also provides extensive guidance and resources for recreational drone users to promote safe flying practices. This includes clear rules about flight altitudes, distances from people, and no-fly zones. These measures are designed to mitigate risks and ensure that recreational drone operations do not compromise public safety. CASA actively participates in international forums and collaborates with organizations such as the International Civil Aviation Organization (ICAO) and the Joint Authorities for Rulemaking on Unmanned Systems (JARUS). This international cooperation is crucial for harmonising drone regulations globally, facilitating cross-border drone operations, and ensuring that Australia's regulatory framework aligns with international standards. Such collaboration helps Australia stay at the forefront of drone technology and regulation.

CASA's approach to regulation is highly inclusive, involving public consultations to gather feedback on proposed regulations. This process ensures that the regulatory framework is reflective of the community's needs and concerns. CASA's efforts in engaging with the public and industry stakeholders help in creating balanced and effective regulations that support both innovation and safety.

CASA is also proactive in addressing emerging technologies through initiatives like the Drone Rule Digitisation and the Emerging Aviation Technology Partnerships Program. These initiatives aim to modernise and streamline regulatory processes, making it easier for businesses to comply with drone laws and for new technologies to be integrated into the national airspace system. This forward-looking approach ensures that Australia remains competitive in the rapidly evolving drone industry.

CASA's comprehensive regulatory framework, international collaborations, and community engagement efforts play a pivotal role in ensuring the safe and responsible integration of drones into Australia's airspace. By fostering an environment that supports innovation while prioritising safety, CASA helps advance the drone industry in Australia and beyond.

China

The Civil Aviation Administration of China (CAAC) is the national authority responsible for regulating civil aviation in China, including the operation of drones. The CAAC develops and enforces regulations to ensure the safety of the national airspace and to promote the responsible use of drones.

China has emerged as a global leader in drone technology, with companies like DJI at the forefront of the market. The CAAC works closely with these industry leaders to develop a comprehensive regulatory framework that supports innovation while maintaining safety. One of the key regulatory frameworks implemented by the CAAC is the "Regulations on the Administration of Civilian Unmanned Aircraft Systems." These regulations outline the requirements for drone registration, pilot certification, and operational limitations. For instance, any drone weighing more than 250 grams must be registered, and operators are required to provide real-name registration, which includes details about the drone and its owner.

A significant aspect of China's regulatory approach is the use of geo-fencing technology to prevent drones from entering restricted areas. This technology is integrated into drones to automatically restrict flights in no-fly zones, such as near airports and sensitive areas like military installations and government buildings.

The CAAC has also implemented the UAS Traffic Management (UTM) system to integrate drones into the national airspace. This system provides real-time tracking and management of drone operations, enhancing both safety and efficiency. The UTM system is crucial for managing the increasing number of drone flights and ensuring they do not interfere with manned aircraft.

China's regulatory framework is further strengthened by international cooperation. The CAAC collaborates with global organizations and participates in international forums to stay updated on the latest developments in drone technology and regulation. This cooperation helps the CAAC align its regulatory framework with global standards and facilitates international drone operations. For instance, China has been actively engaging with the International Civil Aviation Organization (ICAO) and other international bodies to ensure that its regulations are in sync with global best practices.

Overall, the CAAC's comprehensive approach to drone regulation, which includes stringent registration and operational requirements, the implementation of advanced technologies like geo-fencing, and active international cooperation, ensures that China remains a leader in the drone industry while maintaining the highest standards of safety and compliance.

Singapore

The Civil Aviation Authority of Singapore (CAAS) is responsible for regulating civil aviation in Singapore, including the operation of drones. CAAS develops and enforces regulations to ensure the safety of the national airspace and to promote the responsible use of drones. Singapore has been proactive in integrating drones into its urban environment, with initiatives such as the Smart Nation project. CAAS collaborates with industry leaders, government agencies, and research institutions to develop a comprehensive regulatory framework that supports innovation while ensuring safety.

One of the key initiatives in Singapore is the development of the Unmanned Aircraft Systems (UAS) regulatory framework, which sets out the requirements for drone registration, pilot certification, and operational limitations. CAAS also provides guidance and resources for recreational drone users to promote safe flying practices. Singapore's regulatory approach includes the use of the UAS Traffic Management (UTM) system, which integrates drones into the national airspace. The UTM system provides real-time tracking and management of drone operations, enhancing safety and efficiency.

CAAS collaborates with international organizations and participates in global forums to stay informed about the latest developments in drone technology and regulation. This international cooperation helps CAAS to align its regulatory framework with global standards and to facilitate international drone operations.

South Africa

The South African Civil Aviation Authority (SACAA) is the primary body responsible for regulating all aspects of civil aviation, including the operation of drones, in South Africa. SACAA's regulatory framework is designed to ensure the safety of the national airspace while promoting the responsible use of drones. The regulations, initially established in 2015 and updated as recently as 2023, are among the most comprehensive in Africa, balancing innovation with safety and privacy concerns.

South Africa's drone regulations are codified in Part 101 of the Civil Aviation Regulations. These rules distinguish between private and commercial drone operations. Hobbyist or recreational drone operators do not need to register their drones or obtain a Remote Pilot License (RPL), but they must follow strict safety guidelines. These include maintaining visual line of sight (VLOS), flying below 120 meters (400 feet), and staying clear of people, property, and restricted areas such as airports and national parks.

Commercial drone operations require several certifications, including an RPL, an Air Service License, a Remote Pilot Operator Certificate, and an RPA Letter of Approval. These licenses ensure that operators have the necessary skills and knowledge to conduct safe and legal drone activities. The process includes both theoretical and practical training, medical evaluations, and proficiency in English.

The Drone Council South Africa (DCSA) plays a crucial role in advancing the drone industry. The DCSA is engaged in several projects to modernise the regulatory framework and promote the integration of drones into various sectors of the economy. For example, the "Project Operations Catch Up 2023" aims to update the regulatory framework and align it with the latest technological advancements. This initiative involves extensive consultation with stakeholders and the development of a remote pilot operators' certificate regulatory framework to support SMEs.

The DCSA also focuses on education and training, working with higher learning institutions to incorporate drone-related courses into their curricula. This effort is designed to equip students with the necessary skills and certifications to enter the drone industry, supporting the country's broader goals within the fourth industrial revolution. South Africa collaborates with international bodies to ensure its regulations are in line with global best practices. SACAA's engagement with global forums and regulatory bodies helps facilitate the integration of South African drones into international airspace systems. This cooperation is crucial for maintaining high safety standards and supporting the international operation of South African drone technologies.

South Africa's approach to drone regulation is comprehensive and forward-looking, encompassing strict safety guidelines for private use and rigorous certification requirements for commercial operations. Through the SACAA and the DCSA, South Africa is not only ensuring safe drone operations but also positioning itself as a leader in the African drone industry by fostering innovation and supporting the growth of drone technology across various sectors.

India

In India, the Directorate General of Civil Aviation (DGCA) is the primary body responsible for regulating all aspects of civil aviation, including drones. The DGCA has established a comprehensive regulatory framework to ensure the safety and responsible use of drones, which has been significantly updated with the Drone Rules, 2021 and subsequent amendments in 2023.

The regulatory framework categories drones into five classes based on weight: Nano, Micro, Small, Medium, and Large. Each category has specific operational and certification requirements: Nano Drones (weighing up to 250 grams): Do not require pilot licensing or prior permission if flown below 50 feet. Micro Drones (weighing 250 grams to 2 kg): Require permission for flights above 200 feet and within restricted areas. Small, Medium, and Large Drones: Require more stringent regulations, including pilot licensing and flight permissions through the Digital Sky Platform. This Platform is a crucial component of India's drone regulatory ecosystem. It facilitates the online registration of drones, issuance of Unique Identification Numbers (UIN), and permissions for drone operations through a "No Permission, No Takeoff" (NPNT) policy. This platform ensures that every drone flight is logged, and permissions are managed efficiently to prevent unauthorized drone operations.

The DGCA mandates that drone pilots, especially for commercial operations, obtain a Remote Pilot Certificate. This involves training at DGCA-approved flying schools, covering drone operation basics, safety protocols, and regulatory compliance. The certification process ensures that drone pilots are well-equipped to handle the complexities of drone operations safely. India's drone regulations emphasise safety with several technical requirements for drones: GPS and Return-to-Home (RTH) functionality, Anti-collision lights, Identification plates, Flight data logging capabilities and Real-time tracking and geo-fencing technologies. These measures help in maintaining operational safety and accountability, ensuring that drones do not interfere with manned aircraft or enter restricted zones.

The DGCA collaborates with international organizations like the International Civil Aviation Organization (ICAO) to align its regulations with global standards. This international cooperation is crucial for facilitating cross-border drone operations and ensuring a harmonized regulatory environment worldwide. India is also fostering the growth of its drone industry through various initiatives. The Production-Linked Incentive (PLI) scheme aims to boost domestic manufacturing of drones, while the import of foreign drones has been restricted to encourage local production. This policy is expected to significantly increase the annual sales turnover of the Indian drone industry from approximately INR 60 crore in 2020-21 to about INR 900 crore by 2024-25.

The Indian government is actively working on creating drone corridors for cargo deliveries and easing regulations to promote innovation and adoption of drone technology across sectors like agriculture, healthcare, and infrastructure. With a robust regulatory framework and supportive initiatives, India is positioning itself as a significant player in the global drone industry. The DGCA's comprehensive regulatory framework, coupled with initiatives like the Digital Sky Platform and the PLI scheme, ensures that India not only maintains high safety standards but also promotes the growth and innovation of its drone industry. Through international cooperation and continuous updates to regulations, India is well on its way to becoming a leader in the global drone ecosystem. Africa's approach to drone regulation is comprehensive and forward-looking, encompassing strict safety guidelines for private use and rigorous certification requirements for commercial operations.

Nigeria

The Nigerian Civil Aviation Authority (NCAA) is the main regulatory body overseeing all civil aviation activities in Nigeria, including the operation of drones. The NCAA has developed comprehensive guidelines and regulations to ensure the safety of the national airspace while promoting the responsible use of drones. These regulations cover both recreational and commercial drone operations.

The NCAA mandates that all drones weighing more than 250 grams must be registered. Drone operators, both recreational and commercial, need to follow specific safety guidelines which include maintaining visual line of sight (VLOS), not flying above 122 meters (400 feet), and staying at least 9.26 kilometers (5 nautical miles) away from airports. Additionally, drone pilots must be at least 16 years old and obtain a Remotely Piloted Aircraft Systems (RPAS) certificate to legally operate drones in Nigeria.

Nigeria collaborates with international organizations such as the International Civil Aviation Organization (ICAO) to ensure its regulations align with global standards. This cooperation is essential for facilitating cross-border drone operations and maintaining high safety standards. Additionally, Nigeria is exploring the integration of Unmanned Aircraft Systems Traffic Management (UTM) to enhance the management and safety of drone operations within its airspace .

Despite a strong regulatory framework, Nigeria faces challenges such as the enforcement of regulations due to the large number of unregistered drones and the rapid pace of technological advancements. The NCAA, alongside other stakeholders, is working to address these issues through continuous engagement with the public and private sectors, updating regulations, and enhancing enforcement mechanisms.

Nigeria's approach to drone regulation involves a comprehensive framework managed by the NCAA, supported by initiatives from organizations like the DCSA. These efforts aim to ensure safe and responsible drone use while fostering innovation and integration into various sectors. Through international collaboration and ongoing regulatory updates, Nigeria is positioning itself as a leader in the African drone industry.

Collaboration and International Cooperation

International collaborations are essential for the development and implementation of effective regulatory frameworks. These collaborations involve national aviation authorities, industry stakeholders, government agencies, and international organizations working together to harmonise global drone regulations, facilitate cross-border operations, and ensure that regulatory frameworks remain adaptive to technological advancements.

National Aviation Authorities (NAAs) across the globe, such as the Federal Aviation Administration (FAA) in the United States, the Civil Aviation Safety Authority (CASA) in Australia, the European Union Aviation Safety Agency (EASA), and the Civil Aviation Administration of China (CAAC), are key players in forming collaborative frameworks. These authorities engage in regular dialogue with each other and with international bodies like the International Civil Aviation Organization (ICAO) to standardise regulations and ensure global safety standards are met.

For instance, the UAS Integration Pilot Program by the FAA involves partnerships with local and state governments and private sector entities to test and evaluate advanced drone operations. This program has been crucial in developing policies for beyond visual line of sight (BVLOS) operations, nighttime operations, and flights over people, setting precedents for other countries to follow.

The FAA and EASA have established cooperative agreements to streamline the certification processes for drones and their operators. This collaboration ensures that drones certified in one jurisdiction can be recognised in the other, facilitating smoother international operations. Additionally, both agencies participate in joint working groups to address emerging technological challenges and regulatory needs.

India's Directorate General of Civil Aviation (DGCA) has collaborated with the United States FAA and Drone Alliance Europe to align regulatory standards and facilitate international drone operations. These collaborations include joint research programs and the sharing of best practices in drone regulation and safety protocols.

The CAAC's collaboration with international bodies like ICAO and the Joint Authorities for Rulemaking on Unmanned Systems (JARUS) has been instrumental in harmonising China's drone regulations with global standards. This cooperation facilitates the integration of Chinese drones into international airspace systems and supports global commercial drone operations.

The Single European Sky ATM Research (SESAR) Joint Undertaking, in collaboration with EUROCONTROL, is responsible for developing the U-Space framework. U-Space aims to enable the safe and efficient integration of drones into European airspace by providing services for drone traffic management. This initiative includes partnerships with various European nations and industry stakeholders to test and implement new technologies and operational procedures.

India's Digital Sky Platform exemplifies successful collaboration within the drone industry. This platform facilitates drone registration, pilot certification, and real-time monitoring of drone operations, promoting safe and compliant drone usage. The initiative involves partnerships with both domestic and international stakeholders to ensure that the regulatory environment keeps pace with technological advancements.

Japan has been at the forefront of experimenting with BVLOS operations. Collaborations between the Japanese Civil Aviation Bureau and various industry players have led to successful pilot programs for drone deliveries in rural areas. These initiatives provide valuable data and insights that help shape global BVLOS operational standards.

Events such as the International Drone Show in Denmark and Amsterdam Drone Week facilitate knowledge exchange and collaboration among global stakeholders. These forums bring together regulators, industry leaders, and technology experts to discuss advancements, share experiences, and develop standardised protocols for drone operations.

International collaborations are crucial for the advancement of drone technology and the development of effective regulatory frameworks. By working together, national aviation authorities, industry stakeholders, and international organizations can harmonise regulations, facilitate cross-border operations, and ensure that drone regulations remain relevant and adaptive to technological advancements. As drone technology continues to evolve, ongoing dialogue and collaboration among stakeholders will be essential for developing forward-thinking regulatory frameworks that support innovation and safety on a global scale.

In all these countries, collaboration and international cooperation play a crucial role in the development and implementation of drone regulations. National aviation authorities work closely with industry stakeholders, government agencies, and international organizations to create comprehensive regulatory frameworks that support innovation while ensuring safety. These collaborations help to harmonise global drone regulations, facilitate cross-border operations, and ensure that the regulatory frameworks remain relevant and adaptive to technological advancements. As drone technology continues to evolve, ongoing dialogue and collaboration among stakeholders will be crucial for developing effective and forward-thinking regulatory frameworks.

IV

4. Navigating the Regulatory Landscape for Drones

As we stand on the cusp of technological evolution, drones continue to emerge as a critical component of smart and sustainable cities, promising to enhance urban mobility, food production and delivery services. Yet, with this promise comes a plethora of legal and ethical questions that must be navigated carefully. How do we ensure that drone operations comply with existing airspace regulations while fostering innovation? What mechanisms can be put in place to protect the privacy and safety of citizens from potential misuse? How can we harmonise international drone regulations to facilitate cross-border operations and commerce?

This chapter delves into these intricate questions, exploring the multifaceted regulatory landscape that governs drone technology within the context of smart cities. It examine international, regional and national legal frameworks designed to harmonise drone regulations, ensuring safety and facilitating global operations. After analysing regulatory frameworks from European Union, United States, United Arab Emirates, Canada, , and Singapore, this chapter further provides a comprehensive step-by-step guide to navigating the regulatory requirements essential for integrating drones into smart cities.

Moreover, the discussion expands to address the ethical implications of drone use in smart cities. It tackles issues such as transparency in drone operations, accountability for misuse, and the socio-economic impacts of widespread drone adoption. Can drones be used ethically in urban areas without exacerbating social inequalities? What safeguards are necessary to prevent the abuse of drone technology in smart cities?

Finally, this chapter emphasises the importance of establishing robust regulatory frameworks that balance sustainable innovation with safety and privacy concerns. It advocates for legal professionals to actively participate in expert roundtables and think tanks, contributing to the formulation of comprehensive regulations that will shape the future of drones in the age of artificial intelligence. By positioning legal professionals as thought leaders in this evolving field, this chapter aims to create new insight for the informed decision-making processes required to integrate drones seamlessly into smart city infrastructures. Through collaborative efforts and forward-thinking regulatory approaches, we can pave the way for inclusive and accessible smart and sustainable cities, ensuring that the benefits of drone technology are equitably distributed across all segments of society.

Integrating Drones into Smart Cities

The world is more connected than we have ever imagined. In our technologically advanced, and artificially intelligent era, the integration of drones into our daily lives is emerging as a transformative force in smart cities. Drones are rapidly gaining prominence as a key technology reshaping the future of urban planning, logistics, and various other sectors. With their ability to navigate the skies swiftly and efficiently, drones offer immense potential for revolutionising how we move goods, people, and information. However, this rapid expansion brings complex legal and regulatory challenges that must be meticulously addressed. What are the implications of integrating drones into urban airspaces already congested with traditional aircraft?

Integrating drones into urban airspaces presents a unique set of challenges, particularly regarding air traffic management. Unlike traditional aircraft, drones operate at lower altitudes and navigate through tighter spaces. This raises significant safety concerns about potential collisions and interference with manned aircraft. The implementation of Unmanned Traffic Management (UTM) systems is crucial to managing drone traffic. These systems use real-time data to monitor and control drone operations, ensuring they do not pose risks to other airspace users. Collaborative efforts between national aviation authorities and industry leaders are essential to developing these systems. For instance, the FAA's UTM Pilot Program aims to create a comprehensive UTM framework, addressing these very issues.

How then do we protect citizens' privacy and security in an era of pervasive drone surveillance? The proliferation of drones equipped with high-resolution cameras and advanced sensors has sparked significant privacy concerns. How can we ensure that drones used for surveillance or data collection do not infringe on individuals' privacy rights? Legal frameworks must establish clear guidelines on data collection, storage, and usage. Additionally, incorporating geo-fencing technology can restrict drones from entering sensitive areas. The European Union's General Data Protection Regulation (GDPR) provides a robust model for protecting personal data and privacy, which could be adapted for drone operations. Implementing stringent data protection laws and enforcing compliance will be key to addressing these concerns.

Can international cooperation harmonise drone regulations to facilitate seamless cross-border operations? The drone industry is inherently global, with operations often transcending national borders. How can we harmonise international drone regulations to support seamless cross-border operations? Collaboration among international bodies like the International Civil Aviation Organization (ICAO), JARUS, and various national aviation authorities is crucial. For instance, the joint efforts of the FAA and EASA to streamline certification processes exemplify how international cooperation can facilitate smoother global drone operations. These collaborations help establish standardised protocols, ensuring that drones can operate safely and efficiently worldwide.

What are the ethical implications of widespread drone use in urban environments? Widespread drone use in urban environments raises several ethical questions. Can drones be deployed ethically without exacerbating social inequalities? The potential for drones to provide critical services, such as medical deliveries in underserved areas, is immense. However, ensuring equitable access to these benefits requires careful planning and policy-making. Additionally, transparency in drone operations is vital to maintaining public trust. Developing ethical guidelines and accountability frameworks will help mitigate misuse and ensure drones contribute positively to society.

How can regulatory frameworks balance innovation with safety and privacy concerns? Balancing innovation with safety and privacy is a delicate task. Regulatory frameworks must be flexible enough to accommodate technological advancements while maintaining strict safety and privacy standards. The implementation of sandbox environments, where new technologies can be tested under regulatory supervision, allows for innovation without compromising safety. Furthermore, regular updates to regulations, informed by ongoing technological developments and stakeholder feedback, are essential. By fostering an environment of continuous improvement and adaptation, regulatory bodies can support sustainable innovation.

The integration of drones into smart cities presents both immense opportunities and significant challenges. Addressing these complex legal and regulatory questions is crucial for the responsible and sustainable deployment of drone technology. Through international cooperation, robust regulatory frameworks, and ethical considerations, we can harness the full potential of drones to transform urban mobility while ensuring safety, privacy, and equitable access.

Regulatory Frameworks for Drones in Smart Cities

The rapid integration of drones into smart city mobility requires a comprehensive regulatory framework that ensures a balanced approach to innovation, safety, and societal concerns. It is therefore crucial to understand and analyse the existing legal frameworks governing drones and their application in smart mobility. By doing so, we can identify gaps, propose improvements, and advocate for effective regulations that foster the responsible use of this technology.

At the international level, various organisations such as the International Civil Aviation Organization (ICAO) and the European Union Aviation Safety Agency (EASA) have been actively working on establishing harmonized regulations for drones. These efforts aim to ensure consistency across borders and facilitate the safe integration of drones into airspace. Furthermore, national aviation authorities and regulatory bodies have been developing their own guidelines and requirements specific to their jurisdictions.

The ICAO, a specialised agency of the United Nations, plays a crucial role in setting standards and regulations for international civil aviation. In recent years, ICAO has recognised the growing importance of drones and has taken steps to address their unique regulatory challenges. The organization has established the Remotely Piloted Aircraft Systems (RPAS) Panel, which brings together experts from member states to develop standards and guidance for the safe and efficient operation of drones. These efforts aim to ensure a harmonized approach to drone regulations that takes into account factors such as airspace management, licensing and certification requirements, and operational procedures.

Similarly, the EASA, the European Union's aviation safety agency, has been actively involved in shaping the regulatory landscape for drones within the EU. The agency has developed the EU Unmanned Aircraft Systems Regulation, which sets out the rules for drone operations across EU member states. The regulation introduces a risk-based approach that categories drones based on their capabilities and potential risks. It establishes requirements for drone operators, including registration, pilot qualifications, and operational limitations. Key components of the EASA's approach include the Specific Operations Risk Assessment (SORA) and the Light UAS Operator Certificate (LUC), which provide structured methodologies for assessing and mitigating operational risks.

By implementing these regulations, the EASA aims to create a standardised and harmonized framework that ensures the safe integration of drones into European airspace while promoting innovation and growth in the industry. In addition to international organisations, national aviation authorities and regulatory bodies have been playing a crucial role in developing guidelines and requirements specific to their jurisdictions. Recognising the need to address the unique challenges posed by drones, many countries have established dedicated regulatory frameworks or updated existing aviation regulations to accommodate the safe and responsible use of drones. These regulations cover various aspects such as flight restrictions, privacy considerations, and operational limitations.

European Union: The European Union has developed a comprehensive regulatory framework for drones, spearheaded by the European Union Aviation Safety Agency (EASA). This framework, known as the European Union Unmanned Aircraft Systems (EU UAS) Regulation, aims to harmonise drone regulations across EU member states, ensuring consistency in safety standards, operational rules, and certification requirements. The EU UAS Regulation categories drones into three main categories based on their risk levels: open, specific, and certified. Each category has distinct operational limitations and compliance requirements. For instance, drones in the open category are for lower-risk activities and must adhere to specific conditions, such as weight limits and operational guidelines. From January 1, 2024, new drones placed on the market must carry a C classification marking to be used in the open category, ensuring they meet standardised safety and performance criteria.

One of the critical aspects of the EU's regulatory approach is the mandatory registration and remote identification of drones. Starting in 2024, all drones must have remote identification capabilities to enhance airspace safety and accountability. This requirement aims to prevent unauthorized drone operations and ensure that all drones can be tracked and identified in real-time.

The specific category covers operations that involve a higher risk and require authorisation from national aviation authorities. This includes operations beyond visual line of sight (BVLOS) and those involving heavier drones. Operators must conduct a Specific Operations Risk Assessment (SORA) to identify and mitigate potential risks, ensuring safe operations under varying conditions.

The certified category is for the highest-risk operations, such as the use of drones for commercial air transport of goods or passengers. These operations require stringent certification processes for both the drones and their operators, similar to those for manned aviation.

EASA's regulatory framework also includes provisions for ongoing cooperation with international bodies like the International Civil Aviation Organization (ICAO) to ensure global harmonization of drone regulations. This cooperation helps facilitate cross-border drone operations and ensures that European regulations are in line with global best practices.

Through this robust and detailed regulatory framework, the EU aims to foster innovation in the drone industry while maintaining high safety standards. The regulations not only support the integration of drones into the European airspace but also promote their use in various sectors, including logistics, agriculture, and emergency response, ensuring that the benefits of drone technology are maximised in a safe and controlled manner.

United States: In the United States, the Federal Aviation Administration (FAA) regulates drone operations under Part 107 of the Federal Aviation Regulations. This framework outlines specific requirements for drone operations, including pilot certification, operational limitations, and airspace restrictions, ensuring that drones are integrated safely into the national airspace.

Part 107 specifies that drones must be flown below 400 feet, within the visual line of sight of the operator, and away from restricted airspace, such as near airports and sensitive infrastructure. Operators must pass a knowledge test to obtain a Remote Pilot Certificate, which ensures they understand the necessary safety protocols and regulations. This certification must be renewed every two years through a recurrent knowledge test to maintain operational competency.

For operations that exceed the standard limitations, such as flying at night, over people, or beyond visual line of sight (BVLOS), the FAA provides a waiver system. These waivers allow operators to conduct more advanced missions while ensuring that safety standards are met. Recent updates to the regulations under the FAA Reauthorisation Act of 2024 have streamlined the waiver process and expanded the scope of permissible drone activities, facilitating broader commercial applications like drone deliveries and infrastructure inspections.

Another significant development is the implementation of the Remote Identification (Remote ID) rule, which mandates that drones broadcast identification and location information during flight. This rule enhances airspace awareness, allowing for better tracking and management of drone operations, thus mitigating risks associated with expanded drone use. The enforcement of Remote ID has been extended to March 2024 to give operators additional time to comply with these requirements.

Additionally, the FAA's BEYOND program continues to support the integration of drones into national airspace by leveraging private-industry testing and data collection. This program helps inform policy decisions and develop standards for safe drone deployment. The FAA has also established various advisory committees, such as the Drone Advisory Committee (DAC) and the newly formed Unmanned and Autonomous Flight Advisory Committee, to provide expert guidance on drone-related issues, including certification, operational standards, and risk mitigation strategies.

These regulatory frameworks and initiatives underscore the FAA's commitment to promoting innovation in the drone industry while ensuring safety and compliance.

The United Arab Emirates: (UAE) has established a robust regulatory framework for drone operations, overseen by the General Civil Aviation Authority (GCAA). This framework is designed to ensure the safe and secure integration of drones into the national airspace while addressing privacy concerns and operational safety.

The GCAA's regulations, initially laid out in Federal Resolution No. 2 of 2015 and expanded upon in recent years, categorise drones into different types based on their weight and intended use. These categories help streamline the regulatory requirements for both recreational and commercial drone users. For instance, drones weighing up to 5 kilograms are typically allowed for recreational use within designated areas, while those used for commercial purposes, such as aerial photography, surveying, and delivery, must meet more stringent requirements, including registration and operator certification.

One of the key components of the UAE's drone regulations is the mandatory registration of all drones. This process involves obtaining a unique serial number and geo-fencing microchip, which helps authorities track and monitor drone operations. Additionally, commercial drone operators must secure a UAE Operator Authorisation (UOA), which requires thorough documentation, including security clearance and a detailed description of the intended operations. The UAE also places significant emphasis on safety and privacy. Drones are prohibited from flying near airports, heliports, and other sensitive areas. Operators must ensure their drones are equipped with necessary safety features, such as geo-fencing, and are used in compliance with manufacturer instructions. Privacy concerns are addressed by restricting the use of video and image capturing devices in certain areas and requiring special permissions for such activities.

Recently, the UAE has been proactive in incorporating technological advancements to enhance drone regulation. The implementation of the "UAE Drone Fly Zone Map" app allows users to identify approved flying zones and ensure compliance with airspace restrictions. Furthermore, training and certification programs, such as those provided by the Sanad Academy, are mandated for drone operators, ensuring they possess the necessary skills and knowledge for safe operation. The UAE's approach to drone regulation exemplifies a balanced strategy that supports technological innovation while maintaining high standards of safety and privacy. Through continuous updates and international cooperation, the UAE aims to remain at the forefront of drone technology integration, fostering a secure and innovative environment for drone operations.

Singapore: In Singapore, the Civil Aviation Authority of Singapore (CAAS) has established a comprehensive regulatory framework for drones to ensure their safe and secure integration into the national airspace. This framework encompasses registration requirements, operator permits, and strict operational limitations to safeguard public safety and privacy.

Drones weighing more than 250 grams must be registered with the CAAS. Operators of heavier drones, especially those exceeding 1.5 kilograms, are required to obtain a Unmanned Aircraft Pilot License (UAPL). This license mandates passing both a theoretical test and a practical assessment, ensuring that operators are well-versed in safety protocols, air law, and operational procedures. The theoretical component covers various subjects, including general knowledge of unmanned aircraft systems, principles of flight, air navigation, and human factors, while the practical assessment tests the operator's ability to handle the drone safely in different scenarios.

For specific operations that pose higher risks, such as flying beyond visual line of sight (BVLOS) or within five kilometers of an airport, operators must secure additional permits. These include the Operator Permit and the Class 1 or Class 2 Activity Permit, depending on the nature and location of the flight. This regulatory approach ensures that more complex and potentially hazardous drone operations are closely monitored and controlled. Furthermore, the CAAS has implemented a robust system for designating no-fly zones. These zones include areas around airports, military bases, and other sensitive locations. The CAAS provides tools like the OneMap app, which helps drone operators identify permissible flying areas and ensure compliance with regulations.

Singapore is also at the forefront of integrating innovative drone solutions through initiatives such as the trial of Beyond Visual Line of Sight (BVLOS) operations. These trials involve collaborations with industry stakeholders to explore and refine the use of drones for various applications, including logistics and infrastructure inspections.

Overall, Singapore's regulatory framework for drones, characterised by stringent safety and privacy measures, supports the responsible and innovative use of drone technology. The CAAS's continuous updates and collaborative efforts with industry players ensure that the framework remains adaptive to technological advancements while maintaining high standards of safety and security.

Step by Step Guide to Navigating the Regulatory Landscape

To navigate the regulatory landscape of drones for sustainable and smart mobility, taking into account the case studies in the United States, the European Union (EU), and Singapore, the following step-by-step guide can be formulated:

1. **Research and Understand Applicable Regulations**: Begin by thoroughly researching and understanding the regulations governing drone operations in the target jurisdiction(s). This includes studying both national and international regulations, such as those established by the Federal Aviation Administration (FAA) in the United States, the European Union Aviation Safety Agency (EASA) in the EU, and the Civil Aviation Authority of Singapore (CAAS) in Singapore.

2. **Identify Key Regulatory Bodies and Frameworks**: Identify the key regulatory bodies responsible for overseeing drone operations and their associated frameworks in the respective jurisdictions. For example, in the United States, the FAA plays a significant role, while in the EU, EASA provides harmonized regulations, and in Singapore, CAAS oversees the regulatory framework.

3. **Assess Compliance Requirements**: Conduct a comprehensive assessment to identify the specific compliance requirements and standards set by the regulatory bodies. This may include operational limitations, certification processes, safety management systems, pilot qualifications, and documentation requirements, among others.

4. **Analyse Case Studies**: Study case studies such as Wing, the EU, and Singapore to gain insights into practical implications and challenges faced in integrating drones into sustainable and smart mobility systems. Analyse the steps taken by these entities to meet regulatory requirements, including obtaining certifications, establishing safety protocols, and complying with operational guidelines.

5. **Develop Safety Management Systems**: Develop a robust Safety Management System (SMS) tailored to the jurisdiction's regulatory

requirements. This system should address safety risks associated with drone operations, including risk assessments, safety protocols, training programs, incident reporting mechanisms, and ongoing monitoring and evaluation processes.

6. **Establish Operational Plans**: Create detailed operational plans that encompass various aspects of drone operations, such as flight routes, operational procedures, contingency plans, and risk mitigation strategies. Align these plans with the specific requirements outlined by the regulatory bodies in each jurisdiction.

7. **Pilot Certification and Training**: Ensure that drone operators obtain the necessary certifications and qualifications as per the regulatory requirements. This may involve specific training programs, knowledge tests, flight proficiency assessments, and ongoing training to maintain compliance and proficiency.

8. **Documentation and Application Submission**: Prepare the required documentation, including operational plans, safety management systems, pilot qualifications, and any other supporting documents specified by the regulatory bodies. Submit the application for necessary certifications or permits, following the guidelines and procedures outlined by the respective authorities.

9. **Engage with Regulatory Bodies**: Establish communication and engage with the regulatory bodies throughout the application and certification process. Seek clarification on any uncertainties, address any concerns raised, and maintain an open line of communication to ensure compliance and mutual understanding.

10. **Regular Compliance Monitoring**: Once certifications or permits are obtained, establish mechanisms for regular compliance monitoring. This includes adhering to operational limitations, conducting safety assessments, reporting incidents or accidents, and staying updated with any regulatory changes or amendments.

11. **Stay Informed and Adapt:** Continuously monitor and stay informed about evolving regulatory developments, technological advancements, and best practices in the drone industry. Adapt operational plans,

safety management systems, and compliance measures accordingly to align with changing regulations and emerging trends.

By following this step-by-step guide, stakeholders navigating the regulatory landscape of drones can effectively address the complex legal and regulatory challenges. This approach incorporates a comprehensive understanding of applicable regulations, insights from relevant case studies, and the establishment of robust safety management systems and operational plans, leading to responsible and safe integration of drone technology into society.

Safety and Privacy in Drone Regulation

As drones become integral to the infrastructure of smart cities, addressing safety and privacy concerns is paramount to building public trust and ensuring the sustainable development of this technology. What regulatory measures are necessary to ensure drone operations do not compromise airspace safety or individual privacy? How can regulatory bodies balance the innovative potential of drones with the need for stringent safety and privacy safeguards?

Ensuring the safety of drone operations in urban environments involves implementing rigorous regulations pertaining to flight operations, maintenance, and pilot qualifications. Regulatory frameworks must define clear operational rules, including airspace restrictions, maximum flight altitudes, and prohibited areas. For instance, the Federal Aviation Administration (FAA) in the United States mandates that drones must not exceed 400 feet above ground level and must remain within the visual line of sight of the operator.

Advanced collision avoidance systems, robust communication protocols, and effective remote identification mechanisms are crucial for enhancing the safety of drone operations. Collision avoidance systems use sensors and algorithms to detect and avoid obstacles, reducing the risk of accidents. The FAA's Remote Identification rule, which requires drones to broadcast their identification and location information, is an example of how regulatory measures can enhance airspace awareness and safety (FAA, 2024).

The integration of drones into various sectors raises significant privacy concerns, particularly regarding the collection and use of personal data.

Drones equipped with cameras and sensors can capture sensitive information about individuals, their activities, and their surroundings. To protect privacy rights, regulatory frameworks must include provisions that clearly define the scope of permissible data collection and establish strict limitations on data retention. For example, the General Data Protection Regulation (GDPR) in the European Union provides a robust model for protecting personal data, which could be adapted for drone operations.

Guidelines for the use of surveillance technologies embedded in drones, such as facial recognition or advanced imaging systems, are also essential. Regulations should outline specific circumstances under which these technologies can be used, the duration of data retention, and the purposes for which the collected information can be utilised. The European Union Aviation Safety Agency (EASA) has been proactive in addressing these concerns by developing a risk-based framework that categories drone operations based on their potential impact on privacy and safety.

Transparency and accountability are critical in addressing privacy concerns related to drone operations. Regulatory frameworks should require drone operators to provide clear and accessible information about their data collection practices and any surveillance technologies employed. This includes informing the public about the purpose of data collection, the types of data collected, and how long the data will be retained. Mechanisms for individuals to exercise their rights, such as accessing and rectifying their personal data, should also be established and enforced.

The development of privacy-focused regulations should involve stakeholders from legal, technological, and civil society perspectives. Collaborative efforts involving privacy experts, legal professionals, industry representatives, and consumer advocacy groups can help identify potential risks, address societal concerns, and develop robust privacy frameworks. This collaborative approach ensures that the regulations are comprehensive and reflect the diverse perspectives of all stakeholders involved.

Incorporating privacy considerations into the regulatory landscape surrounding drones in smart cities is essential for fostering public trust and acceptance of this transformative technology. Legal professionals specializing in technology law play a crucial role in guiding the development of these regulations. By striking the right balance between innovation, societal benefits, and the protection of individual privacy rights, regulatory frameworks can support the safe and responsible integration of drones into smart city ecosystems.

Ethical and Social Considerations in Drones Regulation

In addition to safety and privacy concerns, the integration of drones in smart cities presents a range of ethical implications that must be carefully examined. It is essential to address these ethical considerations to foster responsible decision-making and shape the future of drones.

The potential displacement of human workers by drone technology raises important ethical considerations that must be addressed within the regulatory framework. While drones can undoubtedly enhance efficiency and reduce costs in various industries, their widespread adoption may lead to job losses, particularly in sectors heavily reliant on manual labor.

To navigate this ethical challenge, it is imperative to implement measures that mitigate the potential negative impacts on the workforce and facilitate a just transition for affected individuals. One approach is to promote the re-skilling and up-skilling of workers to equip them with the necessary skills to adapt to the evolving job market. This could involve providing training programs and educational opportunities that focus on emerging drone technologies and the changing demands of the industry.

For instance, the European Union's Digital Skills and Jobs Coalition initiative provides a framework to support the workforce in acquiring digital skills necessary for new technologies, including drone operation and management. This initiative aims to ensure that workers in sectors likely to be affected by automation, such as logistics and delivery services, can transition into roles that leverage drone technology effectively.

Similarly, in the United States, the Workforce Innovation and Opportunity Act (WIOA) offers funding for training programs that include drone technology courses. These programs are designed to help displaced workers gain new skills in areas like drone piloting, maintenance, and data analysis. By integrating drone technology training into broader workforce development programs, the WIOA helps workers adapt to technological changes and find new employment opportunities within the drone industry.

In Singapore, the SkillsFuture initiative provides citizens with credits that can be used for a wide range of courses, including those related to drone technology. This program encourages continuous learning and skill acquisition, enabling workers to stay relevant in a rapidly changing job market. By offering specialised courses on drone operations, maintenance, and data management, SkillsFuture ensures that workers can transition smoothly into drone-related roles.

Moreover, regulations can mandate that companies using drones in their operations invest in workforce development programs. For example, the UK's Industrial Strategy includes provisions for companies to partner with educational institutions to create apprenticeship programs focused on emerging technologies, including drones. These apprenticeships provide hands-on experience and theoretical knowledge, preparing workers for the new job opportunities created by drone technology.

Another regulatory approach involves establishing public-private partnerships to develop standardised training and certification programs. The Drone Training and Certification Partnership in Australia, a collaboration between the Civil Aviation Safety Authority (CASA) and industry stakeholders, offers nationally recognised qualifications for drone operators. This ensures that the workforce is well-equipped with the necessary skills and knowledge to operate drones safely and effectively, mitigating the risk of job displacement.

By incorporating these specific examples of regulatory measures, we can address the ethical challenges posed by the integration of drones into various industries. Ensuring that workers have access to training and educational opportunities not only facilitates a just transition but also promotes the sustainable and inclusive growth of the drone industry. Through these targeted efforts, we can balance technological innovation with social responsibility, creating a future where both drones and human workers thrive.

Furthermore, policies and regulations can be put in place to encourage the responsible deployment of drones in a manner that complements human labor rather than replacing it entirely. For example, in the delivery services sector, instead of replacing human delivery workers, drones can be used to complement their efforts by facilitating last-mile deliveries or reaching remote locations. This approach ensures that while drones contribute to increased efficiency, human workers continue to play a vital role in the workforce.

Incorporating drones for last-mile delivery can alleviate the workload for human couriers, particularly in urban areas with high delivery volumes or rural areas where access is challenging. This hybrid model leverages the strengths of both drones and human workers, enhancing overall operational efficiency. Policies supporting this model might include tax incentives for companies that adopt drone technology in ways that augment their human workforce or grants for developing drone-human collaborative logistics systems.

Additionally, the regulatory frameworks can mandate that companies deploying drones invest in human resources development, ensuring their workforce is equipped to work alongside advanced technologies. For instance, the UK's Industrial Strategy includes initiatives where companies partner with educational institutions to create apprenticeship programs focused on drone technology. These programs provide hands-on experience and theoretical knowledge, preparing workers for new job opportunities created by drone technology while ensuring that technological advancements do not lead to job losses but rather job transformation and enrichment.

Moreover, collaborative models such as those promoted by public-private partnerships can further facilitate the integration of drones into the workforce. For example, the Drone Delivery Canada (DDC) initiative has collaborated with indigenous communities to create drone delivery routes that complement existing supply chains without eliminating human jobs. This initiative not only brings modern technology to remote areas but also involves local communities in the operation and maintenance of the drone networks, ensuring that technological integration is inclusive and beneficial to all stakeholders.

Another ethical consideration is the potential impact of drone operations on local communities. As drones become more prevalent, it is crucial to address concerns related to noise pollution, visual intrusion, and the potential disruption of everyday activities. Regulatory frameworks should include guidelines and restrictions to minimise these negative effects and ensure that drone operations do not unduly infringe upon the rights and well-being of individuals residing in the vicinity. In addition, incorporating principles of equity and fairness into the regulatory landscape can help mitigate potential social and economic disparities arising from the adoption of drone technology.

To address noise pollution, for instance, regulatory bodies can set operational requirements based on comprehensive noise impact studies. In the UK, drone regulations mandate specific noise level thresholds and operational distances from residential areas to mitigate noise pollution. Studies have shown that maintaining certain distances between drones and buildings can significantly reduce indoor noise levels, ensuring that drone operations do not disrupt the peace and quiet of local communities.

Visual intrusion and privacy concerns can be mitigated by implementing strict flight path regulations and requiring the use of geo-fencing technology to keep drones out of restricted areas. For example, the Civil Aviation Authority (CAA) in the UK requires drone operators to avoid flying over private property without permission, thereby protecting residents from unwarranted visual intrusion and surveillance.

Incorporating principles of equity and fairness into the regulatory landscape can help mitigate potential social and economic disparities arising from the adoption of drone technology. For example, the FAA's Reauthorisation Act of 2024 includes provisions for establishing community engagement programs to ensure that drone deployment does not disproportionately impact marginalised communities. These programs encourage feedback from residents and involve them in the planning and decision-making processes regarding drone operations in their areas.

Furthermore, promoting inclusive use of drones through targeted incentives can ensure that the benefits of drone technology are equitably distributed. The Drone Training and Certification Partnership in Australia, for instance, offers training programs specifically designed for individuals from underserved communities, equipping them with the skills needed to participate in the growing drone industry. This approach helps bridge the digital divide and provides economic opportunities in areas where traditional job markets may be declining.

By incorporating these specific examples of regulatory and industry measures, we can address the ethical challenges posed by the integration of drones into various industries. Ensuring that drones are deployed responsibly and ethically not only protects local communities but also promotes a balanced and sustainable approach to technological innovation. Through these efforts, we can foster an environment where drones enhance, rather than disrupt, the quality of life in smart cities.

Ensuring equal access to drone-related opportunities and benefits, particularly in marginalised communities, can help prevent the exacerbation of existing inequalities. This can be achieved through targeted policies, incentive programs, or partnerships that promote inclusive participation and distribution of benefits.

By considering the ethical implications of drone technology and incorporating measures to mitigate potential negative impacts, legal professionals with expertise in national and international drone laws can play a pivotal role in shaping regulations that foster responsible and inclusive deployment of drones in the context of sustainable and smart mobility. Their insights can help strike a balance between the advantages of technological innovation and the protection of workers' rights, community well-being, and social equity.

As drones are integrated into smart mobility systems, it is important to avoid exacerbating existing social and economic inequalities. The advantages offered by drone technology should not be limited to certain privileged groups but should be accessible to all members of society. To achieve this, measures must be implemented to ensure affordable access to drone-enabled services and consider the specific needs of marginalised communities.

One approach to promoting inclusivity is to establish programs and initiatives that aim to bridge the digital divide. This can involve providing subsidies or incentives to underserved communities, enabling them to access and benefit from the services offered by drones. Additionally, collaborations between public and private entities can be fostered to develop affordable and accessible drone solutions tailored to the unique requirements of different communities.

By prioritising the equitable distribution of drone-enabled services, we can harness the potential of this technology to address societal challenges and promote social cohesion. For example, in transportation, drones can play a crucial role in enhancing mobility options for individuals with limited access to public transportation or those residing in remote areas. By ensuring that these communities are not left behind in the deployment of smart mobility solutions, drones can contribute to reducing transportation inequalities and enhancing overall connectivity. Another crucial ethical aspect that must be addressed is the potential misuse of drones for malicious purposes. With the growing accessibility and affordability of drones, the risk of unauthorized surveillance, privacy breaches, and even physical harm increases. It is essential to establish robust regulations and effective enforcement mechanisms to prevent misuse and hold accountable those who violate the established rules and regulations.

Regulatory frameworks should incorporate measures to address security concerns, including licensing requirements, remote identification systems (RemoteID), and restrictions on prohibited areas or activities. Additionally, public awareness campaigns can play a significant role in promoting responsible drone use and educating individuals about the potential risks and the importance of adhering to ethical guidelines.

Furthermore, collaboration between stakeholders is essential to combating misuse effectively. Public-private partnerships can facilitate information sharing and the development of innovative solutions to mitigate security risks associated with drones. Engaging technology companies, law enforcement agencies, and civil society organisations can foster a collective effort in ensuring the responsible and safe integration of drones into smart mobility systems.

For example, the European Union Agency for Cybersecurity (ENISA) emphasises the importance of public-private partnerships (PPPs) in enhancing cybersecurity and addressing threats posed by new technologies, including drones. ENISA's Good Practice Guide on Cooperative Models for Effective PPPs outlines strategies for establishing and maintaining successful partnerships, which can be adapted to address drone-related security concerns.

In the United States, the Federal Aviation Administration (FAA) collaborates with private sector stakeholders through initiatives like the BEYOND program, which explores advanced drone operations and integrates findings into regulatory frameworks. This partnership not only enhances regulatory oversight but also leverages private sector innovation to improve drone safety and security (FAA, 2024).

Moreover, RAND Corporation's research highlights the importance of a multilayered approach in combating drone threats in correctional facilities. This approach combines drone detection technologies with core correctional practices and partnerships with law enforcement agencies. Such collaborations enable the sharing of critical intelligence and operational best practices, leading to more effective and coordinated responses to security challenges.

The implementation of collaborative models is also evident in the anti-drone industry. Companies like Sentrycs are at the forefront of developing adaptive counter-drone technologies, which are essential for protecting urban environments and sensitive areas. These technologies are often developed in partnership with government agencies and other private sector entities, ensuring a comprehensive approach to drone security.

Public-private partnerships play a crucial role in fostering innovation and addressing the security challenges posed by drones. By combining the resources and expertise of both sectors, these partnerships can develop robust regulatory frameworks and innovative solutions that ensure the safe and responsible use of drone technology in smart cities. Through these collaborative efforts, we can achieve a balanced approach that leverages technological advancements while safeguarding public safety and privacy.

Role of Legal Professionals in Drone Regulation

Legal professionals with expertise in disruptive technologies have a pivotal role in shaping the ethical and regulatory landscape of the drone industry. Their involvement ensures that comprehensive and forward-thinking regulations are developed, addressing social, economic, and security concerns while promoting safe and responsible use of drones.

One key area where legal professionals contribute is in developing privacy and data protection standards. As drones can capture vast amounts of data, ensuring this data is handled responsibly is crucial. Legal experts draft regulations that define clear boundaries for data collection, storage, and use, ensuring compliance with privacy laws such as the GDPR in the European Union. This protects individual privacy and builds public trust in drone technology.

Legal professionals also facilitate the creation of transparency and accountability frameworks. They advocate for regulations requiring drone operators to disclose their data collection practices and ensure the ethical use of surveillance technologies. This transparency is essential for maintaining public confidence and ensuring that drones are used responsibly in various sectors, from delivery services to infrastructure inspections.

Moreover, legal experts are instrumental in crafting policies that encourage equitable access to drone technology. They support initiatives that provide training and entrepreneurial opportunities in underserved communities, bridging the digital divide and promoting social equity. For instance, public-private partnerships can be designed to offer subsidies for training programs, enabling individuals from marginalised backgrounds to enter the drone industry and benefit from its growth.

In addition, legal professionals help shape the security landscape of drone operations. They work with regulatory bodies to develop guidelines and best practices for mitigating security risks. This includes setting standards for drone certification, operational safety, and the use of counter-drone technologies to prevent malicious activities. The RAND Corporation emphasises the importance of multilayered security strategies, involving legal, technical, and operational measures to protect against drone threats.

Finally, legal professionals play a crucial role in international collaboration on drone regulations. By participating in global forums and working with international organizations, they help harmonise regulatory frameworks across borders. This facilitates cross-border drone operations and ensures that safety and ethical standards are maintained globally. Through their expertise, legal professionals ensure that the integration of drones into smart cities is guided by principles of fairness, inclusivity, and societal well-being. Their contributions are vital in balancing technological innovation with the need for comprehensive regulations that address the ethical and practical challenges posed by drone technology.

Expert Roundtables and Think Tanks on Drone Regulation

A strong focus on drones and smart cities requires the active engagement of key stakeholders in expert roundtables and think tanks to contribute to informed decision-making regarding the future of this technology. These platforms provide invaluable opportunities to exchange knowledge, foster collaboration, and shape the regulatory landscape for drones. By participating in these forums, legal professionals can leverage their knowledge and influence to engage policymakers, industry stakeholders, and researchers in discussions that shape regulations, policies, and ethical guidelines for drones.

Expert roundtables serve as a dynamic platform that brings together a diverse range of stakeholders, including legal experts, technologists, policymakers, industry representatives, and academic researchers. These gatherings facilitate collaborative discussions where insights from different perspectives are shared, and ideas are exchanged. By actively participating in expert roundtables, legal professionals can showcase their expertise in technology law and make compelling arguments supported by empirical research and legal analysis.

One of the primary objectives of engaging in expert roundtables is to shape the narrative surrounding drones. Legal professionals can play a vital role in advocating for balanced regulations that consider the unique aspects of drone technology while addressing legal, ethical, and societal concerns. By providing informed perspectives, they can influence the direction of discussions and help draft regulations that strike the right balance between promoting innovation and ensuring safety, privacy, and security.

Furthermore, expert roundtables offer an opportunity to collaborate with other stakeholders, including representatives from the industry, government agencies, and civil society organisations. Through collaborative efforts, legal professionals can contribute to the development of comprehensive and effective regulatory frameworks. By actively engaging with industry stakeholders, they can gain insights into the practical implications of regulations and ensure that the legal framework is aligned with industry realities. This collaborative approach helps bridge the gap between legal requirements and technological advancements, fostering an environment that supports the responsible and sustainable integration of drones into smart mobility systems.

In addition to expert roundtables, participation in think tanks can also provide a platform for legal professionals to contribute to informed decision-making. Think tanks bring together multidisciplinary experts who delve into complex issues and provide in-depth analysis and recommendations. Legal professionals can contribute their expertise in technology law to these forums, sharing insights on legal challenges, proposing innovative approaches, and influencing the development of policies and guidelines. By actively engaging in expert roundtables and think tanks, legal professionals can position themselves as thought leaders in the field of technology law and drones. Their active involvement in these forums allows them to make influential contributions to the ongoing discussions on the regulatory landscape for drones in smart mobility. Through their expertise and persuasive arguments, they can shape the opinions and perspectives of policymakers, industry stakeholders, and researchers, guiding them towards decisions that prioritise safety, privacy, equity, and societal well-being.

Active engagement in expert roundtables and think tanks provides legal professionals with a unique opportunity to contribute to the future of drones in smart cities By leveraging their expertise and presenting strong arguments based on empirical research and legal analysis, they can influence the development of regulations, policies, and ethical guidelines. Through collaboration, knowledge exchange, and advocacy, legal professionals can help shape a regulatory landscape that fosters responsible and inclusive drone integration, while addressing legal, ethical, and societal considerations.

In recent years, the integration of drones into smart cities and the transportation ecosystem has been a major focus of discussions among experts, policymakers, and industry leaders. Numerous roundtables, think tanks, and conferences have been convened to address the regulatory challenges and opportunities presented by drone technology, particularly in the age of artificial intelligence (AI).

One prominent example is the European Drone Summit, an annual event that gathers industry experts, regulators, and researchers to discuss the latest developments in drone technology and regulation. The summit features panels on AI-driven drone innovations, regulatory frameworks, and the integration of drones into urban environments. Expert opinions at these summits emphasise the need for a balanced regulatory approach that fosters innovation while ensuring safety and privacy.

Another significant forum is the International Conference on Unmanned Aircraft Systems (ICUAS), which brings together experts from academia, industry, and government to discuss advancements in drone technology and its regulatory implications. Discussions at ICUAS often focus on the technical challenges of integrating AI into drones, such as machine learning for autonomous navigation and AI-powered image recognition for surveillance. Experts debate the ethical considerations and potential risks, including algorithmic bias and decision-making transparency.

T he European Aviation Safety Agency (EASA) regularly hosts workshops and roundtables on drone regulation. These events provide a platform for stakeholders to discuss the implementation of the EU's harmonized drone regulations and the integration of AI technologies. EASA's discussions highlight the importance of international collaboration and the need for harmonized standards to facilitate cross-border drone operations.

During these roundtables and think tank discussions, several key themes and debates emerge regarding the future of drone integration into smart cities and the transportation ecosystem. One major theme is the balance between innovation and regulation. Experts argue that overly stringent regulations can stifle innovation, while too lenient approaches may lead to safety and security risks. The consensus is that a risk-based regulatory framework, which categories drone operations based on their level of risk, is essential for managing these challenges.

Another critical debate centers on the role of AI in enhancing the capabilities of drones. Proponents highlight the potential of AI to improve safety, efficiency, and scalability. For instance, AI-driven drones can autonomously navigate complex environments, perform real-time data analysis, and execute tasks with minimal human intervention. However, critics raise concerns about the transparency and accountability of AI systems. They emphasise the need for explainable AI (XAI) that can provide clear and interpretable explanations for its decisions.

Privacy and data protection are also major points of discussion. Experts stress the importance of robust privacy regulations to protect individuals from unwarranted surveillance and data collection by drones. They advocate for stringent data protection measures, including anonymization and encryption, to ensure that drone operations comply with privacy laws such as the General Data Protection Regulation (GDPR).

Several innovative regulatory ideas and policies have emerged from these discussions. One notable proposal is the concept of regulatory sandboxes, which allow companies to test new drone technologies and operational models in a controlled environment with relaxed regulatory constraints. This approach enables regulators to observe and evaluate new developments in real-time, providing valuable data and insights to inform future regulations.

Another proposed policy is the implementation of dynamic geofencing and U-space services. Dynamic geofencing involves creating virtual boundaries that drones cannot cross, which can be updated in real-time based on current airspace conditions. U-space services, a set of digital infrastructure and traffic management services, facilitate the safe and efficient integration of drones into the airspace. These services include real-time traffic information, flight authorisation, and conflict detection and resolution.

Collaborative frameworks are also being suggested to enhance regulatory effectiveness. These frameworks involve partnerships between government agencies, industry stakeholders, and research institutions to develop common standards, share best practices, and promote interoperability. Such collaboration can help address the complex challenges of drone regulation and ensure that policies remain relevant and adaptive to technological advancements.

Expert roundtables and think tank discussions provide a rich source of insights and ideas for regulating drones in the age of artificial intelligence. These forums emphasise the need for a balanced approach that fosters innovation while ensuring safety, privacy, and accountability. The proposed policies and regulatory ideas, such as regulatory sandboxes, dynamic geofencing, and collaborative frameworks, offer promising pathways for integrating drones into smart cities and the transportation ecosystem. As drone technology continues to evolve, ongoing dialogue and collaboration among stakeholders will be crucial for developing effective and forward-thinking regulatory frameworks.

V

5. Coexisting Mindsets for Drone Regulations

The rapid evolution and widespread adoption of drone technology have introduced new frontiers of possibilities, yet they also bring significant risks that demand careful regulatory oversight. In response, the European Union has crafted a comprehensive legal framework to address these challenges, focusing primarily on two fundamental principles: safety and privacy. This regulatory intervention is not merely traditional but innovative, establishing a robust set of risk-based rules and guidelines designed to ensure high standards of safety, privacy, and environmental stewardship in the operation of drones, from simple recreational uses to complex commercial applications within European airspace.

The EU's approach reflects a deep understanding of the intricate balance required to foster technological advancement while safeguarding public interest. By setting clear boundaries and operational standards, these regulations aim to mitigate the risks associated with drone technology, ensuring that its benefits can be fully realised without compromising the safety and privacy of European citizens. This framework not only protects individuals and communities but also paves the way for sustainable and responsible growth in the drone industry.

The harmonised EU drone rules and guidelines entered into force on the 1st of January, 2021 and forms a significant part of the Single European Sky Strategy from September, 2023. These rules and guidelines offer a broad range of solutions deeply rooted in a risk-based approach with rules for drone airspace integration and standards for the design of law-abiding drones.

Risk-based approach to regulation describes, develops and applies suitable codes and requirements that can be traced to, and are proportionate with, the degree of operational risk posed by a given drone system. Risk-based approach in the context of aviation regulatory processes involves specification of regulations (rule-making), assessment against regulations (compliance assessment) and the decision making process used to judge compliance (compliance finding).

Furthermore, the risk-based approach to regulation recognises that drones come in various sizes, shapes, and capabilities, and that different types of operations pose varying levels of risk to people, property, and the environment. The EU drone rules and guidelines have classified drones into categories based on their weight, performance, and intended use. The categories are open, specific, and certified, and they correspond to different regulatory requirements and obligations. The open category includes drones that are considered low-risk, such as those that weigh less than 250g and those that operate in limited areas. The specific category includes drones that operate in more complex environments or that pose higher risks, and they require a risk assessment and authorisation from the national aviation authority. Finally, the certified category includes drones that are used for commercial operations and that require a certificate of airworthiness and compliance with strict standards and regulations.

The risk-based approach to regulation also recognises that the integration of drones into the airspace requires a collaborative effort from all stakeholders, including regulators, drone manufacturers, operators, and users. The EU drone rules and guidelines promote the concept of "shared responsibility," which means that all stakeholders are responsible for ensuring the safe and responsible use of drones. For instance, drone manufacturers are required to design and produce drones that meet the safety and performance standards set by the regulatory authorities. Operators are required to conduct a risk assessment before each flight, to ensure that the operation does not pose a danger to people, property, or the environment. Users are required to comply with the rules and guidelines, and to respect the privacy and security of others.

EU Legal Principles applicable to commercial drones

The regulation of commercial drones in the European Union extends beyond public law and encompasses principles derived from EU private law. A critical principle from EU private law applicable to commercial drones is liability. Under EU law, drone operators and manufacturers can be held liable for damages caused by drone operations. This principle ensures that there is accountability and that victims of drone-related incidents can seek compensation. For instance, the EU's regulations mandate that drone operators carry adequate insurance to cover potential liabilities, ensuring financial responsibility and protection for third parties (EASA, 2024).

Another important aspect is contract law, which governs agreements between various parties in the drone ecosystem, including service providers, operators, and clients. These contracts often include clauses that delineate responsibilities, service levels, and dispute resolution mechanisms, providing a clear legal framework for commercial transactions involving drones. Contract law principles are applicable in various aspects of commercial drone operations, such as the purchase and sale of drones, service agreements between drone operators and clients, or agreements between manufacturers and suppliers. Contractual obligations, warranties, and liabilities are governed by contract law principles, ensuring legal certainty and protecting the rights and interests of all parties involved. The enforcement of these contracts under EU private law principles ensures that all parties adhere to their obligations, fostering trust and reliability in the drone industry.

Intellectual property rights (IP) also play a significant role. As drone technology advances, protecting innovations through patents, trademarks, and copyrights becomes crucial. EU private law provides robust mechanisms for protecting intellectual property, encouraging innovation by ensuring that creators can reap the benefits of their inventions. This protection not only incentives technological advancements but also ensures that companies can compete fairly in the market.

Consumer protection is another critical principle. EU private law includes stringent consumer protection regulations that apply to drone sales and services. These regulations ensure that consumers are informed about the capabilities and limitations of drones, receive adequate product safety information, and have avenues for redress in case of defects or non-compliance. By upholding high standards of consumer protection, the

EU fosters a market where consumers can trust the products and services they purchase.

Conflict of laws is also relevant, especially given the cross-border nature of drone operations. EU private international law principles help determine which jurisdiction's laws apply in cases involving multiple countries. This is particularly important for resolving disputes and ensuring legal clarity in international drone operations, which are becoming increasingly common as the technology evolves (European Commission, 2024).

Tort law principles play a crucial role in determining liability for harm caused by commercial drones. Under EU private law, individuals or entities that operate drones can be held liable for damages resulting from their negligent actions or omissions. The principles of negligence, duty of care, and causation are instrumental in assessing liability and ensuring accountability in cases of drone-related incidents.

Private international law principles govern the resolution of legal disputes involving cross-border drone operations. In cases where a drone operator, manufacturer, or victim resides in a different EU member state, private international law principles guide the determination of the applicable jurisdiction, choice of law, and recognition and enforcement of judgments. These principles facilitate the harmonization and coordination of legal proceedings across different jurisdictions within the EU.

The EU legal framework places significant importance on protecting fundamental rights, including privacy, data protection, and property rights, which are relevant in the context of commercial drone operations. The principles enshrined in the EU Charter of Fundamental Rights and other human rights instruments provide a foundation for safeguarding individuals' rights and balancing them against the legitimate interests served by drones.

Understanding the interplay between these EU private law principles and the public law framework is crucial for comprehensively regulating commercial drone operations. The application of private law principles helps address legal disputes, allocate liability, and protect the rights of individuals and entities affected by drone activities. In the subsequent chapters, we will delve deeper into the specific application of these EU private law principles and their implications for commercial drone operations. By examining case studies, legal analyses, and emerging trends, we aim to provide a comprehensive understanding of the complex regulatory landscape surrounding commercial drones in the European Union.

EU Common Rules in the Field of Civil Aviation

In addition to the principles derived from EU private law, the European Union has established common rules in the field of civil aviation that are pertinent to the regulation of commercial drones. This section provides an overview of these common rules and their significance in shaping the regulatory framework for drone operations in the EU.

1. **Unmanned Aircraft Systems (UAS) Regulation:** The EU has adopted specific regulations for unmanned aircraft systems, which encompass drones. The UAS Regulation sets out the common rules and procedures for the operation of drones in the EU, including requirements for registration, certification, remote identification, and operational limitations. These regulations aim to ensure the safe and responsible operation of drones, addressing concerns related to privacy, security, and environmental impact.

2. **Rules for the Operation of Unmanned Aircraft:** The EU has also developed rules specifically tailored to the operation of unmanned aircraft, taking into account their unique characteristics and potential risks. These rules cover areas such as flight restrictions, airspace integration, pilot qualifications, and insurance requirements. By establishing comprehensive operational rules, the EU seeks to minimise the risks associated with drone operations and promote their safe integration into the existing aviation ecosystem.

3. **Risk-Based Approach:** The common rules in the field of civil aviation adopt a risk-based approach to drone regulation. This approach entails assessing the level of risk associated with different drone operations and applying appropriate regulations accordingly. Factors such as the weight and capabilities of the drone, the type of operation, and the location of flights are taken into consideration to determine the regulatory requirements. This approach allows for a more nuanced and proportional regulation of drone activities.

4. **Harmonization of Standards:** The EU has developed a harmonized set of technical standards and requirements for drones, ensuring a consistent level of safety and performance across member states. These standards cover aspects such as drone design, construction, maintenance, and operation. By establishing common rules, the EU

aims to facilitate the free movement of drones within the internal market while ensuring a high level of safety.

By implementing common rules in the field of civil aviation, the EU aims to create a unified regulatory framework that ensures the safe and efficient integration of drones into the airspace. These rules provide a foundation for member states to develop their national regulations, while also facilitating cross-border drone operations and harmonising safety standards throughout the EU. In the subsequent chapters, we will delve deeper into the specific provisions of the EU common rules in the field of civil aviation and analyse their practical implications for the regulation of commercial drones. Through case studies, legal analysis, and industry insights, we aim to provide a comprehensive understanding of the evolving regulatory landscape and its impact on the sustainable and responsible use of drones in the EU.

Unmanned Aircraft System Regulation

The European Union has developed specific regulations and rules that govern the operation of unmanned aircraft systems (UAS), including drones. These regulations, collectively known as the Unmanned Aircraft Systems Regulation and the Rules for the Operation of Unmanned Aircraft, play a crucial role in ensuring the safe and responsible use of drones within the EU. This section provides an overview of these regulations and their key provisions.

1. **Unmanned Aircraft Systems Regulation:** The Unmanned Aircraft Systems Regulation sets out the common rules and procedures for the operation of drones in the European Union. It encompasses various aspects of drone operations, including registration, certification, remote identification, operational limitations, and privacy protection. These rules aim to establish a harmonized framework that ensures the safety, security, and privacy of drone operations while promoting innovation and facilitating the integration of drones into the European airspace.

2. **Operational Limitations:** The regulations define operational limitations that drone operators must adhere to when conducting flights. These limitations include restrictions on flight altitude, distance from people and buildings, operation in controlled airspace,

and adherence to no-fly zones or prohibited areas. By imposing these operational limitations, the regulations aim to mitigate potential risks associated with drone operations and ensure the safety of people and property on the ground.

3. **Remote Identification:** The regulations require drones to be equipped with a remote identification system. This system enables the identification of drones in real-time by authorities, allowing for enhanced safety and security oversight. Remote identification helps prevent unauthorized drone operations, facilitates law enforcement actions when necessary, and promotes accountability and responsible behaviour among drone operators.

4. **Authorisation and Certification:** The regulations establish procedures for obtaining authorisation and certification for specific drone operations. Depending on the nature of the operation, drone operators may need to obtain specific authorisations, such as operating in certain airspace or conducting specialised activities like aerial work or commercial deliveries. Additionally, certain categories of drones may require certification to ensure compliance with technical and safety standards.

These regulations reflect the EU's commitment to fostering a favourable environment for drone operations while ensuring the highest level of safety and accountability. They address critical aspects of drone operations, including registration, operational limitations, remote identification, and authorisation procedures. By implementing these regulations, the EU aims to strike a balance between enabling the innovative potential of drones and mitigating potential risks and concerns associated with their use.

In the upcoming chapters, I will delve deeper into the specific provisions of the Unmanned Aircraft Systems Regulation and the Rules for the Operation of Unmanned Aircraft. Through case studies, legal analysis, and practical insights, we will explore the implications of these regulations for drone operators, manufacturers, and other stakeholders. Our goal is to provide a comprehensive understanding of the regulatory framework and its impact on the sustainable and responsible integration of drones into the European airspace.

Current and Future perspectives in Drone Regulations

How can regulatory frameworks evolve to address the persistent inefficacies inherent in traditional normative interventions? It is well understood that even stringent regulations often fail to entirely prevent unwanted behaviours. For instance, murder is universally prohibited and subject to severe penalties, yet it continues to occur. This reality underscores the limitations of conventional regulatory regimes, which are never entirely foolproof. However, this does not imply that regulators should abandon rule-making altogether. Instead, it calls for the development and implementation of more innovative and effective forms of regulation that complement traditional approaches. By exploring adaptive regulatory strategies, leveraging technology, and fostering collaborative governance, we can enhance the efficacy of regulatory interventions and better manage societal risks.

The regulatory landscape for drones, particularly in the European Union, presents a unique set of challenges that mirror the complexities seen in other areas of law and policy. The rapid advancement of drone technology, especially with the integration of AI, demands dynamic regulatory responses that can adequately address both current and emerging issues. While traditional normative regulatory frameworks provide a foundation, they often struggle to keep pace with technological innovation. This results in a regulatory environment that may not fully capture the nuances of drone operation and its implications across various sectors.

One significant challenge is the scalability of regulations to accommodate an increasing variety of drone applications. From commercial deliveries and agricultural monitoring to emergency response and surveillance, the diverse use cases for drones require differentiated regulatory approaches. A one-size-fits-all model proves inadequate, leading to potential over-regulation in some areas while under-regulating in others. This necessitates a more granular approach to rule-making, where regulations are tailored to specific uses and risks associated with different types of drone operations.

Furthermore, cross-border operations of drones present another complex regulatory challenge within the EU. As drones can easily traverse national boundaries, consistent and harmonized regulations across member states are essential to prevent regulatory arbitrage and ensure safety and privacy standards are maintained throughout the Union. This requires not only cooperation among EU countries but also a central framework that can reconcile national legal systems with overarching EU policies.

How can future regulatory frameworks for drones in the EU evolve to dynamically leverage technology for better enforcement and compliance? The perspective on drone regulations in the EU is increasingly leaning towards adaptive frameworks that integrate advanced technologies. For example, employing blockchain technology could maintain a tamper-proof record of drone activities, ensuring data integrity and traceability. Additionally, artificial intelligence (AI) could be utilised to monitor drone operations in real-time, ensuring adherence to designated flight paths and compliance with no-fly zones. Such technologies not only enhance enforcement capabilities but also increase transparency and accountability in drone operations, paving the way for a more robust and responsive regulatory environment.

In addition, the concept of "co-regulation" is gaining traction. This approach involves stakeholders, including drone manufacturers, operators, and user communities, in the regulatory process. By doing so, regulators can harness the expertise of industry participants and technology developers to craft more practical and effective regulations. This collaborative approach can also foster innovation while ensuring safety, privacy, and security standards are met.

By addressing these challenges and embracing forward-looking regulatory strategies, the EU can create a robust framework that not only mitigates the risks associated with drone technologies but also maximizes their societal benefits. Such a progressive stance is essential as drones continue to evolve and become increasingly embedded in the fabric of European socio-economic activities.

How can the European Union address the persistent challenges in drone regulations to ensure their future effectiveness? While the EU has made significant strides in developing comprehensive drone regulations, several obstacles remain that require ongoing evaluation and adaptive strategies. As I delve into some of these key challenges and outline potential future directions for enhancing drone regulations in the EU, I assess the possibility of updating regulatory frameworks, integrating new technologies, and fostering international cooperation. I argue that by adopting adaptive, dynamic, and sophisticated regulatory frameworks that leverage advanced technologies, such as blockchain for secure and transparent record-keeping and AI for real-time monitoring and compliance, the EU can effectively navigate the normative complexities of drone operations. Furthermore, through continuous evaluation and international collaboration, the EU can maintain and even elevate a high standard of safety, privacy, and accountability, ensuring that drone technology is harnessed responsibly and beneficially.

How can the European Union create a cohesive and adaptable regulatory environment for drone operations amidst diverse national implementations, evolving technological landscapes, and varying public perceptions? As the EU strives to refine its drone regulations, it must address several key challenges that impede the creation of a unified and effective regulatory framework. These challenges encompass issues such as the lack of uniformity in national implementation, significant privacy and data protection concerns, the complexities of urban integration and airspace management, the rapid pace of technological advancements coupled with safety considerations, and the fluctuating public perception and acceptance of drone technology. Understanding and addressing these multifaceted challenges is crucial for the EU to maintain high standards of safety, privacy, and accountability in the burgeoning field of drone operations. These challenges include:

1. **Lack of Uniformity in National Implementation:** Although the EU has introduced harmonized drone regulations, the implementation and enforcement of these regulations can vary across member states. This lack of uniformity poses challenges for drone operators, particularly those conducting cross-border operations. Harmonising the interpretation and application of regulations among member states is crucial to ensure consistency and streamline operations within the internal market. How can the EU effectively align national enforcement practices to create a truly unified regulatory framework that accommodates both local specificities and the broader goals of safety, innovation, and cross-border operational efficiency?

2. **Privacy and Data Protection Concerns:** The use of drones raises significant privacy and data protection concerns. Drones equipped with cameras and sensors can collect vast amounts of personal data, potentially infringing on individuals' privacy rights. Ensuring compliance with existing data protection regulations, such as the General Data Protection Regulation (GDPR), and establishing guidelines specifically tailored to drone operations are essential for

safeguarding individuals' privacy while facilitating the legitimate use of drone technology. How can the EU develop and enforce a regulatory framework that effectively balances the innovative potential of drones with the stringent requirements of privacy and data protection, particularly in the context of rapidly evolving technological capabilities?

3. **Urban Integration and Airspace Management:** As drones become more prevalent in urban areas, the integration of drones into existing airspace management systems poses considerable challenges. Ensuring safe and efficient integration, avoiding conflicts with manned aircraft, and managing the increased traffic density require advanced technological solutions and coordination between drone operators, air traffic management authorities, and urban planners. How can the EU develop and implement a comprehensive urban airspace management strategy that not only leverages cutting-edge technology for conflict avoidance and traffic management but also ensures seamless cooperation among diverse stakeholders to support the dynamic and safe operation of drones in densely populated areas?

4. **Technological Advancements and Safety Considerations:** The rapid evolution of drone technology brings both opportunities and challenges in terms of safety. Emerging technologies such as artificial intelligence, autonomous systems, and beyond-visual-line-of-sight (BVLOS) operations present new possibilities for drone applications but also require robust safety mechanisms and regulations. Striking a balance between promoting technological innovation and ensuring safety is crucial to harness the full potential of drones. How can the EU craft a regulatory framework that not only fosters innovation and accommodates cutting-edge advancements but also implements stringent safety standards and anticipatory measures to mitigate the inherent risks associated with these rapidly evolving technologies?

5. **Public Perception and Acceptance:** Public perception and acceptance of drones play a vital role in shaping the regulatory landscape. Concerns about privacy, noise pollution, and safety incidents can influence public opinion and policymakers' decisions. Engaging in public dialogue, raising awareness about the benefits

and responsible use of drones, and addressing concerns through transparent and participatory processes are essential for building public trust and acceptance. How can the EU effectively engage the public and stakeholders in the regulatory process to ensure that the evolving drone technology is not only accepted but also actively supported by the community, balancing innovation with the legitimate concerns of privacy, safety, and environmental impact?

Looking ahead, several future perspectives lies in the ability to address these complex challenges and the questions they pose. The answers will guide the development of EU drone regulations of the future, ensuring they remain relevant and effective in a rapidly evolving technological landscape. Therefore, it is essential for drone regulations to embody continuous adaptation and flexibility. As drone technology advances and new use cases emerge, regulations must evolve to keep pace. This requires regular updates and revisions, informed by stakeholder feedback, technological developments, and societal needs. Such a dynamic approach ensures that the regulatory framework remains future-proof and responsive to the challenges and opportunities presented by the ever-changing drone industry.

International cooperation and standardisation are also crucial. Collaboration with international partners and standardisation bodies helps establish common frameworks that promote interoperability of drone operations across borders. Aligning EU regulations with international best practices and fostering global cooperation can facilitate the harmonisation of rules, creating a conducive environment for cross-border drone operations. This alignment is essential for the seamless integration of drones into global airspace, supporting international commerce and enhancing safety standards worldwide.

Balancing safety with innovation through innovation-friendly regulations is another key perspective. Encouraging responsible innovation within the regulatory framework is vital for the growth of the drone industry. Implementing regulatory sandboxes and experimentation programs can provide controlled environments where new technologies and applications can be tested and validated without compromising safety. These initiatives can help foster technological advancements while ensuring that safety standards are rigorously maintained.

Enhanced enforcement and compliance mechanisms are paramount to address safety risks and prevent unauthorised operations.

Strengthening these mechanisms involves effective monitoring, surveillance, and imposing penalties for non-compliance. Such measures can deter unsafe practices and promote responsible drone operations. Ensuring that drone operators adhere to regulations is critical for maintaining public trust and safety in the skies.

Stakeholder collaboration plays a vital role in developing inclusive and effective regulations. Engaging all relevant stakeholders, including drone operators, manufacturers, industry associations, regulatory bodies, and the public, is essential. Collaborative approaches that foster dialogue, knowledge sharing, and multi-stakeholder participation can lead to well-informed and balanced regulations. These regulations should reflect the diverse perspectives and interests involved, ensuring that the needs and concerns of all parties are addressed.

By addressing these challenges and adopting future-oriented perspectives, the EU can build an effective regulatory framework that enables the safe, sustainable, and innovative use of drones across various sectors. The next chapters of this book will delve further into specific aspects of EU drone regulations, examining their effectiveness and exploring potential strategies for improvement and advancement.

From Normative to Non-Normative Dimensions

How will the evolution of regulatory frameworks adapt to the increasing reliance on technological tools over traditional normative legal interventions? According to Roger Brownsword, future regulators will be more inclined to use technology to ensure compliance because it promises greater effectiveness. Whether the legal approach to regulating technology and human life will fully transition from normative to non-normative dimensions remains uncertain.

Notwithstanding, Brownsword argues that the legal approach needs to be re-imagined—not as a precise set of rules (i.e., 'the law') but within a broader regulatory environment that incorporates non-normative forms of regulation. He refers to this as an era of technology management and defines technological management as a regime that relies on technologies like code, design, and architecture to guide behaviour directly, rather than through the indirect means of normative commands and prohibitions. This shift towards a technologically managed regulatory framework can enhance compliance by embedding regulatory controls within the technologies themselves, thus reducing the reliance on human judgment and enforcement.

Brownsword's perspective underscores the necessity of integrating non-normative regulatory mechanisms to address the dynamic and complex nature of technological advancements. By embedding regulatory principles within the design of technological systems, we can create more adaptive and responsive regulatory environments that can better manage the risks and opportunities presented by new technologies. This approach not only promises greater effectiveness but also aligns regulatory practice with the realities of an increasingly digital and automated world.

Brownsword defines the technological management regime in detail as follows: "Technological management means the use of technologies-typically involving the design of products or places, or the automation of processes-with a view to managing 'specific' types of risks by excluding (i) the possibility of certain actions which, in the absence of this strategy, might be subject only to a rule regulation, or (ii) human agents who otherwise might be implicated (whether as rule-breakers or as innocent victims of rule-breaking) in the regulated activities".

Brownsword's perspective on the evolution of regulation within the realm of technology and human life raises critical points regarding the efficacy and nature of traditional legal frameworks. As we transition into an era where technology permeates every aspect of our lives, including how drones are operated and managed within the European Union, the appeal of technological management as a regulatory tool becomes increasingly evident. This method promises to enhance regulatory effectiveness by directly integrating control mechanisms into technology itself, thus preemptively managing risks rather than merely responding to breaches.

The concept of technological management as outlined by Brownsword suggests a shift from purely normative methods, which rely on the establishment and enforcement of general rules, to systems where technology itself restricts or enables certain behaviours. This approach could be particularly effective in drone regulation, where rapid advancements in technology and the wide range of applications make traditional rule-making slow and sometimes ineffective. For instance, geofencing technology can be employed to automatically prevent drones from entering restricted airspace, thereby eliminating the reliance on operators to manually avoid these no-fly zones.

Moreover, the automation of compliance processes through AI and machine learning can further ensure that drones operate within regulatory requirements without constant human oversight. Such systems can monitor compliance in real-time, dynamically adapting to new data or regulatory changes—something that traditional regulatory approaches cannot easily accommodate. This not only reduces the burden on regulatory bodies but also decreases the likelihood of human error.

The advocacy for a broader regulatory environment resonates with the need for an ecosystem where legal, technological, and social regulations coexist and complement each other. In the context of drone regulation, this could mean a layered approach where traditional laws set the broad parameters within which technology-specific rules and mechanisms operate. For instance, while traditional laws can govern privacy and data protection in drone operations, technological management can provide the tools to enforce these laws through secure data transmission protocols and automated data handling processes that respect privacy by design.

However, the shift towards technological management also raises important questions about transparency, accountability, and the potential for technology to embed new forms of bias or error. It is essential that these systems are designed in a way that allows for oversight and intervention by human regulators and that they are subject to rigorous testing and review to ensure they do not inadvertently create new risks.

The concept of technological management in the regulation of drones represents a promising path forward. It calls for a re-imagining of legal frameworks as part of a larger, more dynamic regulatory ecosystem that can better address the complexities introduced by modern technologies. This approach not only enhances the effectiveness of regulation but also ensures that it remains adaptable and responsive to the continuous evolution of technology.

In essence, technological management is an adaptive and sophisticated approach which involves designing structures that render non-compliance with the rules practically impossible. For example, to eradicate the risk of commercial drones crashing and causing physical harm, regulators may require or encourage regulatees to include fail safe or back-to-home configuration into commercial drones by design. Such technology instrument which enables drones to simply returns to take-off point in the case of a malfunction is considered more effective form of regulation.

Technological management, by embedding compliance directly into the design of technology, offers a proactive approach to regulation that can significantly enhance safety and efficiency, especially in complex systems like commercial drone operations. This design-oriented approach to regulation not only makes it difficult for users to deviate from established rules but also reduces the dependency on post-incident enforcement and penalties, which are often less effective in preventing accidents and breaches in the first place.

The inclusion of fail-safe or return-to-home features in drones is a prime example of how technological management can be implemented. These features are engineered to activate automatically under certain conditions, such as loss of signal, low battery, or system failure. By ensuring that drones return to their point of origin during malfunctions, the risk of accidents, such as collisions with other aircraft or injuries to people on the ground, is substantially mitigated. This approach not only protects public safety but also preserves the integrity of drone operations, enhancing trust among the public and regulatory bodies alike.

Expanding this concept further, regulators could also require that drones incorporate advanced detection and avoidance technologies that enable them to identify and evade obstacles autonomously. By integrating sensors and real-time processing capabilities, drones can dynamically navigate around potential hazards, further reducing the risk of collision. This level of technological sophistication could extend to weather adaptation capabilities, allowing drones to alter their flight parameters in response to changing weather conditions, thereby avoiding weather-related incidents.

The broader application of technological management could extend to how drones are used in specific sectors. For example, in agricultural drone use, technology could ensure that drones disperse pesticides or fertilisers in precise amounts, adhering strictly to environmental regulations. Similarly, in urban areas, drones could be designed to operate only within specified noise levels to comply with local noise ordinances, using technology that monitors and adjusts their propulsion systems accordingly.

Although, technological management holds significant promise, it is crucial to consider its limits and the importance of maintaining a balanced regulatory approach, I hereby argue that dependence solely on technology could lead to over-reliance and potential gaps in situations where technology fails or is unable to handle unexpected scenarios.

Therefore, it's essential that such technological solutions are part of a broader, multi-faceted regulatory framework that includes robust human oversight, regular updates to technological standards based on new research and developments, and ongoing training for operators to handle emergency situations effectively.

While technological management significantly enhances the ability to regulate complex systems like drone operations effectively, it should be integrated thoughtfully within a comprehensive regulatory strategy that combines technology, human oversight, and adaptive legal frameworks to address both current needs and future challenges.

How can we ensure that the integration of advanced technological instruments within regulatory frameworks effectively mitigates the risks associated with commercial drone operations, such as crashes over public or private property due to technical faults or pilot negligence? The technology fix to a technology problem reduces the practical option of commercial drones crashing over public or private property and causing harm due to a technical fault or negligence of the pilot. Considering the true promise of technological instruments as a regulatory strategy to eliminate the risk of non-compliance, the questions is to what extent the regulatory environment for commercial drones facilitates the compliance by design approach?. By embedding compliance mechanisms directly into the drone technology itself, such as through real-time monitoring systems, autonomous navigation controls, and automatic fail-safes, regulators can ensure a layer of guarantee that drones operate within established safety parameters, thereby significantly reducing the potential for accidents and enhancing overall public safety.

What necessitates the shift in regulatory approach to manage the disruptions caused by emerging technologies like drones, blockchain, and cryptocurrencies? The need for such change in approach is premised on the disruption posed to the existing legal system by the emergence of transformational technologies such as drones, blockchain and cryptocurrencies. Brownsword argues that there are three separate but coexisting mindsets about the role of law in the management of disruptive technologies. These three co-existing mindsets are framed as follows: Law 1.0, Law 2.0 and, Law 3.0.

Law 1.0 mindset refers to the traditional normative approach to the law, which involves the application of legal principles to factual situations. This mindset is synonymous with lawyerly thinking. It is the default mindset that lawyers apply when advising a client or judges when justifying their decisions.

Law 1.0 represents the foundational methodology upon which the legal profession is built. This approach is characterised by its reliance on established legal principles and precedents to interpret and resolve factual situations. It encapsulates the essence of lawyerly thinking—methodical, principle-driven, and predominantly deductive. This mode of legal reasoning is not just prevalent; it is the bedrock upon which legal advice is given, and judicial decisions are made.

At its core, Law 1.0 is about applying a set of predefined rules to a given set of facts. Lawyers are trained to dissect and analyse cases within the framework of these rules, crafting arguments that align with legal precedents. Similarly, judges use these principles to frame their judgments, ensuring that their decisions are rooted in law and consistent with earlier rulings. This process is fundamental to maintaining the rule of law, as it ensures consistency and predictability in legal proceedings, which in turn reinforces the legal system's integrity.

However, this traditional approach also hinges on the stability and continuity of legal doctrines. It assumes a relatively static legal landscape where changes are incremental and predictable. Lawyers and judges are experts in navigating this landscape, adept at linking the specific facts of a case with the broad principles that govern the relevant legal area. This expertise is crucial in areas of law where precedent and principle are paramount, such as contract law, torts, and property law.

Despite its strengths, the traditional normative approach faces challenges in the face of rapidly evolving fields like technology, where new scenarios often emerge faster than the law can adapt. In these areas, Law 1.0 can sometimes lag, struggling to apply old paradigms to new problems. For example, issues like digital privacy, cybersecurity, and artificial intelligence present novel legal challenges that do not neatly fit into existing legal frameworks. The pace at which technology evolves requires a legal approach that can anticipate and adapt to new developments without the latency typically associated with legislative processes.

Moreover, the traditional approach tends to be reactive rather than proactive. It focuses on resolving disputes after they have arisen rather than preventing them. This aspect of Law 1.0 can be seen as a limitation in today's fast-paced world where the economic and social costs of litigation, and the slow pace of judicial proceedings can be significant. While Law 1.0 provides a critical foundation for the practice and application of law, its traditional mechanisms and pace may not always align with the needs of an increasingly complex and fast-moving global society. As such, the legal field continues to evolve, integrating new technologies and methodologies to better serve both the public and the principles of justice. This leads us to Law 2.0

Law 2.0 mindset results from the arrival of new technologies. The classic application of rules in a Law 1.0 sense are disrupted, and jurisdictions begin asking whether existing rules are fit for the purpose. The onus of articulating compliance to new rules and frameworks to regulate innovative technology shifts from the courts to the regulator as performed by the legislative and executive branches. This approach is regularly adopted in the enacting of a new regulation to govern the use of disruptive technologies like commercial drones.

Law 2.0 represents a paradigm shift in the legal landscape, precipitated by the rapid development and integration of new technologies that challenge the traditional norms and applications of law. As technologies evolve at an unprecedented pace, the classic Law 1.0 application of static rules to static scenarios becomes increasingly inadequate. This inadequacy is particularly evident in fields such as cyber law, intellectual property rights surrounding digital content and software, and regulations governing emerging technologies like drones, autonomous vehicles, and AI.

How can jurisdictions worldwide effectively reevaluate and adapt their existing legal frameworks to address the unique complexities and rapid changes characteristic of technological advancements? In response to these challenges, jurisdictions worldwide find themselves reevaluating the suitability of existing legal frameworks. This evaluation often reveals that many traditional rules are ill-equipped to address the unique complexities and rapid changes characteristic of technological advancements. As a result, the focus shifts from a purely judicial interpretation of law to a more dynamic regulatory approach. This shift places greater responsibility on legislative and executive branches to develop and implement laws that are specifically designed to manage the impact and integration of disruptive technologies.

This proactive legislative approach involves crafting new regulations that not only address the immediate implications of technologies but also anticipate future developments and challenges. For example, in the realm of commercial drones, regulators are tasked with balancing innovation and public safety, crafting rules that govern everything from airspace access to privacy concerns and liability issues. These regulations must be flexible enough to adapt to advancements in drone technology while robust enough to protect public interest.

Moreover, Law 2.0 encourages a more collaborative regulatory process, involving a wide range of stakeholders including technologists, industry experts, public interest groups, and government agencies. This inclusive approach ensures that the regulations are comprehensive and consider multiple perspectives, which is vital for the successful integration of new technologies into society. For instance, the development of guidelines for drone operations may involve consultations with aviation authorities, privacy advocates, commercial drone operators, and local communities to address the varied implications of drone use.

Additionally, Law 2.0 utilises advanced tools such as data analytics, machine learning, and blockchain to enforce and monitor compliance. These technologies provide regulators with powerful mechanisms to track the implementation of laws in real-time, predict potential issues, and ensure that regulations are being followed. For example, blockchain could be used to create a secure, immutable record of drone flight paths and operational data, helping to enforce flight restrictions and protect sensitive airspace.

In essence, Law 2.0 is about creating a legal framework that is as adaptive, responsive, and forward-looking as the technologies it seeks to regulate. It represents a significant shift from the reactive, precedent-based approach of Law 1.0 to a more proactive, regulatory-driven approach that is better suited to the complexities of the modern technological landscape. This evolution is crucial for ensuring that the law not only keeps pace with technology but also facilitates its development in a way that maximizes benefits and minimises risks to society. This leads us to Law 3.0.

Law 3.0 mindset evolves from the disruption of the regulatory mindset. As new technology increasingly presents themselves as possible regulatory tools, jurisdictions move from the regulatory mindsets to question whether technologies could be employed to limit, mitigate or entirely prevent certain behaviours. In this non-normative approach, regulators adopt a technology mindset, alongside black letter law, to discourage, limit or eliminate harmful, antisocial, or criminal behaviour.

In this technology mindset, jurisdictions adopt technological instruments to make the practical option of non-compliance impossible. Law 3.0 represents an even further evolution in the legal landscape, wherein technology is not only the subject of regulation but also becomes an integral part of the regulatory process itself. As technology advances, it opens up new possibilities not just for enabling activities, but also for preventing undesired outcomes. Jurisdictions are increasingly exploring how technologies can be directly implemented as regulatory tools to preclude, mitigate, or outright prevent harmful or undesirable behaviours. This shift marks a significant departure from traditional regulatory approaches that typically involve setting rules and then reacting to breaches.

In the realm of Law 3.0, the concept of using technology as a regulatory mechanism is known as "regulatory technology" or "RegTech." This approach leverages technological innovations to achieve regulatory objectives more effectively and efficiently than traditional methods allow. For example, advanced surveillance technologies equipped with facial recognition can be used to enhance public safety by identifying and tracking individuals who are under surveillance orders. Similarly, IoT (Internet of Things) devices can be programmed to monitor environmental conditions and automatically enforce regulations by controlling pollution outputs from factories in real-time, ensuring compliance with environmental standards without the need for human oversight.

Law 3.0 entails the use of technological instruments to complement or substitute traditional regulatory interventions. The goal of Law 3.0 is to render non-compliance with the rules practically impossible. This technological management regime is a new frontier in regulatory interventions, and it can be seen in the increased reliance on technical solutions to manage risks in various sectors, including aviation. By focusing on the potential of technological solutions, Law 3.0 could be an effective strategy for regulating commercial drones.

This technology-driven regulatory approach also extends to influencing social behaviours. For instance, smart city technologies can manage traffic flows based on real-time data and usage patterns, reducing congestion and pollution, and encouraging the use of public transport options. In the digital realm, algorithms can be used to monitor online platforms and automatically detect and remove content that violates hate speech regulations or other legal standards, thereby mitigating the spread of harmful content.

Moreover, Law 3.0 involves embedding compliance into the design of products and systems—a concept known as "privacy by design" in the data protection field. This approach ensures that technologies are built from the ground up to comply with legal standards, thus making non-compliance practically impossible. For example, smart home devices can be designed in a way that they automatically comply with data protection laws, collecting only the data necessary for their function and encrypting it to protect user privacy.

However, the transition to Law 3.0 raises significant ethical and practical concerns. There is the risk of over-reliance on technology, which may not always be foolproof or could lead to new forms of abuse. Furthermore, the use of technology as a regulatory tool requires careful oversight to ensure that it does not infringe on individual rights or lead to discrimination. Regulators must maintain a delicate balance, ensuring that while technology can be employed to enforce and facilitate compliance, it does not replace the need for human judgment and accountability.

At its core, Law 3.0 represents a transformative phase in regulatory approaches, where technology becomes a central element not just in facilitating but in enforcing compliance. This non-normative approach seeks to integrate technological solutions into the fabric of legal and regulatory frameworks, making the practical option of non-compliance increasingly challenging and setting a new standard for how laws are enforced in an increasingly digital world.

However, a significant challenge with the adoption of technological management is the risk of unanticipated consequences. The application of technological solutions to regulate behaviour may not always result in the desired outcome. For instance, designing commercial drones to automatically return to the take-off point in case of a malfunction may increase the risk of drones colliding with each other in crowded airspace. Thus, while technological management holds promise, regulators must be mindful of the potential for unintended outcomes and consider ways to mitigate such risks.

How can the adoption of technological management be effectively balanced with comprehensive risk assessments and mitigation strategies to avoid unintended consequences in complex, real-world scenarios? The adoption of technological management, while promising, indeed requires careful consideration of potential unintended consequences, which can arise when deploying technology in complex, real-world scenarios. As noted, the automated features designed to enhance safety, like the return-to-home function in drones, could inadvertently create new hazards, such as increased mid-air collision risks in crowded airspace. This example underscores a broader issue within Law 3.0: the need for comprehensive risk assessment and mitigation strategies when integrating technological solutions into regulatory frameworks.

To address these complexities, regulators must adopt a multi-faceted approach that includes robust scenario planning and continuous monitoring of technological deployments. This involves conducting thorough impact assessments prior to the implementation of new technologies to understand potential negative outcomes and develop strategies to mitigate these risks. Such assessments should consider various scenarios, including those that may seem unlikely at first glance, to ensure a comprehensive understanding of the possible implications of regulatory technologies.

Moreover, the establishment of feedback mechanisms is crucial. These systems allow for the ongoing collection and analysis of data regarding the performance of technology-based regulatory tools. By analysing this data, regulators can identify patterns or instances of unintended consequences and adjust regulatory measures accordingly. For instance, if data reveals that automated drone returns are leading to congestion in certain areas, regulators could update flight path algorithms or adjust operational restrictions to alleviate these issues.

In addition to these technical measures, there is also a need for legal and ethical oversight to ensure that the use of regulatory technologies aligns with broader societal values and legal principles. This includes ensuring transparency in how these technologies operate and making provisions for appeals or redress in cases where technology-driven decisions negatively impact individuals or groups. Establishing governance frameworks that include ethical guidelines for the use of regulatory technologies can help maintain public trust and ensure that these tools are used responsibly.

Furthermore, regulators should foster an environment of collaboration among technology developers, legal experts, and other stakeholders. This collaborative approach can facilitate the sharing of knowledge and experiences, leading to better-designed regulatory technologies that are more likely to achieve their intended outcomes without undesirable side effects. For example, engaging with drone manufacturers and airspace management experts could lead to more innovative solutions that effectively balance safety with efficiency in drone regulation.

In essence, while technological management introduces a new horizon of possibilities for regulatory compliance, it also brings challenges that require careful management. By adopting a holistic approach that includes risk assessment, continuous monitoring, ethical oversight, and stakeholder collaboration, regulators can harness the benefits of technological solutions while minimising the potential for unintended consequences. This proactive and thoughtful approach is essential in navigating the complexities of Law 3.0 and ensuring that technology serves as a beneficial tool in the pursuit of regulatory objectives.

Another challenge with the technological management approach is the issue of accountability. When technological instruments are employed to prevent non-compliance, the responsibility for compliance shifts from the individual to the technology. This shift in accountability can make it challenging to determine who is responsible in case of a failure or an unintended outcome. The use of technological management, therefore, requires careful consideration of issues of accountability and responsibility.

When technological instruments are used to enforce compliance, the blurred lines of accountability present significant challenges for regulators and legal frameworks. As the responsibility for certain actions shifts from individuals to technology, determining liability in the case of failures or unintended outcomes becomes complex. This raises pivotal legal and ethical questions: If an automated system fails, who is to be held accountable—the designer, the operator, the manufacturer, or the technology itself?

For instance, consider a scenario where a drone equipped with automatic collision avoidance technology fails to prevent an accident. Determining who is liable involves dissecting layers of responsibility: Was the technology designed appropriately? Were there manufacturing defects? Did the operator override or improperly maintain the system? Or was there a failure in the software algorithm that led to incorrect decision-making? Each of these questions reflects the multifaceted nature of accountability in technological management systems.

To address these complexities, there is a growing consensus on the need for creating clear legal frameworks that delineate responsibilities among all parties involved in the lifecycle of a technological product. Such frameworks would need to establish guidelines for liability that consider the roles of various stakeholders—manufacturers, software developers, users, and others—in maintaining the safety and reliability of technology. This could involve adopting models of shared accountability, where liability is distributed across multiple parties according to their degree of control and involvement in the technological process.

Moreover, it is crucial to integrate robust mechanisms for tracing back failures to their sources within technological systems. This could involve mandating that all autonomous technologies have built-in auditing and logging capabilities that can record operational decisions and the inputs influencing those decisions. Such transparency would not only aid in post-incident investigations but also in preemptive monitoring, thereby enhancing overall accountability.

How can regulatory bodies effectively monitor and ensure the compliance of emerging technologies with safety standards and regulatory requirements while fostering a culture of ethical development and accountability among technologists and companies? Regulatory bodies might also consider establishing independent oversight committees dedicated to the monitoring and assessment of these technologies. These committees could perform regular evaluations and audits to ensure compliance with safety standards and regulatory requirements. Their role could be crucial in providing an objective analysis of technological failures and recommending policy adjustments or improvements to prevent future incidents. Additionally, fostering a culture of ethical development and deployment among technologists and companies is vital. Encouraging the adoption of ethical guidelines that emphasise responsibility and accountability in the development of autonomous systems can lead to more thoughtful and safer technological innovations. This ethical commitment needs to be supported by educational initiatives that raise awareness about the potential impacts of technology and the importance of incorporating accountability in every stage of development.

Essentially, addressing the accountability issues within technological management requires a comprehensive approach that includes clear legal definitions of liability, enhanced transparency mechanisms, proactive regulatory oversight, and an ethical framework guiding the development and use of technology. Such measures are essential to ensure that while technological innovations continue to advance, they do so in a manner that is safe, responsible, and aligned with societal values.

The adoption of technological management in regulating commercial drones also raises questions about the role of the regulator. While traditional regulatory interventions are enforced by regulatory bodies, the use of technological instruments requires a different approach. Regulators may need to collaborate with technology developers and manufacturers to ensure that the necessary technological solutions are integrated into commercial drones. This collaboration is necessary to ensure that the technological instruments are effective in managing risks and that they comply with regulatory standards.

How can regulatory bodies transition from being mere enforcers to active participants in the development and implementation of drone technology, ensuring they possess both the legal and technical knowledge necessary to effectively oversee and influence these technologies? The necessity for collaboration between regulators, technology developers, and manufacturers underlines a significant shift in the role of regulatory bodies from mere enforcers to active participants in the development and implementation of technology. This transition demands that regulators not only understand the legal implications of drone technologies but also acquire a deep technical knowledge that allows them to effectively oversee these technologies. As such, regulators must become well-versed in the technical aspects of drone operations and the underlying technologies to ensure they can adequately assess and influence the design and functionality of these systems.

This collaborative approach also requires a change in the regulatory framework itself. It suggests moving towards a more proactive form of regulation that is iterative and adaptive. Regulators will need to engage in continuous dialogue with technology developers to stay updated on advancements and ensure that emerging technologies can be integrated smoothly into the regulatory framework without compromising public safety or efficiency. Such engagement might involve regular consultations, joint research projects, and the establishment of pilot programs to test new technologies under controlled conditions. These initiatives can help both regulators and technologists identify potential issues early and refine technologies to meet both commercial and regulatory needs.

Moreover, the collaboration extends to standard-setting. Regulators, in partnership with industry stakeholders, are tasked with developing standards that dictate how technologies should be implemented to meet safety, privacy, and efficiency benchmarks. These standards must be clear, measurable, and enforceable, with built-in flexibility to accommodate future technological developments. For instance, standards could specify requirements for electronic visibility, control systems redundancy, data security measures, and fail-safe mechanisms in drones.

This integrated regulatory approach also implies a need for transparency and accountability in the decision-making process. Stakeholders including the public, industry experts, and advocacy groups should be invited to participate in discussions about regulatory policies affecting drone technology. This inclusivity helps ensure that the regulatory framework reflects a wide range of interests and concerns, balancing innovation with public interest and societal values.

Finally, the success of this collaborative regulatory model depends on the ability of regulatory bodies to enforce compliance effectively. This requires not only the traditional tools of regulation such as inspections and penalties but also innovative enforcement mechanisms that leverage technology itself. For example, regulators could use automated systems to monitor drone operations remotely, detecting violations of flight regulations or unauthorized use of airspace in real time. Such technological enforcement tools, when integrated effectively into the regulatory strategy, can enhance compliance while also reducing the burden on regulatory resources.

Fundamentally, the adoption of technological management in drone regulation necessitates a reimagined role for regulators—one that blends legal oversight with technical expertise and collaborative governance. This approach can lead to more effective and adaptive regulations that ensure the safe and responsible integration of advanced drone technologies into society.

Critically, the adoption of technological management in the regulation of commercial drones requires careful consideration of the implications for privacy. For instance, requiring commercial drones to include cameras and other surveillance technologies as a means of managing risks may raise significant concerns about privacy. Regulators must, therefore, consider the privacy implications of technological management solutions and ensure that they comply with privacy laws and regulations.

The integration of surveillance technologies such as cameras and sensors in commercial drones, while enhancing regulatory compliance and safety, indeed heightens concerns surrounding privacy and data protection. These technologies can collect a vast amount of data, much of which may be personal or sensitive, leading to potential infringements on individual privacy rights. As such, it is crucial for regulators to devise strategies that safeguard privacy while still harnessing the benefits of technological management.

One essential approach involves ensuring that all technological solutions comply with established privacy laws, such as the General Data Protection Regulation (GDPR) in the European Union. These regulations mandate strict guidelines on data collection, processing, and storage, requiring that data be handled in a way that respects privacy and ensures security. For instance, drones equipped with cameras should have systems in place that limit data collection to necessary information only, anonymise collected data where possible, and secure data against unauthorized access and breaches.

Furthermore, privacy by design should be a core principle in the development and deployment of drone technologies. This approach requires that privacy considerations be integrated into the technology from the earliest stages of design and throughout the product lifecycle. It involves embedding data protection features directly into the design of drones and their operational procedures, ensuring that privacy is maintained without sacrificing functionality. For example, drones could be programmed to avoid recording in sensitive areas or during times when privacy risks are particularly high.

Regulators also need to establish clear guidelines for the permissible uses of drone-collected data, specifying what data can be collected, how it should be used, and who can access it. These guidelines should be transparent and enforceable, with strict penalties for violations to deter misuse of data. Public awareness and consent are also vital; individuals in areas subject to drone surveillance should be informed about the presence of drones, the nature of the data being collected, and the purposes for which it will be used.

To address the dynamic challenges posed by technological advances, regulatory frameworks may also need to be adaptive. This could involve regular reviews and updates to privacy regulations as drone technology and data use practices evolve. Additionally, creating avenues for public input and feedback on drone policies can help ensure that regulations remain relevant and are responsive to citizens' privacy concerns.

Moreover, cross-sector collaboration can enhance privacy protections. By working together, regulators, drone manufacturers, technology developers, privacy advocates, and legal experts can develop innovative solutions that balance effective risk management with robust privacy safeguards. This collaborative approach can lead to the development of advanced cryptographic techniques, more effective anonymization processes, and smarter data minimisation practices that protect individual privacy without impeding the practical applications of drones.

In essence, while the adoption of technological management in drone regulation offers promising benefits for compliance and safety, it must be approached with a conscientious focus on privacy. By embedding privacy considerations into every aspect of drone regulation and operations, creating clear and enforceable guidelines, and fostering a collaborative regulatory environment, privacy concerns can be effectively managed. This approach ensures that drone technologies are used responsibly and ethically, aligning with broader societal values and legal norms.

The emergence of transformational technologies such as commercial drones and artificial intelligence calls for a re-imagining of the legal approach to regulation. The traditional normative approach to the law, Law 1.0, is not always effective in managing risks posed by disruptive technologies. The use of technological management, Law 3.0, offers a promising approach to managing risks in various sectors, including aviation. However, the adoption of technological solutions in the regulation of commercial drones requires careful consideration of issues of accountability, collaboration with technology developers and manufacturers, and privacy implications. Overall, a balance between normative and non-normative regulatory interventions is necessary to ensure effective regulation of commercial drones while promoting innovation and growth in the industry.

The rapid evolution and widespread application of technologies such as commercial drones and artificial intelligence necessitate a dynamic and flexible legal framework that transcends traditional normative approaches. Law 1.0, with its reliance on static, predefined rules and regulations, often struggles to keep pace with the fast-moving nature of technology, where new capabilities and applications are constantly emerging. In contrast, Law 3.0, which embraces technological management, presents an adaptive model that integrates technology directly into the regulatory process, offering a more proactive and responsive approach to governance.

This shift toward Law 3.0 involves a comprehensive reevaluation of regulatory strategies to ensure they are capable of addressing the unique challenges posed by advanced technologies. Key to this transformation is the need for greater accountability mechanisms within the frameworks of technological management. As technology takes on more responsibility for enforcement and compliance, clear guidelines must be established to delineate responsibilities among developers, operators, and regulators. This includes creating protocols for when technology fails or when unintended consequences arise, ensuring that there are effective remedies and that those accountable can be easily identified and held responsible.

Ultimately, the successful regulation of disruptive technologies like commercial drones and artificial intelligence will depend on developing a regulatory ecosystem that is both innovative and grounded in legal tradition. This ecosystem should be capable of adapting to new developments while providing a stable and predictable legal environment that supports growth and innovation in the industry. By embracing both Law 3.0 and traditional legal approaches, regulators can create a flexible, effective framework that meets the challenges of today's rapidly evolving technological landscape.

In general, I build on Brownsword's idea of the three coexisting mindsets and employ the technology mindset as the lens to view the existing regulatory environment for commercial drones in the EU particularly in the age of artificial intelligence. I argue that non-normative rules are considered sophisticated because they are able to effectively balance liability and regulation by steering innovation towards 'precluding the practical option of non-compliance' more importantly in the face paced age of AI. Therefore, I prescribe that the level of sophistication of regulatory rules for drones is determined by the extent to which the regulation excludes the possibility of certain actions, which in absence of a technology mindset, might be subject to a rule, or the extent to which regulation excludes human or non-human agents from being implicated as rule breakers in the regulatory regime.

Expanding upon Brownsword's concepts, I further argue that adopting a Law 3.0 or technology mindset to coexist with the traditional mindsets not only reflects a sophisticated regulatory strategy but also represents a necessary evolution in the governance of emerging technologies like commercial drones. The sophisticated approach leverages non-normative tools to directly embed compliance into the technological framework of drones, thereby diminishing the chances of non-compliance to near zero. This mindset is particularly effective in the context of drones incorporated with artificial intelligence, where the speed of technological development and its deployment will most certainly outpace traditional regulatory mindset.

In evaluating the sophistication of drone regulations, I considers the degree to which these regulations integrate technological solutions to enforce compliance preemptively. Such integration involves the use of AI to monitor drone operations continuously, automatically enforce no-fly zones through geo-fencing technologies, and manage traffic without human intervention. By limiting the physical capacity of drones to operate outside designated parameters, the regulation effectively precludes the possibility of certain non-compliant actions before they occur.

Furthermore, this approach shifts the focus from punitive measures post-violation to preventive measures that anticipate and neutralise potential breaches. This shift is crucial in an era where the capabilities of drones are expanding rapidly, making traditional enforcement methods less effective and more resource-intensive. For instance, AI-enabled drones can be programmed to perform self-checks and maintenance diagnostics to predict and prevent failures that could lead to non-compliance with safety standards.

The sophistication of such regulatory frameworks is also evident in their ability to dynamically adapt to new challenges and technological advancements. This adaptability is facilitated by the continuous feedback loop between the operational data collected by AI technologies and the regulatory parameters set by governing bodies. Such a system allows for real-time updates to regulatory measures based on actual drone behaviours and emerging trends, ensuring that the regulatory environment remains relevant and effective.

Additionally, the technology mindset fosters a collaborative regulatory environment where manufacturers, operators, technology developers, and regulatory bodies work together to ensure compliance. This collaboration is instrumental in crafting regulations that are not only enforceable but also promote innovation within safe and legal boundaries.

In essence, I posit that the true measure of regulatory sophistication in the age of AI and commercial drones lies in the ability to preclude non-compliance through integrated technological solutions. By adopting a Law 3.0 approach, EU drone regulations can more effectively manage the complex interplay between rapid technological advancement and the need for robust, enforceable governance. This approach not only enhances compliance and safety but also supports sustainable innovation, ultimately benefiting both the industry and society at large. In this context, I ask the following question

Are the rules that presently apply to commercial drone operations in the European Union sufficiently adaptive and sophisticated in their regulatory strategy to effectively preclude the practical option of non-compliance and exclude human and non-human agents from being implicated as rule breakers in the fast evolving age of artificial intelligence?

To answer this question, I undertake a critical evaluation of the sophistication and efficacy of the EU's regulatory framework for commercial drones, especially in light of the advancements and integration of artificial intelligence technologies. I delve into the EU Common Rules in the Field of Civil Aviation and the Unmanned Aircraft Systems Regulation, along with the Rules for the Operation of Unmanned Aircraft, I assess whether these regulations are adequately designed to keep pace with the rapid evolution of drone technology and AI.

The juxtaposition of these rules against the broader backdrop of the EU Artificial Intelligence Act allows for a comprehensive analysis of how well drone regulations integrate with overarching AI policies. The EU Artificial Intelligence Act, which aims to set standards for AI development and usage that ensure safety and respect for existing laws and values, serves as a critical benchmark for assessing drone regulations. This synthesis not only highlights areas where drone regulations are aligned with AI policies but also identifies gaps where current drone regulations may fall short.

A sophisticated regulatory strategy for drones in the age of AI should not only enforce compliance effectively but also anticipate and mitigate the risks associated with AI integration. This involves understanding the capabilities of AI to enhance operational efficiency and safety but also recognising potential pitfalls such as biased decision-making processes or the manipulation of AI systems. I scrutinise whether current EU regulations have the capacity to adapt to the technological landscape where drones are not merely tools but active agents capable of autonomous decision-making.

Moreover, I analyse the extent to which mechanisms are in place for accountability and enforcement. I critically examine whether the regulations sufficiently detail the accountability of human operators, manufacturers, and the AI itself. With AI's potential to act independently, it is imperative that regulations clearly delineate responsibilities to prevent ambiguities that could lead to compliance failures.

Additionally, I explore the preventive measures stipulated in the regulations and whether they specifically address the risks associated with AI technologies in drones. These include requirements for transparency in AI decision-making processes, mandates for AI robustness and accuracy, and protocols for emergency management when AI systems malfunction. The effectiveness of these preventive measures is pivotal in determining the sophistication of the regulatory framework.

In framing the discussion around these points, I sets the stage for further research into the intersection of drone law and AI regulation, including exploring potential future scenarios as AI becomes more embedded in drone technology. Through this extensive discussions and the questions it raises, this book creates a deeper understanding of the adaptive and sophisticated mindset that drone regulations in the EU require to remain robust and effective in the fast-evolving landscape of disruptive technologies.

VI

6. AI-Enhanced Drone Capabilities

Artificial Intelligence (AI) refers to the branch of computer science that is concerned with building smart machines capable of performing tasks that typically require human intelligence. In the context of drone technology, AI is revolutionising how drones are operated and utilised, marking a significant shift from manually operated models to autonomous systems. This transition allows drones to perform complex tasks independently, without the need for constant human oversight. AI-driven drones integrate various forms of machine learning and computer vision technologies, enabling them to understand and interact with their environment in real-time.

One of the primary functionalities of AI in drones is advanced navigation. Unlike traditional drones that require manual control or pre-programmed routes, AI-enabled drones can dynamically adjust their flight paths, recognize and avoid obstacles, and navigate through challenging terrains autonomously. This is achieved through the use of sophisticated sensors and onboard processing capabilities that analyse environmental data in real-time. This autonomous navigation not only increases the safety of drone operations by reducing the likelihood of collisions but also enhances their efficiency in completing tasks.

Another crucial AI functionality in drones is obstacle avoidance. This feature is particularly important as it directly impacts the operational safety and reliability of drones, especially in crowded or complex environments. By utilizing AI algorithms, drones can detect and analyse obstacles in their immediate vicinity and make split-second decisions to manoeuvre around them safely. This capability is supported by technologies such as LIDAR (Light Detection and Ranging), radar, and cameras that feed data into the drone's AI system, which processes this information to guide its movements.

Additionally, AI enhances drones' ability for sophisticated data collection and real-time decision-making. Drones equipped with AI can be used for a variety of data-intensive applications, such as agricultural monitoring, where they collect data on crop health and soil conditions, or in infrastructure inspection, where they assess the integrity of structures like bridges and buildings without human intervention. In these applications, not only do drones collect data, but they also analyse it on-the-fly, using AI to identify patterns, anomalies, or critical conditions that need immediate attention. This capability allows for more informed decision-making in fields ranging from environmental monitoring to emergency response, showcasing the expansive potential of AI in enhancing drone functionalities.

How has the integration of AI into drone technology enhanced their efficiency, accuracy, and versatility, particularly in tasks requiring repetitive or extensive monitoring and precise data collection? The integration of AI into drone technology has notably enhanced their capabilities, making them far more efficient, accurate, and versatile. AI enables drones to perform tasks more quickly and with greater precision than manual or less sophisticated automated systems. This improvement in efficiency is particularly evident in tasks that require repetitive or extensive monitoring, where AI-driven drones can operate continuously without fatigue, covering vast areas in a fraction of the time it would take humans. Additionally, the accuracy of data collection with AI drones is significantly enhanced by advanced sensors and analytic capabilities, allowing for precise measurements and detailed environmental assessments. This level of precision is crucial in applications that depend on fine-grained environmental data.

Moreover, the ability of AI drones to operate in diverse environments has broadened their applicability across various sectors. For example, these drones can navigate difficult terrains in agricultural or natural settings to monitor wildlife or crop health. Their robustness and adaptability make them ideal for use in conditions that are typically challenging and hazardous for humans.

One specific example where AI drones are making a significant impact is in agricultural monitoring. Farmers are using AI-equipped drones to assess crop health, monitor soil hydration, and manage pests and diseases. These drones can quickly analyse vast fields and provide data-driven insights that help farmers make informed decisions about irrigation, fertilisation, and pest control, leading to improved crop yields and reduced waste.

In disaster response, AI drones are transforming how emergencies are managed by providing real-time data and imagery from affected areas. This capability allows for rapid assessment of damage, identification of safe and unsafe zones, and efficient allocation of resources. For instance, following a natural disaster like an earthquake or flood, AI drones can be deployed immediately to survey the affected region, providing rescue teams with critical information that can guide their efforts and save lives.

Additionally, in healthcare delivery, AI drones are being used to transport medical supplies and vaccines to remote areas that are difficult to reach via traditional means. This use of drones is particularly impactful in regions with poor infrastructure, where timely delivery of medical aid can mean the difference between life and death. Drones equipped with AI are able to navigate to remote locations with little to no human guidance, ensuring that medical supplies are delivered quickly and safely.

AI technologies in drones, while offering substantial benefits, also introduce a range of complexities and risks that must be carefully managed. One significant concern is algorithmic bias, which can occur if the data used to train AI models is not fully representative of diverse conditions or populations. This can lead to skewed or unfair outcomes, such as a drone not properly recognising certain obstacles or misinterpreting data based on biased inputs. In drone applications, such as surveillance or data collection, this could lead to inaccuracies that have real-world consequences, affecting everything from environmental monitoring to public safety.

Another critical issue is decision-making transparency. AI systems, particularly those using complex algorithms like deep learning, often operate as "black boxes," where the decision paths are not easily understandable by humans. This lack of transparency can be problematic in drone operations, where understanding the basis of decisions is crucial, especially in incidents involving accidents or unintended intrusions into private spaces. The challenge of ensuring that AI-driven drones make decisions that are interpretable and justifiable is essential for legal and ethical accountability.

The unpredictability of AI behaviours in uncontrolled environments further compounds these risks. AI-driven drones may encounter scenarios that were not anticipated or covered during their training, leading to unexpected behaviours. For instance, a drone might react inadequately to unforeseen whether conditions or unexpected obstacles, posing risks to public safety and property. This unpredictability makes it difficult to fully trust AI systems in high-stakes situations without robust safeguards and continuous learning mechanisms in place.

The implications of these AI-related risks for privacy, security, and safety are profound. In terms of privacy, drones equipped with AI can collect and process vast amounts of data, including personal information, often without the explicit consent of individuals. This raises significant concerns about surveillance and data privacy, particularly if data handling practices are not transparent or do not comply with stringent regulations like the GDPR.

From a security perspective, the autonomous nature of AI drones makes them potential targets for hacking or misuse. If a drone's AI system is compromised, it could be used to conduct unauthorized surveillance, or worse, be repurposed as a tool for malicious activities. Ensuring the security of AI systems in drones is therefore critical to prevent data breaches and unauthorized access.

The safety risks associated with AI in drones cannot be underestimated. Unpredictable behaviours could lead to accidents, causing injury or damage. Ensuring the safety of these systems involves rigorous testing and validation under diverse conditions, continuous monitoring, and rapid response mechanisms to deal with failures as they occur. Addressing these complexities and risks is essential for the sustainable integration of AI technologies in drones. This involves not only improving the robustness and reliability of AI systems but also implementing comprehensive regulatory and ethical frameworks to guide their development and deployment.

As AI technologies become increasingly integrated into drone operations, existing drone regulations often struggle to fully address the unique challenges posed by AI. These regulations, typically designed for more traditional, manually-operated or less sophisticated automated drones, may not adequately cover the complexities introduced by AI, such as the need for transparency in decision-making, accountability for autonomous actions, and mitigation of algorithmic bias. One of the primary shortcomings of current regulations is their lack of specificity regarding the oversight and validation of AI algorithms. This includes how data is collected, processed, and used by AI systems within drones, raising concerns over privacy and security that are not sufficiently addressed by traditional aviation laws.

How can regulatory frameworks keep pace with the rapid development of AI technologies in drones, ensuring that these systems remain safe, transparent, and compliant with fundamental rights despite the slower legislative processes? The rapid development of AI technologies outpaces the slower legislative processes, leading to gaps in regulatory frameworks that might not foresee the rapid advancement and application of AI in drones. For example, current regulations do not have provisions for the real-time monitoring of AI systems in operation, which is crucial for ensuring that these systems do not behave unpredictably or deviate from approved operational parameters. Recognising these gaps, the European Union has been proactive in discussing and proposing new regulations and amendments specifically tailored to address the integration of AI into drones. One significant initiative is the ongoing development of the EU framework for trustworthy AI, which seeks to establish criteria and standards that AI systems, including those used in drones, must meet to be considered safe, transparent, and compliant with fundamental rights. This framework emphasises ethical guidelines for AI, including accountability, data governance, transparency, and the robustness of AI systems.

In addition to the broader AI framework, there are specific discussions within the EU aimed at updating the U-space regulations— the framework intended for managing drone traffic in European skies. These discussions focus on integrating AI safety standards, ensuring that AI-driven drones can coexist safely with manned aircraft and other airspace users. The proposed regulations are expected to include more stringent requirements for AI testing and certification, ensuring that drones can reliably perform under the diverse conditions encountered across Europe. These ongoing discussions and proposals reflect an understanding within the EU that as drones become more autonomous through AI, regulations must evolve not just to manage but also to foster innovation while protecting public safety and privacy. The EU's approach serves as a model for how regulatory bodies can anticipate technological advancements and create flexible, adaptive regulations that accommodate the future landscape of drone operations and AI integration.

The integration of AI into drone technology exemplifies a dual-edged sword, offering tremendous benefits while introducing significant risks that must be carefully managed. On one hand, AI enhances drone capabilities, enabling them to perform complex tasks with greater efficiency and precision, from precision agriculture and critical infrastructure inspections to rapid emergency response and healthcare delivery in remote areas. These advancements promise not only to boost economic productivity but also to improve societal well-being across various sectors. On the other hand, the autonomous nature of AI-powered drones raises critical concerns about privacy, security, and safety. The potential for algorithmic bias, decision-making opacity, and unpredictable behaviours in uncontrolled environments presents complex challenges. These risks necessitate rigorous regulatory oversight to ensure that drones operate safely and ethically, protecting individuals' rights and societal norms.

This balance between leveraging the benefits of AI in drones and mitigating the associated risks underscores the need for a nuanced understanding of drone regulations. Current regulations may fall short in addressing the rapid advancements in AI, highlighting the urgent need for updated or new regulations that specifically consider the unique aspects of AI integration. The complexities introduced by AI technologies demand that regulatory frameworks evolve continuously to keep pace with innovation while safeguarding public and airspace safety.

AI Technologies Integrated into Drones

The integration of artificial intelligence (AI) into drone technology has significantly expanded the capabilities and applications of drones across various industries. AI technologies such as machine learning algorithms for navigation, AI-powered image recognition for surveillance, and autonomous decision-making processes have transformed drones from simple remote-controlled devices into sophisticated autonomous systems. This section discusses these specific AI technologies in detail, highlighting their functionalities, benefits, and the challenges they pose.

Machine Learning Algorithms for Navigation

Machine learning (ML) algorithms play a pivotal role in enhancing the autonomous navigation capabilities of drones. These advanced algorithms enable drones to learn from their surroundings and make real-time adjustments to their flight paths, ensuring navigation that is both safe and efficient. Through the integration of ML, drones are equipped with several critical functionalities that enhance their operational capabilities.

One of the primary functionalities is obstacle detection and avoidance. ML algorithms allow drones to detect and avoid obstacles in their flight path by utilising data from various sensors, including LiDAR, ultrasonic sensors, and cameras. These algorithms can identify objects in the environment and calculate the best path to avoid collisions, significantly reducing the risk of accidents. For example, in an urban setting, a delivery drone can autonomously navigate around buildings, trees, and other potential hazards, ensuring that it reaches its destination safely.

Another essential capability provided by ML algorithms is terrain mapping. Drones equipped with these algorithms can create detailed maps of their surroundings by analysing data from their sensors. This enables them to identify terrain features and adjust their altitude and route accordingly. Terrain mapping is particularly useful in applications such as agriculture, where drones can survey fields, monitor crop health, and identify areas that require attention. By accurately mapping the terrain, drones can navigate safely and efficiently, even in complex environments.

Path planning is another critical aspect of autonomous navigation facilitated by ML algorithms. These algorithms enable drones to select the optimal route based on real-time environmental conditions and mission requirements. Dynamic path planning is essential for various applications, such as delivery services, where drones must navigate complex urban environments. By continuously analysing data from their sensors, drones can adapt their flight paths to avoid traffic, weather conditions, and other obstacles, ensuring timely and safe delivery of packages.

The integration of machine learning algorithms into drone navigation systems represents a significant advancement in the field of autonomous technology. These algorithms not only enhance the safety and efficiency of drone operations but also expand their potential applications across various industries. As ML technology continues to evolve, the capabilities of drones will further improve, opening up new possibilities for innovation and practical applications. The integration of machine learning (ML) algorithms into drone technology has brought about significant advancements, particularly in terms of safety, efficiency, and adaptability. However, these benefits are accompanied by several challenges that must be addressed to fully realise the potential of this technology.

One of the primary benefits of using ML algorithms in drones is improved safety. Drones equipped with these algorithms have the capability to detect and avoid obstacles, significantly reducing the risk of collisions. This enhances the overall safety of drone operations, making them more reliable and reducing the likelihood of accidents that could result in damage or injury. For instance, drones used for delivery services can navigate through urban environments, avoiding buildings, trees, and other obstacles, ensuring that packages are delivered safely and efficiently.

Operational efficiency is another key advantage. Autonomous navigation, enabled by ML algorithms, allows drones to perform tasks with minimal human intervention. This not only increases the speed and accuracy of operations but also reduces labour costs. For example, in logistics and delivery services, drones can autonomously transport packages, optimise delivery routes, and drop off items at designated locations without requiring constant oversight from a human operator. This level of efficiency can lead to significant cost savings and higher productivity.

Adaptability is a crucial benefit offered by ML algorithms. These algorithms enable drones to adapt to changing environmental conditions, making them suitable for a wide range of applications and environments. Whether it's flying in varying weather conditions or navigating different terrains, drones can adjust their flight paths and operational parameters based on real-time data. This adaptability makes them highly versatile tools for industries such as agriculture, construction, and environmental monitoring, where conditions can change rapidly and unpredictably.

Despite these benefits, there are several challenges associated with integrating ML algorithms into drones. One of the primary challenges is the requirement for large datasets to train these algorithms. Acquiring and processing the necessary data can be a significant hurdle, particularly in ensuring that the data is representative and comprehensive enough to cover all potential scenarios that a drone might encounter. This challenge is compounded by the need for high-quality, annotated data, which can be time-consuming and expensive to obtain. Another challenge is the computational power required for real-time processing of sensor data and running ML algorithms. Drones need to process vast amounts of data quickly to make real-time decisions, which requires significant computational resources. This can be a limitation for smaller drones that have limited onboard processing capabilities. Ensuring that drones have the necessary hardware to support these demands without compromising their flight performance is a critical consideration.

Ensuring the robustness and reliability of ML algorithms in diverse and unpredictable environments is also a significant challenge. Drones must be able to operate effectively in a variety of conditions, from urban landscapes to rural areas, and in different weather conditions. Developing ML algorithms that can perform reliably across these varied scenarios requires extensive testing and validation. The algorithms must be able to handle unexpected situations and continue to function correctly even when faced with unforeseen challenges.

While the integration of ML algorithms into drone technology offers substantial benefits in terms of safety, operational efficiency, and adaptability, it also presents several challenges. Addressing the data requirements, computational power needs, and ensuring the robustness of these algorithms is essential for fully harnessing the potential of drones. As the technology continues to evolve, ongoing research and development will be crucial in overcoming these challenges and realising the full capabilities of AI-driven drones.

Image Recognition for Surveillance

AI-powered image recognition has dramatically transformed the use of drones in surveillance and monitoring applications. By enabling drones to analyse visual data in real-time, these technologies can identify and track objects of interest, providing enhanced surveillance capabilities that were previously unimaginable.

AI-powered image recognition algorithms are designed to detect specific objects within a drone's camera feed, making them particularly useful for security monitoring. These algorithms can identify unauthorized personnel or vehicles, ensuring that potential threats are quickly recognised and addressed. Advanced image recognition technologies can also perform facial recognition, identifying individuals in real-time. This capability is invaluable for law enforcement and security applications, where accurately identifying suspects or missing persons can be crucial. Moreover, image recognition algorithms can analyse human activities and behaviours, detecting suspicious or unusual movements. This functionality is particularly useful in applications such as crowd monitoring and border security, where understanding and responding to human behaviour is essential.

The benefits of AI-powered image recognition in drones are significant. Enhanced surveillance capabilities allow drones to perform continuous and comprehensive monitoring over large areas, identifying potential threats quickly and efficiently. Real-time analysis of visual data enables immediate responses to detected incidents, improving security and response times. Additionally, drones equipped with AI-powered image recognition can perform surveillance tasks that would otherwise require extensive human resources, thereby reducing operational costs.

However, the deployment of AI-powered image recognition in drones also presents several challenges. Privacy concerns are paramount, as the use of facial recognition and continuous surveillance raises significant issues related to data protection and regulatory compliance. Ensuring the accuracy of image recognition algorithms and mitigating biases in the data are critical challenges that impact the reliability of surveillance outcomes. Furthermore, real-time image analysis requires substantial processing power and storage capabilities, which can be limiting for some drone platforms.

Autonomous decision-making processes represent another significant advancement in drone technology, enabling drones to make complex decisions without human intervention. This capability allows drones to perform tasks independently and adapt to unforeseen circumstances, greatly enhancing their utility and efficiency.

Autonomous Decision Making Algorithms

Autonomous decision-making algorithms enable drones to plan and execute missions autonomously. These algorithms can select optimal routes, manage resources such as battery life, and adjust plans based on real-time data. This dynamic response capability allows drones to adapt to changes in their environment, such as altering their flight path in response to weather conditions or obstacles. Autonomous decision-making also enables drones to perform specific tasks, such as inspecting infrastructure, waste management, delivering packages, or conducting search and rescue operations, based on predefined criteria and real-time data analysis.

The benefits of autonomous decision-making in drones are substantial. Increased autonomy reduces the need for constant human oversight, allowing drones to operate in remote or hazardous environments where human intervention is impractical. Drones can make quick decisions and execute tasks more efficiently than humans, particularly in time-sensitive situations such as emergency response. The ability to operate autonomously also allows for the deployment of large fleets of drones, scaling operations without a proportional increase in human resources.

Despite these benefits, the deployment of autonomous decision-making in drones presents several challenges. Ethical and legal considerations are significant, particularly regarding accountability and liability in the event of an accident or failure. Developing robust and reliable autonomous decision-making algorithms is complex and requires extensive testing and validation. Additionally, ensuring seamless integration of autonomous drones with existing air traffic management systems and other infrastructure is essential for safe and coordinated operations.

The integration of AI technologies into drones has significantly enhanced their capabilities, enabling advanced navigation, sophisticated surveillance, and autonomous decision-making. These technologies have expanded the potential applications of drones across various industries, from security and logistics to agriculture and disaster response. However, the deployment of AI in drones also presents challenges, including data requirements, computational power, privacy concerns, and ethical and legal considerations. Addressing these challenges requires a balanced approach that fosters innovation while ensuring safety, privacy, and regulatory compliance. As AI technologies continue to evolve, they will play an increasingly critical role in shaping the future of drone operations, driving advancements in efficiency, safety, and functionality.

Risks Associated with AI Technologies in Drones

While AI technologies offer significant benefits to drone operations, they also introduce several potential risks that must be carefully considered. These risks include algorithmic bias, decision-making transparency, and the challenge of unpredictable AI behaviours in uncontrolled environments. Understanding and mitigating these risks are crucial to ensuring the safe, ethical, and effective deployment of AI-driven drones.

Algorithmic bias

Algorithmic bias refers to the systematic errors that occur in AI systems due to biased data or flawed algorithmic design. This bias can lead to unfair or discriminatory outcomes, which is particularly concerning in applications like surveillance and decision-making. Bias can arise from several sources. Data bias occurs when AI algorithms are trained on large datasets that are unrepresentative or contain inherent biases. For instance, if a facial recognition system is trained primarily on data featuring a specific demographic, it may perform poorly in identifying individuals from other demographics. Design bias can also occur during the development phase. If developers inadvertently incorporate their own biases or make assumptions that favour certain groups over others, the resulting AI system will reflect these biases.

The implications of algorithmic bias are significant. In surveillance applications, biased algorithms could result in unfair targeting or misidentification of certain groups, leading to violations of privacy and civil liberties. Operational inefficiency is another concern, as biased navigation algorithms might preferentially avoid certain areas, leading to suboptimal routing and resource usage. Moreover, algorithmic bias raises serious legal and ethical concerns. Ensuring fairness and non-discrimination in AI systems is critical for maintaining public trust and compliance with anti-discrimination laws. To mitigate algorithmic bias, it is essential to use diverse and representative datasets for training. Continuous data auditing and updates are necessary to maintain dataset quality. Regularly testing AI algorithms for bias and performance across different demographic groups can help identify and address biases before deployment. Transparent development practices, including clear documentation and open-source collaboration, can also help identify and correct biases during the design phase.

Decision-making transparency

Decision-making transparency refers to the ability to understand and explain how AI systems arrive at their decisions. In the context of drones, transparency is crucial for accountability, safety, and public trust. However, several challenges impede decision-making transparency. Many AI models, especially those based on deep learning, operate as "black boxes" with decision-making processes that are not easily interpretable. This lack of transparency makes it difficult to understand why a drone made a specific decision. Additionally, there is often a trade-off between the explainability of an AI model and its performance. More complex models tend to perform better but are harder to interpret, whereas simpler models are easier to explain but may not be as effective. The lack of transparency has significant implications. Accountability issues arise if an AI-driven drone makes an error, such as misidentifying a target or choosing an unsafe route, because it is challenging to determine the cause and assign responsibility. Unexplained decisions can lead to safety risks, especially in critical applications like search and rescue or the delivery of medical supplies. Operators need to trust and understand the AI system to manage and mitigate risks effectively. Furthermore, transparency is essential for gaining public trust in AI technologies. Without clear explanations of how AI systems work and make decisions, public scepticism and resistance to drone deployment may increase.

Enhancing decision-making transparency involves several strategies. Explainable AI (XAI) techniques can help make AI decision-making processes more understandable by creating models that provide clear, interpretable explanations for their decisions. Implementing regulatory requirements for transparency can ensure that AI systems include mechanisms for logging and explaining their decisions. Regulators can mandate documentation and regular audits of AI systems. Additionally, designing user interfaces that present AI decisions in an understandable format can help operators and stakeholders comprehend and trust the system. Visualisations and summaries can aid in interpreting complex data and decisions, thereby improving transparency and trust in AI-driven drones.

Unpredictable AI Behaviours

Unpredictable AI behaviours refer to unexpected actions taken by AI systems, particularly when operating in uncontrolled or dynamic environments. These behaviours can pose significant risks to safety and reliability, necessitating careful consideration and mitigation strategies.

One major factor contributing to unpredictable behaviours is environmental variability. Drones often operate in diverse and changing environments, and AI systems may encounter situations or obstacles they were not trained to handle. For instance, a drone operating in an urban setting might encounter unexpected obstacles such as construction equipment or dynamic changes in traffic patterns, leading to unforeseen responses from the AI system.

Incomplete training data also plays a critical role. AI algorithms rely on extensive training data to learn how to behave in different scenarios. If the training data does not encompass all possible situations, the AI system may act unpredictably when faced with unfamiliar conditions. For example, a drone trained predominantly in rural environments might struggle when navigating complex urban landscapes.

Furthermore, complex interactions in uncontrolled environments introduce additional challenges. Drones may interact with various entities, such as humans, animals, or other drones. These interactions can create complexities that AI systems might find difficult to manage, leading to erratic or unexpected behaviours. An AI system might misinterpret the presence of a flock of birds or a group of pedestrians, resulting in sudden and potentially unsafe manoeuvres.

The implications of unpredictable AI behaviours are significant. Safety hazards are a primary concern, as these behaviours can lead to accidents, collisions, or other dangerous situations. For instance, a drone might misinterpret a shadow as an obstacle and make abrupt, unsafe manoeuvres. Operational disruptions are another consequence, as inconsistent behaviour can reduce the efficiency and reliability of drone missions. This is particularly problematic in critical applications such as logistics or emergency response, where reliable performance is essential. Additionally, regulatory compliance requires predictable and reliable AI behaviours.

Unpredictable actions can lead to violations of airspace regulations and operational guidelines, complicating regulatory adherence. To mitigate unpredictable behaviours, several strategies are essential. Robust training is crucial, ensuring that AI systems are trained on diverse and comprehensive datasets to handle a wide range of scenarios. Using simulated environments to expose AI systems to rare or extreme conditions can further enhance their preparedness. Implementing safety mechanisms, such as fail-safes and redundancy systems, can also help manage unpredictable behaviours. For instance, if an AI system encounters an unfamiliar situation, it could switch to a pre-programmed safe mode or return to a known location to prevent unsafe actions. Continuous monitoring and updates are vital as well. Regularly updating AI algorithms based on new data and real-world experiences can improve their robustness and reliability. Feedback loops that incorporate real-world performance data into ongoing training help address gaps and enhance the predictability of AI behaviours.

While AI technologies significantly enhance drone capabilities, they also introduce challenges related to unpredictable behaviours. Addressing these challenges through robust training, safety mechanisms, and continuous monitoring is essential for ensuring the safe, reliable, and effective deployment of AI-driven drones.

Case Study: AI-Driven Drone Incidents in South Korea

In South Korea, AI-driven drones have been increasingly used for various applications, including surveillance and delivery. In 2020, an incident involving an AI-driven surveillance drone raised significant concerns about the limitations and risks of AI in real-world applications. The drone, equipped with AI-powered facial recognition software, was deployed for security monitoring in a busy urban area. During its operation, the drone misidentified several individuals as persons of interest, leading to unnecessary and intrusive questioning by law enforcement. The AI system's facial recognition algorithms had been trained on a dataset that did not adequately represent the diverse population of the area, resulting in biased and inaccurate identifications.

In response to this incident, South Korean regulators took immediate action to address the shortcomings of AI-driven drone technologies. The government mandated more stringent testing and validation processes for AI algorithms used in public safety applications. This included requiring diverse and representative training datasets to minimise bias and improve the accuracy of AI systems. Additionally, regulators introduced policies for greater transparency and accountability, requiring developers to document their AI systems' decision-making processes and performance metrics. The incident underscored the critical need for comprehensive regulatory oversight and continuous monitoring of AI-driven technologies. It highlighted the potential for algorithmic bias to lead to real-world consequences and the importance of rigorous testing and validation to ensure AI systems are reliable and fair.

This case illustrates the varied responses to incidents involving drones and AI-driven technologies. The Gatwick Airport incident led to significant regulatory tightening and the adoption of advanced detection technologies, while the AI-driven drone incident in South Korea prompted stricter testing and transparency requirements for AI systems

Regulatory responses to drone incidents, whether involving AI or not, typically focus on enhancing safety, accountability, and transparency. Key regulatory measures include expanding no-fly zones, mandatory registration and training for operators, and implementing advanced technological solutions to detect and mitigate drone threats. For AI-driven drones, additional measures are necessary to address specific challenges such as algorithmic bias and decision-making transparency. These incidents underscore the importance of a proactive regulatory approach that anticipates and mitigates potential risks associated with AI-driven drone technologies. Ensuring that AI systems are rigorously tested, transparent, and accountable is crucial for maintaining public trust and safety. As AI technologies continue to evolve, regulators must remain vigilant and adaptive, continuously updating policies and frameworks to keep pace with technological advancements and emerging risks.

The integration of AI technologies into drones offers significant advancements in navigation, surveillance, and autonomous decision-making. However, these technologies also introduce potential risks, including algorithmic bias, decision-making transparency, and unpredictable behaviours in uncontrolled environments. Addressing these risks requires a comprehensive approach that includes diverse and representative training data, explainable AI techniques, robust safety mechanisms, and continuous monitoring and updates. By proactively managing these challenges, stakeholders can ensure the safe, ethical, and effective deployment of AI-driven drones, balancing innovation with responsibility and public trust.

In the next sections of this book, we will delve deeper into the specific regulatory frameworks and principles that govern drone operations in the EU. We will examine how these regulations are adapting to the challenges posed by AI, exploring both the current landscape and the forward-looking measures that are being considered to ensure that drones contribute positively to our future. This discussion will provide a comprehensive overview of how drone regulations are being shaped by technological advances, setting the stage for a more sophisticated and informed regulatory approach in the age of artificial intelligence.

VII

7. Risk-Based Approach to Drone Regulation in the EU

Building upon the previous discussions on the regulatory mindset for managing drone technology risks, referred to as Law 2.0, I will now delve into several critical questions that arise within this context: What is the nature of EU competence and the legal basis to intervene in the commercial drone industry? What are the harmonized risk-based regulatory rules, standards, and principles applicable to commercial drone operations in the EU? How are drone rules placed in the context of risk-based measures, precautionary measures, retributive measures, and distributive measures? How does the regulatory mindset, synonymous with rules and frameworks performed by legislative and executive branches, address risks associated with disruptive technological innovations? What is the essence of the EU harmonized drone rules, and how are they typified in a manner familiar to private law? To what extent do these risk-based harmonized rules make non-compliance an impossible option?

EU Competence in Drone Operations

The competence of the EU to intervene in the commercial drone industry is derived from the Treaty on the Functioning of the European Union (TFEU) which prescribes 'Shared competence between the Union and the Member States applies in the area of transport'. This means that both the EU and Member States are able to legislate and adopt legally binding acts in the area of transport. Therefore, EU countries may exercise their competence where the EU does not, or has decided not to exercise its competence.

The EU legal basis for intervention in the commercial drone industry is derived from the TFEU which provides that 'The European Parliament and the Council may lay down appropriate provision for sea and air transport. They shall act after consulting the Economic and Social Committee and the Committee of the Regions'. In addition, legal basis for harmonisation measures concerning the EU's internal market is also be derived form the TFEU. 'The European Parliament and the Council shall adopt the measures for the approximation of the provisions laid down by law, regulation or administrative action in Member States which have as their object the establishment and functioning of the internal market'.

The Commission is further required to take as a base a high level of protection of health, safety, environmental protection and consumer protection and to take into account new developments based on scientific facts. The Commission, in exercise of this competence, adopted in 2014 and 2015 respectively a Communication on a new era for civil aviation and the Aviation Strategy for Europe which highlighted that safety is crucial to the successful integration of drones in the airspace and the development of this industry and the services and applications enabled by drones. Notwithstanding, there are other issues beyond safety that must also be addressed in order to ensure the social acceptance of drones in Europe. Therefore, the need for a regulatory framework under which the aviation industry can thrive and remain competitive on the global market, including new business models and technologies such as drones was prioritised.

The Aviation Strategy sets the objective to establish a basic legal framework for the safe development of drone operations in the EU and to prepare more detailed rules that allow drone operations and the development of industry standards. This regulatory framework is now in place. In 2019, the Commission adopted Delegated Regulation

2019/945 on unmanned aircraft systems and on third-country operators of unmanned aircraft systems and Implementing Regulation 2019/947 on the rules and procedures for the operation of unmanned aircraft.

The EU's regulatory environment covers current and future drone operations, creating maximum harmonised rules for commercial drones in the EU with the mindset for the highest level of safety possible with some flexibility for member states in certain aspects. EU drone rules also facilitate the enforcement of privacy rights and addresses security issues and environmental issues to a limited extent. Notwithstanding, a recent study shows that national legislators are now faced with the challenging task of replacing their national regulations with EU rules.

Understanding the Risk-Based Approach

The risk-based approach is a fundamental paradigm in the regulation of drones, providing a framework for addressing potential risks and ensuring safety in their operations. What is the concept and what are the objectives of risk-based regulation? How does risk-based regulation guide the development of drone regulations? How does risk-based regulation shape the regulatory landscape? What are the principles and rationale behind the risk-based approach? How does the risk-based approach promote responsible and effective drone regulation?

The concept and objectives of risk-based regulation are critical to understanding how the EU shapes its regulatory landscape for commercial drones. This approach aims to create a dynamic and adaptive framework that responds effectively to the varying levels of risk associated with different drone operations. Risk-based regulation involves categorising drone operations into different risk levels and tailoring regulatory requirements accordingly. EASA has implemented such a framework by defining three categories of drone operations: open, specific, and certified. Each category corresponds to a different level of risk, with "open" being the lowest risk and "certified" the highest. This categorisation allows for proportionate regulatory responses that ensure safety without imposing unnecessary burdens on lower-risk operations.

One of the primary objectives of risk-based regulation is to enhance safety by focusing regulatory efforts on higher-risk activities. For example, operations within the "specific" category, which involve moderate risks, require operators to conduct a Specific Operations Risk Assessment (SORA). This assessment identifies potential hazards, evaluates risks, and implements mitigation strategies to ensure that the operation can be conducted safely. By requiring such detailed assessments, regulators can ensure that appropriate safety measures are in place for more complex and riskier drone operations.

Additionally, risk-based regulation aims to promote innovation and flexibility within the drone industry. By not applying a one-size-fits-all regulatory approach, the EU allows for more tailored and innovative uses of drone technology. This flexibility is crucial for fostering the development of new applications and encouraging investment in the sector. For instance, the adoption of advanced technologies such as AI and autonomous navigation is facilitated under this framework, as long as the associated risks are adequately managed.

Another key objective is to ensure that regulatory frameworks remain adaptive and responsive to technological advancements. The rapid pace of innovation in drone technology necessitates a regulatory approach that can evolve quickly to address new risks and opportunities. The EU's use of risk-based regulation allows for continuous monitoring and adjustment of regulatory measures to keep pace with technological changes and emerging threats.

Risk-based regulation in the EU is designed to balance safety, innovation, and flexibility. By categorising drone operations based on risk and tailoring regulatory requirements accordingly, the EU can ensure that safety is prioritised while also fostering an environment conducive to technological advancement and industry growth. This approach not only addresses current challenges but also positions the regulatory framework to adapt to future developments in the drone industry.

Benefits and Limitations of the Risk-Based Approach

What are the benefits and limitations of adopting a risk-based approach to drone regulation, and how does this approach impact flexibility, efficiency, and regulatory focus on high-risk areas while addressing concerns about subjectivity, resource allocation, and the need for ongoing risk assessment?

The risk-based approach to drone regulation offers several notable benefits, but it also presents certain limitations that must be carefully considered. One of the primary advantages of this approach is its inherent flexibility. By tailoring regulations based on the level of risk associated with different drone operations, regulators can create more adaptable and responsive frameworks. This flexibility allows for the accommodation of a wide range of drone applications, from low-risk recreational activities to high-risk commercial operations, ensuring that regulatory measures are proportionate to the potential hazards involved.

Furthermore, the risk-based approach promotes efficiency by concentrating regulatory efforts on areas that pose the greatest risks. By identifying and prioritising high-risk operations, resources can be allocated more effectively, ensuring that critical safety and compliance issues are addressed without imposing unnecessary burdens on low-risk activities. This targeted allocation of resources helps streamline regulatory processes and reduces the administrative workload for both regulators and drone operators.

However, the risk-based approach is not without its challenges. One significant limitation is the potential for subjectivity in risk assessments. Determining the level of risk associated with specific drone operations can be complex and may vary depending on the criteria and methodologies used. This subjectivity can lead to inconsistencies in regulatory enforcement and may undermine the perceived fairness and reliability of the regulatory framework.

Another concern is the allocation of resources required to implement and maintain a risk-based regulatory system. Continuous monitoring, data collection, and risk assessment are essential components of this approach, demanding significant investment in both technology and personnel. Ensuring that regulators have the necessary resources to perform these tasks effectively is crucial for the success of the risk-based approach.

Additionally, the dynamic nature of drone technology and its rapid evolution necessitate ongoing risk assessment and adaptation of regulations. The regulatory framework must be able to evolve in response to new developments and emerging risks, requiring a proactive and forward-looking approach. This need for constant vigilance and adaptability can strain regulatory bodies and may pose challenges in keeping pace with technological advancements.

By examining the benefits and limitations of the risk-based approach, we gain valuable insights into its practical implications for drone regulation. While this approach offers significant advantages in terms of flexibility and efficiency, it also demands careful consideration of subjectivity in risk assessments, resource allocation, and the need for continuous adaptation. Balancing these factors is essential for developing a regulatory framework that effectively promotes the safe and responsible use of drones while fostering innovation and growth in the industry.

Integration of Risk Assessment in Drone Regulations

How can the integration of risk assessment methodologies into drone regulations ensure effective safety measures and proportional regulatory requirements? The effective implementation of a risk-based approach to drone regulations necessitates the integration of detailed risk assessment methodologies. One widely adopted methodology is the Specific Operations Risk Assessment (SORA), which provides a structured framework for evaluating the risks associated with drone operations and developing appropriate mitigation strategies. SORA, developed by the Joint Authorities for Rulemaking on Unmanned Systems (JARUS), is designed to offer a standardised approach that can be applied internationally.

The process of conducting risk assessments for drone operations involves several key steps. Initially, operators must determine both the ground and air risk classes. The ground risk assessment considers factors such as the drone's size, operational environment, and population density of the area where the drone will be used. Mitigations to reduce ground risk might include technical containment systems and emergency response plans. Similarly, the air risk assessment evaluates the likelihood of encountering manned aircraft in the operating airspace and applies strategic and tactical mitigations to minimise collision risks.

The culmination of these assessments results in a Specific Assurance and Integrity Level (SAIL), which defines the robustness of the required mitigations. A high SAIL indicates a higher-risk operation that necessitates stringent safety measures, while a low SAIL corresponds to lower-risk operations with less stringent requirements. This nuanced approach allows regulators to tailor their oversight to the specific risks associated with each drone operation, ensuring that safety measures are proportionate to the identified risks.

Moreover, integrating risk assessments into drone regulations enhances regulatory effectiveness by focusing resources on higher-risk areas, thereby optimising safety and compliance efforts. It also promotes transparency and accountability, as operators are required to document and justify their risk mitigation strategies, which are then reviewed by regulatory authorities.

By understanding and applying these risk assessment methodologies, regulators can ensure that safety considerations are effectively addressed. This approach not only fosters innovation within the drone industry but also maintains high safety standards, thus enabling the responsible integration of drones into commercial and public applications across the EU and beyond.

Assessing and Managing Drone Risks

What methodologies and approaches are used to assess and manage risks in the context of drones, and how do these methods identify and categorise risks while implementing effective risk mitigation strategies? Assessing and managing risks in drone operations is critical for developing robust regulatory frameworks that effectively protect against potential hazards. This involves a multi-step process that includes identifying risks, categorising them, and implementing strategies to mitigate these risks.

The first step in risk assessment is to identify the various risks associated with drone operations. This includes evaluating potential hazards such as equipment failure, loss of control, adverse weather conditions, and collisions with other aircraft or objects. For instance, SORA framework, provide a structured approach to classify and evaluate risks based on the type of operation and the environment in which it takes place.

Risks are generally categorised into ground risk and air risk. Ground risks consider the potential impact on people, properties, and critical infrastructures, while air risks assess the likelihood of mid-air collisions and interactions with manned aircraft. These assessments take into account factors such as population density, type of airspace, and the complexity of the operation.

Once risks are identified and categorised, the next step is to develop and implement risk mitigation strategies. These strategies can include both corrective and preventive actions. Corrective actions address immediate hazards, such as modifying flight paths to avoid populated areas. Preventive actions, on the other hand, involve long-term measures like regular maintenance protocols, training programs for drone operators, and the implementation of advanced technological solutions such as geofencing and real-time monitoring systems. For instance, regular inspections and maintenance are essential to ensure that drones remain in optimal condition, thereby reducing the risk of mechanical failures. Training programs are crucial to equip operators with the necessary skills to handle various operational scenarios and to understand the regulatory requirements.

Moreover, adopting a systematic approach to documenting the entire risk management process is vital. This includes maintaining detailed logs of all flights, safety incidents, and mitigation actions. Such documentation not only ensures continuous safety assurance but also provides valuable data for ongoing risk assessment and improvement of safety measures.

Proactive measures play a key role in reducing risks and enhancing safety in drone operations. These measures include using AI-driven systems for real-time risk assessment, developing redundancy plans to handle emergencies, and ensuring compliance with privacy laws to protect sensitive data. Engaging with local communities and stakeholders is also important to address public concerns and to foster a positive perception of drone operations.

Assessing and managing drone risks involves a comprehensive process that integrates identification, categorisation, and mitigation of risks. By adopting structured frameworks like SORA, implementing robust maintenance and training programs, and leveraging advanced technologies, regulators and operators can significantly enhance the safety and reliability of drone operations. This proactive approach not only mitigates potential hazards but also supports the sustainable growth of the drone industry.

Identification and Categorisation of Drone Risks

How are drone risks identified and categorised in the EU regulatory framework? What factors are considered in determining the proportionality of rules and procedures for drone operations? How do the operational characteristics of drones and the characteristics of the area of operation influence the risk assessment process? What are the specific criteria used to establish the three categories of drone operations (open, specific, and certified)? How are proportionate risk mitigation requirements applied according to the level of risk involved?

The identification and categorisation of drone risks are fundamental steps in the risk assessment process within the EU regulatory framework. The process involves evaluating various factors that could potentially pose hazards during drone operations. The EU uses a structured approach to ensure that these risks are systematically identified and classified. This helps in understanding the potential impact of different drone operations and tailoring regulations accordingly. By systematically identifying and categorising risks, the EU can implement targeted mitigation strategies to enhance overall safety.

In determining the proportionality of rules and procedures, the EU regulatory framework considers the nature and risk level of the drone operation, as well as the specific operational characteristics of the drone. Additionally, characteristics of the area of operations, such as population density, surface characteristics, and the presence of buildings, are crucial. These factors help regulators ensure that the rules are appropriately scaled to the potential hazards, thus preventing over-regulation of low-risk operations while imposing stricter controls on high-risk activities.

The operational characteristics of drones, including their size, weight, and performance capabilities, directly influence the risk assessment process. Additionally, the characteristics of the area of operation, such as whether the area is densely populated or has significant infrastructure, also play a critical role. For instance, flying a small drone in a remote, sparsely populated area involves different risks compared to operating a large drone over a crowded urban center.

By considering these variables, regulators can tailor safety measures to the specific context of each operation.

The EU regulatory framework categories drone operations into three distinct categories based on risk level: open, specific, and certified. The 'open' category covers low-risk operations that do not require prior authorisation, such as recreational flights. The 'specific' category involves medium-risk operations that necessitate a risk assessment and operational authorisation. Finally, the 'certified' category includes high-risk operations similar to manned aviation, which require certification of the drone, operator, and sometimes the remote pilot. These categories help streamline regulatory requirements according to the complexity and potential hazards of the operation

It is important to reiterate that identifying and categorising risks is a critical part of assessing drone operations, particularly within the regulatory framework of the European Union. The EU's regulations are tailored to match the nature and risk level of each operation, taking into account the specific characteristics of both the drone and the environment in which it operates. Factors such as population density, surface features, and the presence of buildings influence these assessments.

To effectively manage these risks, operations are classified into three main categories: open, specific, and certified as conceptualised in the diagram below. Each category has its own set of risk mitigation requirements, proportionate to the level of risk and the operational characteristics involved. For example, the "open" category is for low-risk operations with minimal regulatory requirements, while the "specific" category covers medium-risk operations that require a risk assessment and authorisation. The "certified" category applies to high-risk operations, necessitating rigorous certification and compliance with stringent safety standards.

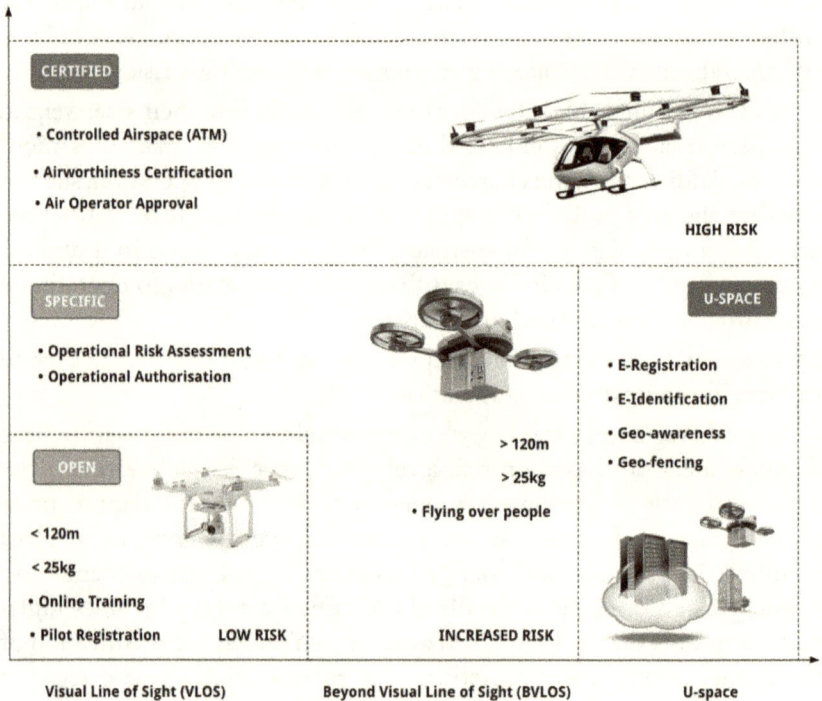

This structured approach ensures that drone operations are conducted safely and responsibly, with appropriate measures in place to mitigate potential risks based on the specific context and conditions of each operation. The classification system and corresponding risk mitigation strategies are depicted in detailed regulatory diagrams provided by the EU aviation authorities, ensuring clarity and compliance across all drone activities. The figure also incorporates the U-space requirements which is an essential cloud-based infrastructure for drone operational coordination and risk mitigation.

Open Risk Category

Drones in open category presents the lowest risk. Operations are in the open category only where they have a maximum take-off mass (MTOM) of less than 25kg. The remote pilot of drones flown in open category must ensure that the drone is kept at a safe distance from people and that is not flown over assemblies of people. The remote pilot must also ensure that the drone is flown in Visual Line of Sight (VLOS) at all times except when flown in follow-me mode or when using a drone observer. The remote pilot of a drone flown in open category must maintain the drone within 120 meters from the closest point of the surface of the earth, except when flying over an obstacle and during the flight, the drone must not carry dangerous goods or drop any material.

How are remote pilots regulated within the open category of drone operations, and what are the specific training and certification requirements for these pilots to ensure safety and competency? The minimum age for remote pilots operating a drone in open category is 16 years. Drones flown in open category are not subject to any prior operational authorisation, nor to any operational declaration by the drone operator before the operation takes place. Notwithstanding, remote drone pilots are required, depending on the subcategory, to complete an online training course and obtain a Certificate of remote pilot competency. The certificate of remote pilot competency is valid for five years and renewal is subject to demonstration of competencies in relation to the subcategory flown.

How do regulations for drone operations in the open category ensure operational safety and compliance, including the use of Remote ID systems, and what responsibilities do drone operators have regarding pilot competence and the use of operational procedures? Regulations for drone operations in the open category in the European Union are designed to ensure operational safety and compliance through a combination of stringent guidelines and technological requirements. One of the critical components of these regulations is the implementation of Remote ID systems. As of January 1, 2024, drones operating in the open category (C1, C2, and C3 classes) and the specific category must be equipped with Remote ID technology. This system allows for the remote identification of drones, providing crucial information such as the operator's registration number, the drone's serial number, its geographical position, and flight trajectory. This technology enhances transparency and accountability, helping to prevent unauthorized drone activities and ensuring that operators adhere to safety regulations.

Drone operators have specific responsibilities to maintain safety and compliance. They must develop operational procedures tailored to the type of operational risk and ensure that these procedures effectively support the efficient use of radio spectrum to avoid harmful interference. Additionally, operators must ensure that remote pilots possess the necessary competencies for the subcategory of drones they intend to fly. This includes completing online training courses and obtaining a Certificate of Remote Pilot Competency, which is valid for five years and requires periodic renewal to demonstrate continued proficiency.

Furthermore, operators must ensure that all personnel involved in drone operations are familiar with the user manuals provided by the drone manufacturers. This requirement is crucial for maintaining high safety standards and ensuring that all operational procedures are followed correctly. The combination of these regulatory measures, technological advancements, and operator responsibilities forms a comprehensive framework aimed at promoting safe, responsible, and efficient drone operations within the EU. Drone operators in open category are required to develop operational procedures to adapt the type of operational risk and ensure that all operations effectively use and support the efficient use of radio spectrum in order to avoid harmful interference. Drone operators must ensure that remote pilots have the appropriate competence in the subcategory intended. Drone operators must ensure that the remote pilot and all other personnel performing a task in support of the operation are familiar with the users manual provided by the manufacturer.

Manufacturers of drones in open category must affix a class identification label on it and ensure compliance with class requirements. Drone manufacturers from Third countries entering the internal market must ensure that appropriate conformity assessment procedures have been carried out and that CE marking and technical documentation are provided for inspection to the competent national authorities. Drones can only be placed on the market if they meet product requirements and do not endanger the health or safety of persons, animals and property. Member States are not allowed to prohibit, restrict or impede the making available on the market of products that are in conformity with harmonised standards. Member States are required to designate notifying authorities responsible for the setting up and carrying out the necessary procedures for conformity assessment. Member states are required to organise and perform market surveillance of products that are placed on the internal market.

Specific Risk Category

Drone operations in the specific category covers operations presenting a higher risk therefore thorough risk assessment must be conducted to indicate which requirements are necessary to ensure safe operations. Drones in specific category require operational authorisation from the competent authority or declaration to the competent authority in case of standard scenarios. Standard scenarios in the specific category means scenarios for which precise list of mitigation measures has been identified in such a way that the competent authority can be satisfied with declarations in which such operators declare that they will apply the mitigating measures when executing the specified type of operation.

For operations within the Specific Risk Category, drone operators must perform thorough risk assessments to determine the unmitigated ground risk and evaluate the operational environment. Factors such as flying beyond visual line of sight (BVLOS), population density in overflown areas, and the dimensions and characteristics of the drone are crucial in this assessment. The European Union has developed the Specific Operations Risk Assessment (SORA) methodology, which helps operators identify the risks and implement appropriate mitigation measures.

SORA is a comprehensive framework designed to evaluate and mitigate risks associated with drone operations, particularly those that require specific permissions from regulators, such as beyond visual line of sight (BVLOS) flights. Developed by the Joint Authorities for Rulemaking on Unmanned Systems (JARUS), SORA provides a standardised methodology for assessing the risks and determining the necessary safety measures to ensure secure and compliant drone operations. SORA encompasses various documents and guidelines aimed at identifying potential risks and outlining mitigation strategies. The key components include:

1. **Concept of Operations (ConOps):** A detailed document that provides all relevant technical, operational, and system information needed to assess the risks associated with the proposed drone operation. It includes the flight plan, the environment, and how the operation will be executed.

2. **Ground Risk Class (GRC) Determination:** Evaluates the risk of a drone impacting people or property on the ground. This assessment considers the drone's size, speed, type of flight (e.g., VLOS or BVLOS), and the operational area. Mitigating measures such as emergency parachutes or active geofencing can reduce the GRC.

3. **Air Risk Class (ARC) Determination:** Assesses the probability of the drone encountering crewed aircraft. Factors influencing ARC include whether the flight is in controlled or uncontrolled airspace, proximity to airports, and whether it flies over urban or rural areas. Strategic and tactical mitigations can lower the ARC.

4. **Specific Assurance and Integrity Levels (SAIL) Determination:** Provides a confidence level for the flight operation by integrating ground and air risk analyses. SAIL levels range from 1 to 6, each specifying objectives and supportive activities.

5. **Operational Safety Objectives (OSO):** Based on SAIL levels, OSOs outline requirements for the drone, its operator, and the operating organization. These include standards for operator knowledge and skills, as well as technical assessments of the drone and its equipment.

The time required to complete a SORA can vary depending on the complexity of the operation and the thoroughness of the risk assessment process. Typically, it involves multiple steps and may take several weeks to complete. The information needed includes detailed descriptions of the drone system, operational environment, proposed flight paths, risk mitigation measures, and compliance with safety standards. SORA is not just a static document; it is an iterative process that may involve multiple exchanges with regulatory authorities to refine and approve the risk assessment. The framework ensures that all potential hazards are addressed, and appropriate safety measures are in place before the operation is authorised. SORA is an essential tool for ensuring safe and compliant drone operations, particularly for high-risk activities. It involves a detailed and structured assessment of risks, backed by robust mitigation strategies, to safeguard people, property, and other airspace users.

In addition to the SORA, operators must obtain operational authorisation from their National Aviation Authority (NAA) unless they are operating under a predefined risk assessment (PDRA) or a standard scenario (STS). Standard scenarios provide a set of predefined mitigation measures that, if adhered to, simplify the authorisation process. For example, the STS-01 scenario involves operations over a controlled ground area in a populated area, whereas STS-02 pertains to operations in sparsely populated areas with extended BVLOS capabilities.

Mitigation measures are tailored to the specific risks identified during the assessment process. These measures can include the implementation of advanced technical solutions such as geofencing, automated emergency landing systems, and real-time tracking. Additionally, operators must ensure that all personnel involved, including remote pilots and observers, are adequately trained and certified according to the specific requirements of the operation Drone operators in the Specific Category must also maintain detailed documentation, including an Operations Manual, which outlines the procedures and protocols for safe operation. This documentation must be submitted to the NAA for review and approval. Failure to provide comprehensive documentation or to comply with regulatory requirements can result in the rejection of authorisation applications or additional scrutiny.

The European Union continues to refine its regulatory framework to address the evolving landscape of drone technology. Future amendments may include more sophisticated risk assessment tools and increased integration of AI and machine learning to enhance safety and efficiency. By continuously updating the regulatory environment, the EU aims to ensure that drone operations remain safe, responsible, and aligned with technological advancements.

200 / Drone Law 3.0

Certified Risk category

What are the requirements and considerations for operating drones within the certified risk category in the European Union, and how do regulatory bodies ensure compliance and safety in these high-risk operations? The certified risk category for drone operations in the EU is designed to address high-risk activities that involve significant potential hazards. These operations typically include flying over large assemblies of people, transporting passengers, or carrying dangerous goods. Given the heightened risks, stringent regulations are imposed to ensure safety and compliance.

Operations in the certified category are principle subject to rules on certification of the operator, and the licensing of remote pilots. Drone operations fit into the certified category if they involve flying over assemblies of people, involve the transport of people or involve the carriage of dangerous goods, that may result in high risk for third parties in the case of an accident. Operations in the certified category require comprehensive certification processes for the drone, the operator, and in some cases, the remote pilot. This involves obtaining operational authorisations from national aviation authorities (NAAs) and complying with specific airspace restrictions and safety protocols. For instance, the EASA mandates detailed operational risk assessments to identify potential hazards and implement appropriate risk mitigation measures.

To support these requirements, member states must establish accurate registration systems for certified drones and their operators. This ensures that all parties involved in high-risk drone operations are properly documented and can be monitored for compliance with safety standards and regulatory requirements. Competent authorities may also classify a drone operations certified where, based on the risk assessment provided, it is considered that the risk of the operation cannot be adequately mitigated without the certification of the drone and of the drone operator and, where applicable, without the licensing of the remote pilot.

Drone operations in the certified category must provide operational risk assessment that identifies the risk of the operation and possible risk mitigation measures. When conducting operations in the certified category, drone operators must follow a comprehensive risk assessment framework to identify potential risks and implement effective mitigation measures. The primary methodology used for this purpose is the Specific Operations Risk Assessment (SORA) for Certified operations, which provides a structured approach to evaluating both ground and air risks associated with drone operations.

Key Steps in the SORA for Certified Operations Process:

1. **Concept of Operations (ConOps):** Describe the operational context, including the type of drone, the qualifications of the remote pilot, the intended flight path, and the operational environment. This step involves detailing who will be involved, what equipment will be used, and where the operation will take place.

2. **Determining the Ground Risk Class (GRC):** Assess the risk posed to persons and property on the ground. This involves evaluating factors such as population density, the operational scenario (e.g., over populated or unpopulated areas), and the potential impact of a drone crash. Mitigation measures might include emergency response plans, the use of parachutes, or geofencing technologies to reduce the ground risk.

3. **Determining the Air Risk Class (ARC):** Evaluate the risk of collision with manned aircraft. This is done by analysing the airspace where the drone will operate, including whether the airspace is controlled or uncontrolled, proximity to airports, and the type of terrain (urban vs. rural). Implementing strategic and tactical mitigations, such as restricted flight times and areas, can help reduce air risk.

4. **Operational Safety Objectives:** Identify and implement safety objectives and limitations for the operation. This includes training requirements for personnel, technical specifications for the drone, and operational procedures to ensure safety.

5. **Submission and Approval:** Compile the risk assessment and mitigation measures into an operations manual. This manual, along with other relevant documentation, is submitted to the National Aviation Authority (NAA) for review and approval before commencing operations.

6. **Continuous Monitoring and Reporting:** After receiving authorisation, operators must continuously monitor the operation and report any incidents or deviations from the approved plan. This ensures ongoing compliance and helps improve safety protocols over time.

7. **Risk Mitigation Strategies and Best Practices:** Preventive Measures: Implementing robust testing, quality control, and risk assessment procedures to preemptively address potential issues. Using advanced technologies such as AI for real-time data analysis, automated reporting systems, and emergency response mechanisms. Engaging with regulatory bodies, insurers, and other stakeholders to develop standardised guidelines and share best practices. Ensuring that all personnel involved in drone operations are adequately trained and aware of the latest safety protocols and regulations.

Gathering sufficient data on drone operations and incidents is crucial for accurate risk assessment but can be challenging due to the diverse nature of drone applications. Implementing standardised data collection practices and utilizing automated tools can help streamline this process. Maintaining comprehensive and up-to-date documentation is essential for compliance but can be burdensome. Utilizing document management systems can facilitate easier documentation and ensure that all regulatory requirements are met. By adhering to these guidelines and leveraging the SORA methodology, drone operators in the certified category can effectively manage risks and ensure safe and compliant operations.

U-space Drone Traffic Management

U-space services means a service relying on digital services and automation of functions designed to support safe, secure and efficient access to U-space services for a large number of drones to operate. What is the regulatory framework for U-space in the EU and how does it ensure the safe integration of Unmanned Aircraft Systems (UAS) into the aviation system? What are the rules and procedures laid out for U-space airspace, and how do they facilitate safe drone operations? Why is U-space so crucial for the future of drone operations, and what technologies and services does it deploy to ensure compliance and safety?

The regulatory framework for the U-space (drone traffic management platform) was adopted on 22 April 2021. The Regulation lays down rules and procedures for the safe operations of UAS in the U-space airspace and for the safe integration of UAS into the aviation system and for the provision of U-space services. The U-space airspace is defined as a drone geographical zone designated by Member States, where drone operation are only allowed to take place with the support of U-space services.

U-space services provides geo-awareness, flight authorisation, electronic registration and electronic identification of drones throughout the duration of flight. SESAR has been mandated to lead the development of the Drone Traffic Management System (U-Space) which will enable the use of fully automated drones in the lower level aviation airspace. In addition, U-space has been established with a technology mindset which emphasises the deployment of automation technologies aims to remove or limit the practical option of non-compliance with drone rules. The automation technologies to be deployed includes detect and avoid, tactical deconfliction, dynamic geofencing and a collaborative interface.

Moreover, U-space is particularly important because it supports the development and deployment of advanced automation technologies aimed at ensuring compliance with drone regulations. Technologies such as detect and avoid systems, tactical deconfliction, dynamic geofencing, and collaborative interfaces play a pivotal role in mitigating risks associated with drone operations. Detect and avoid systems, for example, enable drones to identify and steer clear of potential obstacles autonomously, significantly reducing the likelihood of collisions. Tactical deconfliction helps in real-time management of airspace to prevent conflicts between drones and other aircraft, while dynamic geofencing creates virtual boundaries that drones cannot cross, ensuring they remain within safe operational areas.

The involvement of SESAR (Single European Sky ATM Research) in leading the development of U-space underscores its importance. SESAR's mandate includes creating a Drone Traffic Management System that will facilitate the use of fully automated drones in lower-level airspace. This initiative is aimed at integrating drones seamlessly with manned aviation, promoting a harmonious coexistence that is crucial for the broader adoption of drone technology.

Essentially, U-space is indispensable for the safe and efficient integration of drones into the aviation ecosystem. By leveraging automation and digital services, it ensures that drone operations are conducted within a secure and regulated framework, thus fostering innovation while maintaining high safety standards.

Precautionary measures for drone risk management

Drone operators and remote pilots must ensure that they are adequately informed about applicable EU law and national rules relating to the intended operations, in particular with regard to safety, privacy, data protection, liability, insurance, security and environmental protection. In case of areas considered sensitive to drone operations, Member States may lay down national rules to make subject to certain conditions the operations of drones for reasons falling outside the scope of this Regulation, including environmental protection, public security or protection of privacy and personal data in accordance with the Union law.

Member States can impose regulations to protect environmentally sensitive areas from drone activities. This includes national parks, wildlife reserves, and other ecologically significant regions where drone flights could disturb wildlife or damage natural habitats. For instance, specific flight restrictions or no-fly zones can be established to minimise the impact of drones on the environment.

To ensure public safety, national rules may restrict drone operations in areas where there are heightened security concerns. This includes critical infrastructure such as power plants, water treatment facilities, and government buildings. In some cases, temporary restrictions may be applied during public events or in areas with large gatherings to prevent potential security threats posed by drones. Privacy concerns are paramount, especially in densely populated areas where drones could capture personal data without consent. Member States may establish rules that restrict drone operations in residential zones or near private properties to protect individuals' privacy. These rules could mandate certain altitudes for flights, require prior notification to local authorities, or even restrict the use of drones with specific capabilities like high-resolution cameras.

For example, Germany has strict regulations regarding drone flights in residential areas to protect privacy. Operators must obtain explicit permission to fly over private property and are prohibited from using drones to capture images or videos of individuals without their consent.

France has also designated specific no-fly zones over nuclear power plants and military installations to enhance public security. Additionally, drones are not allowed to fly over certain urban areas and historical sites to protect public safety and privacy.

Further, drone operators in specific category must provide a statement confirming that the intended operation complies with any applicable Union and national rules relating to it, in particular, with regard to privacy, data protection, liability, insurance, security and environmental protection. Drone pilots in open category must keep the drone in VLOS and maintain a thorough visual scan of the airspace surrounding the unmanned aircraft in order to avoid any risk of collision with any manned aircraft. The remote pilot shall discontinue the flight if the operation poses a risk to other aircraft, people, animals, environment or property.

Competent authorities in Member States may define certain areas as UAS geographical zones for safety, security, privacy or environmental reasons. UAS geographical zone means a portion of airspace established by the competent authority that facilitates, restricts or excludes UAS operations in order to address risks pertaining to safety, privacy, protection of personal data, security or the environment, arising from drone operations. Competent authorities may restrict or prohibit specific or all categories of drones based on precautionary measures. They are required to maintain registration systems for drones whose design are subject to certification and for UAS operators whose operation may present a risk to safety, security, privacy, and protection of personal data or the environment.

Safety obligations

Safety rules are essential requirements to in the regulatory framework for drones in the EU. How do safety rules within the EU regulatory framework ensure the airworthiness, noise control, emissions standards, and environmental compliance of drones, and what responsibilities do drone manufacturers, operators, and service providers have in maintaining these standards?

Drone manufacturers, operators and service providers are obliged to ensure that drones comply with airworthiness, noise, emissions and environmental requirements set in the regulations. Manufacturers must ensure that drones are airworthy, which means they must be built and maintained to specific safety standards. This includes structural integrity, reliability of control systems, and overall performance. Compliance with these standards is verified through certification processes, where drone designs are rigorously tested and evaluated. Noise pollution and emissions are significant concerns in urban areas. Drones must comply with noise regulations to minimise their impact on communities. This involves using quieter propulsion systems and incorporating noise-reducing technologies. Emission standards are also crucial, especially for larger drones that may use combustion engines. Compliance ensures that drones do not contribute significantly to air pollution. Environmental protection is another key aspect. Drones operating in the EU must adhere to regulations that limit their environmental impact, which includes considerations for wildlife disturbance and habitat protection. Manufacturers must ensure that their products do not harm the environment, aligning with broader EU environmental policies.

Drone designers and manufacturers in the EU are obliged to apply for certification and shall be issued an approval upon compliance with essential requirements set out in the regulations. Personnel involved in release and maintenance of drone products must be licensed to ensure that they comply with essential safety requirements. Drone manufacturers and operators are advised to have detailed understanding of the essential requirements for operating in the applicable classification.

Privacy obligations

Drones are integrated with cameras and are able to collect significant amount of personal data during their operation. Data collected and retained have far reaching consequences on the right to privacy. Commercial drone manufacturers and operators must have clear understanding of the impact of drones on fundamental rights, as protected by the European Convention on Human Rights, the EU Charter of Fundamental Rights, and the General Data Protection Regulation (GDPR).

GDPR applies by default to processing of personal data via drones, either by private or public entities for purposes other than law enforcement. Operators that intend to share images or videos captured with a drone must ensure that there are no private images or data that may identify a subject. As a precautionary measure, all footage captured should be treated as personal data. Drone manufacturers and operators are advised to adopt anonymization technology to make subjects unrecognisable.

How do EU regulations under the GDPR accommodate the

activities of drone operators in media and journalism, and what specific derogations are allowed to reconcile the right to privacy with the freedom of expression and information? EU regulations under GDPR provide specific derogations for drone operators engaged in media and journalism activities. These derogations allow EU Member States to make exceptions to certain parts of the GDPR, provided they are necessary to reconcile the right to privacy with the rules of freedom of expression and information. This means that when drone operators use drones for journalistic purposes or for academic, artistic, or literary expression, they may be subject to different regulatory requirements compared to other drone users.

The purpose of these derogations is to ensure that the processing activities carried out by drone operators in these fields do not unduly infringe on privacy rights while still allowing for the essential functions of media, journalism, and other expressive activities. For example, drone operators in journalism may need to capture aerial footage of public events or areas of interest that would typically be restricted under standard GDPR regulations. The provided exceptions recognize the societal value of such activities and aim to balance it against privacy concerns.

This regulatory flexibility is crucial for the functioning of a free press and the dissemination of information, as it allows journalists and media professionals to use advanced technology like drones to gather news and provide coverage that might not be possible otherwise. However, it also places a responsibility on drone operators to ensure that their use of drones is justified, proportionate, and respectful of individuals' privacy.

By allowing these derogations, the EU ensures that the rights to privacy and data protection are balanced with the need for freedom of expression and information, acknowledging the unique role that media and journalism play in democratic societies. This approach helps maintain the integrity and effectiveness of journalistic work while upholding fundamental privacy rights.

How do regulators ensure that drone operators integrate necessary features and functionalities to mitigate risks related to privacy and protection of personal data, while complying with principles of privacy by design and by default? To address risks pertaining to privacy and protection of personal data, regulators may require drone operators to incorporate specific features and functionalities into their operations. These mandated features and functionalities are designed to adhere to the principles of privacy by design and by default, ensuring that privacy and data protection considerations are integrated into the development and operation of drones from the outset.

One critical requirement is that drones must be equipped with mechanisms that allow for easy identification of the aircraft, as well as the nature and purpose of the operation. This might include features such as remote identification systems, which enable authorities and other stakeholders to identify and monitor drones in real-time, thereby enhancing accountability and traceability.

Privacy by Design (PbD) is a fundamental principle in ensuring that privacy and data protection are embedded throughout the entire lifecycle of drone technology—from initial design to final deployment and commercialisation. This proactive approach aims to incorporate privacy measures into the core functionality of drones, ensuring that data protection is not an afterthought but a fundamental component of the technology.

Privacy by Design means considering privacy and data protection from the earliest stages of drone design. This involves identifying potential privacy risks and implementing technical and organizational measures to mitigate these risks before the drones are brought to market. For example, the Belgian Data Protection Authority (DPA) advises drone operators to incorporate features and functions that are directly proportional to the drone's intended use. This ensures that data collection is limited to what is necessary and relevant, adhering to the principle of data minimisation.

Drone features should be proportionate to their specific purposes. For instance, if a drone is used for agricultural monitoring, it should not be equipped with high-resolution cameras capable of identifying individuals unless absolutely necessary. This targeted approach helps to limit the scope of data collection, reducing the risk of infringing on individuals' privacy unnecessarily.

To further protect privacy, drones can be equipped with anonymization software that obscures the identities of individuals captured during data acquisition. This can include techniques such as blurring faces or using algorithms that prevent the identification of individuals in recorded footage. Additionally, data-erasing software can be implemented to automatically destroy data once it is no longer needed or if the drone is lost or stolen. These measures ensure that personal data is not exposed or misused, even in the event of unforeseen circumstances.

The Belgian DPA provides guidance on implementing Privacy by Design for drone operators. They recommend that all integrated features and functionalities should be necessary and proportionate to the drone's operational purpose. This includes not only technical specifications but also procedural safeguards to protect data throughout the drone's operational lifecycle

Under the GDPR, the concept of Privacy by Design is codified, requiring that data protection measures are integrated into the processing activities and the technology itself. This regulation mandates that data controllers and processors implement appropriate technical and organizational measures designed to implement data protection principles effectively.

One of the primary challenges in implementing Privacy by Design in drones is balancing the need for advanced functionalities with robust privacy protections. As drones become more sophisticated, ensuring that privacy measures do not compromise operational capabilities is crucial. Continuous innovation and collaboration between technologists and privacy experts are essential to achieving this balance.

As drone technology evolves, so do the potential privacy threats. Regulators and drone manufacturers must stay ahead of these developments by regularly updating privacy measures and ensuring that new threats are promptly addressed. This might involve periodic reviews and updates to the privacy features embedded in drones.

Privacy by Design is critical in ensuring that drones operate within the boundaries of privacy laws and ethical standards. By integrating privacy measures from the ground up, drone operators can build trust with the public and regulators, paving the way for broader acceptance and utilization of drone technology in various sectors. Ensuring privacy is not merely a regulatory requirement but a fundamental aspect of responsible drone innovation and deployment.

The privacy by design guide has been recognised as official privacy guidance by the European Union Aviation Safety Agency (EASA). The comprehensive guide developed by Drone Rules Pro project and sponsored by the European Commission, offers the drone industry practical tools to meet their privacy and data protection obligations. Manufacturers have been advised to build active sensor mode into drones to indicate to people on the ground that the drone's sensors are operating and collecting data.

Environmental obligations

Drones must be designed to minimise noise and emissions to the greatest extent possible. This involves the use of quieter propulsion systems and more efficient engines to reduce the acoustic footprint of drones, particularly in urban and wildlife-sensitive areas. Electric drones, for instance, offer a quieter and cleaner alternative to traditional internal combustion engine models, significantly reducing both noise pollution and greenhouse gas emissions.

The European Union Aviation Safety Agency (EASA) mandates that drones comply with environmental protection standards as part of their essential requirements. Drones must obtain a certificate of airworthiness and a noise certificate where certification is required. EASA's regulations aim to prevent significant harmful effects on climate, environment, and health by addressing emissions and noise

Despite these measures, specific provisions for environmental considerations relating to wildlife are currently lacking. The noise and presence of drones can disturb animals, particularly in protected areas. For instance, wildlife monitoring using drones must balance the need for data collection with the potential stress caused to animals. Establishing no-fly zones and flight altitude restrictions in ecologically sensitive regions could help mitigate these impacts.

The drone industry must also address the environmental impact of drone manufacturing and disposal. This involves adopting sustainable materials and manufacturing processes, as well as implementing end-of-life recycling programs. The use of eco-friendly materials and the development of drones designed for easy disassembly and recycling are critical steps towards reducing the environmental footprint of drone production.

To further enhance the environmental sustainability of drone operations, it has been proposed that common EU rules and specifications on environmental sustainability of drones be established. These regulations could cover topics such as sustainable logistics and operational systems for delivery drones, aiming to minimise their impact on wildlife and promote broader sustainability in the drone value chain. For example, delivery drones could be required to follow specific flight paths that avoid sensitive ecological areas, and to operate at times that minimise disturbance to both human and wildlife populations.

Operational best practices also play a crucial role in reducing the environmental impact of drones. This includes optimising flight paths to reduce energy consumption, scheduling flights to avoid peak wildlife activity times, and using drones for environmental monitoring and data collection that supports conservation efforts. Additionally, the integration of advanced technologies such as AI and machine learning can enhance the efficiency and precision of drone operations, further reducing their environmental footprint.

In essence, while drones offer significant benefits across various industries, their environmental impact must be carefully managed. This involves not only adhering to existing noise and emission regulations but also developing new standards and practices that specifically address the ecological challenges posed by drones. By promoting sustainable design and operation, the drone industry can contribute to achieving broader environmental and sustainability goals, ensuring that the benefits of drone technology do not come at the expense of the planet's health and biodiversity.

Retributive measures

Drone operators and remote pilots are required to ensure that they are adequately informed about applicable EU and national rules relating to the liability of the intended drone operations. How does the statement of compliance required from drone operators impact their liability in the event of accidents or injuries. The requirement for drone operators to provide a statement of compliance confirming adherence to EU and national rules on liability implies a significant responsibility on their part. This statement ensures that operators acknowledge their duty to comply with regulations, thereby accepting liability in the case of bodily injury or damage caused by their drones. The compliance statement can act as a legal document that may be used in court to establish the operator's accountability. If an operator fails to meet these compliance standards, they may be held liable for accidents, thus highlighting the importance of rigorous adherence to regulatory requirements.

Under what circumstances could a drone manufacturer be held liable for a defective product, and how does this differ from operator liability? Liability for defective products can arise if damage or bodily harm is caused by a drone due to manufacturing defects or failures beyond the operator's control. This type of liability falls on the manufacturer rather than the operator. If a drone crashes due to a defect that was present at the time of production, the manufacturer could be subject to product liability claims. This differs from operator liability, where the operator is held responsible for mishaps resulting from their operation of the drone. Establishing causation is key in determining whether the defect is attributable to the manufacturer or if the incident was due to operational errors.

What challenges exist in implementing safety-by-design features in drones, and how do these challenges affect liability? Implementing safety-by-design features such as propeller protection and parachute drags presents challenges primarily due to the need for drones to remain lightweight and maintain their flight efficiency. These safety features, although beneficial, add weight and complexity, potentially affecting the drone's performance and battery life. The difficulty in integrating these features effectively means that manufacturers and operators must find a balance between safety and functionality. The absence of adequate safety features could increase liability for both manufacturers and operators in the event of accidents, as it might be argued that insufficient safety measures were in place.

How can drone operators mitigate liability risks associated with data loss or theft during operations? Drones can be lost or stolen during operations, posing significant liability risks, especially if sensitive data is compromised. Operators can mitigate these risks by implementing robust data security measures such as encryption and data-erasing features that activate if a drone is lost or stolen. These measures ensure that even if the drone falls into the wrong hands, the data it collected cannot be accessed or misused. Additionally, regular audits and updates to data security protocols can help maintain high standards of protection against cyber threats, reducing the risk of liability from data breaches.

What role does remote identification play in managing liability and ensuring accountability in drone operations? Remote identification is crucial for managing liability and ensuring accountability in drone operations. It allows for the real-time broadcasting of drone data, including the operator's registration number, which helps identify the operator and establish liability in the event of incidents. This feature is particularly useful for law enforcement and investigators as it provides immediate access to information about the drone's operator and flight activities, facilitating faster and more accurate investigations. The integration of remote identification systems also deters unlawful activities and encourages responsible operation by making operators easily traceable.

How does mandatory registration of drone operators in their Member State enhance regulatory compliance and liability management?

Mandatory registration of drone operators in their Member State ensures that operators are accountable and can be easily identified and contacted. This registration process requires operators to keep their information accurate and up-to-date, fostering a transparent and accountable operational environment. By preventing operators from registering in multiple Member States simultaneously, the regulation simplifies the tracking of operator activities and enhances the enforcement of compliance. This system supports liability management by ensuring that there is a clear record of who is responsible for each drone, aiding in the swift resolution of any legal issues that arise from drone operations.

Distributive measures

Liability risks may be covered by insurance solutions from specialised insurers with drone manufacturers that adopt rigorous testing, quality control, and risk assessment procedures. How can insurance solutions be tailored to effectively cover the unique liability risks associated with drone operations? Tailoring insurance solutions to effectively cover the unique liability risks associated with drone operations requires a deep understanding of the specific risks and challenges involved. Unlike traditional aviation, drones operate in a diverse range of environments and for various purposes, which introduces different risk profiles.

Specialised insurers need to develop policies that address these specific risks, including coverage for accidents, property damage, bodily injury, cybercrime, and data breaches. One approach is to conduct rigorous testing, quality control, and risk assessment procedures to better understand the likelihood and impact of potential incidents. Insurers can use this data to create more accurate and relevant insurance products for drone operators. Furthermore, as the drone industry evolves, continuous data collection and analysis will be crucial in refining these insurance solutions to ensure they remain relevant and comprehensive.

What are the limitations of current drone liability insurance, and how do these limitations affect operators and manufacturers? Current drone liability insurance solutions are limited due to the lack of comprehensive data on accidents and damages. Most available insurance premiums are benchmarked to traditional civil aviation, which may not adequately reflect the specific risks associated with drones. This can result in either insufficient coverage or excessively high premiums that do not align with the actual risk profile of drone operations. For operators and manufacturers, these limitations mean they might face significant financial risks in the event of an incident. Without tailored insurance solutions, they may be underinsured, leading to substantial out-of-pocket expenses or legal liabilities if an accident occurs. This gap in insurance coverage underscores the need for the development of specialised drone insurance products that consider the unique aspects of drone usage and risks.

How does the EU's regulation on compulsory insurance for aircraft impact drone operators, particularly regarding the MTOM classification? The EU's regulation on compulsory insurance for aircraft, which includes drones with a Maximum Takeoff Mass (MTOM) of 20 kg or more, mandates a third-party liability insurance coverage of at least approximately €930,179.96. This regulation ensures that victims of drone accidents can claim compensation, providing a safety net for affected parties. However, for drones under 20 kg, it can be argued that the regulation does not explicitly apply, creating a potential gap in mandatory insurance coverage. This gap could leave victims of accidents involving lighter drones at risk of inadequate compensation. The MTOM classification as the sole criterion for insurance requirements has been criticised as inadequate, as it does not account for other relevant factors such as the pilot's experience or the nature of the drone operation. A more nuanced approach that includes these factors could provide a more accurate assessment of the necessary insurance coverage.

What are the benefits of drone operators subscribing to higher insurance coverage, and what specific risks should these policies address? Subscribing to higher insurance coverage provides drone operators with a greater level of protection against various risks. These policies should cover multiple exposures, including physical damage to the drone, third-party liability for bodily injury or property damage, invasion of privacy, cybercrime liability, and data theft or hacking claims. By opting for comprehensive insurance policies, operators can mitigate the financial impact of incidents and ensure they are better prepared to handle any legal or compensation claims that arise. Higher insurance coverage can also enhance the credibility and reliability of drone operators, potentially leading to more business opportunities and partnerships. Additionally, comprehensive insurance policies that address specific risks associated with drone operations can provide peace of mind to both operators and their clients, fostering a safer and more responsible industry.

How can legislation address the current lacuna in drone insurance requirements to ensure adequate protection for all stakeholders? To address the current gaps in drone insurance requirements, legislation should evolve to encompass a broader range of factors influencing drone operations. This could include mandating insurance coverage based not only on the MTOM but also considering the operator's experience, the type of operation, and the specific environments in which drones are used. Legislators could also establish minimum insurance requirements for all drones, regardless of weight, to ensure that even lighter drones have adequate coverage. Furthermore, creating standardised guidelines for insurance policies tailored to drones can help insurers develop more relevant products. Enhanced collaboration between regulatory bodies, insurers, and industry stakeholders is essential to develop a comprehensive framework that provides robust protection for all parties involved in drone operations.

Monitoring and Reporting Obligations

Monitoring and reporting obligations play a vital role in ensuring compliance with risk-based regulations and facilitating ongoing safety assessment. How do monitoring and reporting obligations ensure compliance with risk-based regulations in drone operations? Monitoring and reporting obligations are essential for ensuring that drone operators adhere to risk-based regulations, which are designed to mitigate potential hazards associated with drone use. These obligations require operators to systematically track and report incidents, performance metrics, and operational data. By maintaining a comprehensive database of drone-related incidents, regulatory bodies can identify trends, pinpoint common issues, and develop targeted regulations to address specific risks. This ongoing process not only ensures compliance but also helps in refining and enhancing regulatory frameworks based on empirical data.

What specific incident reporting requirements are imposed on drone operators, and how do these contribute to overall safety in the drone industry? Drone operators are typically required to report incidents such as crashes, near-misses, and any operational anomalies that could impact safety. These reports must include detailed information about the nature of the incident, the circumstances leading up to it, and any measures taken to mitigate the impact. The detailed reporting enables regulatory bodies to analyze incidents comprehensively and develop strategies to prevent recurrence. This systematic reporting contributes to overall safety by creating a culture of transparency and accountability, where operators are continuously encouraged to adhere to best practices and safety protocols.

How do data collection and performance monitoring obligations impact the operational efficiency and safety of drone activities? Data collection and performance monitoring are critical components of drone operations that impact both efficiency and safety. By continuously monitoring performance metrics such as battery life, flight stability, and system health, operators can identify potential issues before they lead to failures. This proactive approach helps in maintaining high operational standards and reducing the likelihood of accidents. Moreover, the data collected can be used to optimise flight paths, improve maintenance schedules, and enhance overall drone performance, leading to more efficient and safer operations.

What role do regulatory bodies play in overseeing compliance with monitoring and reporting obligations, and how do they ensure that operators adhere to these requirements? Regulatory bodies play a crucial role in overseeing compliance with monitoring and reporting obligations. They establish the standards and guidelines for incident reporting and data collection, ensuring that all operators follow a consistent approach. Regulatory agencies also maintain and analyze comprehensive databases of reported incidents and operational data. Through regular audits, inspections, and evaluations, they can verify that operators comply with the established requirements. Additionally, regulatory bodies may provide feedback and recommendations based on the analysis of collected data, helping operators to improve their practices and enhance safety measures.

How can the establishment of a comprehensive database of drone-related incidents promote transparency, accountability, and continuous improvement in the drone industry? Establishing a comprehensive database of drone-related incidents promotes transparency by making information about drone operations and incidents publicly available. This openness encourages accountability as operators are aware that their actions and any incidents will be scrutinised. Furthermore, the data collected in such databases provides invaluable insights into the causes and frequency of incidents, allowing for continuous improvement in both regulatory frameworks and operational practices. By analysing this data, industry stakeholders can identify best practices, develop new safety measures, and share lessons learned across the industry, leading to a safer and more efficient drone ecosystem .

What challenges do drone operators face in meeting monitoring and reporting obligations, and how can these challenges be addressed to ensure effective compliance? Drone operators face several challenges in meeting monitoring and reporting obligations, including the technical complexity of data collection, the administrative burden of reporting, and the need for real-time data transmission and analysis. To address these challenges, operators can leverage advanced technologies such as automated reporting systems and AI-driven analytics to streamline data collection and reporting processes. Training and education programs can also help operators understand the importance of these obligations and how to fulfil them effectively. Regulatory bodies can support operators by providing clear guidelines, user-friendly reporting tools, and ongoing support to ensure effective compliance.

8. Understanding Liability in Drone-Related Violations

As we venture deeper into the complex and rapidly evolving domain of drone technology, it is imperative to scrutinise the traditional legal frameworks that govern our societal norms and behaviours. Here, I set out to critically examine the applicability and adequacy of what we term Law 1.0—the established traditional norms of private law—when faced with the dynamic and often unpredictable nature of commercial drone operations. This exploration aims to help understand how existing legal principles adapt to or fall short against the backdrop of technological disruption.

The foundational principles of Law 1.0, encompassing default rules, standards, and general legal doctrines, have long served as the bedrock of legal reasoning. These principles, entrenched in the fabric of our legal system, are designed to offer predictability, fairness, and a structured approach to resolving disputes. However, as we confront the advent of drones—unmanned aircraft systems (UAS) capable of autonomous operations across diverse environments—we must question the sufficiency of these traditional frameworks.

What are the existing legal principles that govern liability in commercial drone operations within the EU? In exploring this question, I turn to the foundational principles of private law,. This body of law encompasses default rules, standards, and general legal doctrines, which have traditionally governed civil liabilities. Key principles include negligence, strict liability, and product liability. Negligence, a cornerstone of tort law, requires establishing a duty of care, a breach of that duty, and a resultant harm. In contrast, strict liability imposes responsibility without fault, particularly relevant to inherently dangerous activities where the operator's caution may not suffice to mitigate risk. Product liability, on the other hand, addresses the accountability of manufacturers and sellers for defective products that cause harm. While these principles provide a starting point, their application to the autonomous and pervasive nature of drone technology poses significant challenges.

However, before delving deeper into the specific liability principles, it is crucial to comprehend the unique characteristics and challenges associated with drone-related violations. Their autonomous capabilities, ability to operate in diverse environments, and potential to interact with humans and property present distinct considerations when assessing liability. Understanding these aspects will provide a solid foundation for analysing the applicable drone liability frameworks.

What unique characteristics and challenges do drones present when assessing liability? The autonomous nature of drones affects liability in several ways. First, there are legal implications when a drone makes independent decisions that lead to damage or injury. The operational environments of drones are varied; they can function in urban areas, where the risk of collision with buildings or interference with people is high, or in remote locations, where the challenges include difficult terrain and limited oversight. Moreover, drones' interactions with people and property can result in unique legal challenges. For instance, determining liability in cases where drones cause personal injury or property damage involves assessing the drone's autonomy, the operator's control, and the manufacturer's responsibility for any technological failures.

In addressing these questions, I lay the groundwork for a comprehensive analysis of how existing legal principles can adapt to and adequately regulate the rapidly evolving field of drone technology. By examining case studies and legal precedents, I aim to highlight the strengths and limitations of current legal frameworks, proposing enhancements that ensure robust regulation and accountability in the age of aerial autonomy. This chapter argues that while Law 1.0 offers a foundational approach, it may lack the precision and flexibility needed to address the unique attributes and risks associated with drone technologies. Through this analysis, we aim to uncover the gaps, propose modifications, and anticipate future legal developments in drone law. This is not only essential for legal professionals but also for operators, manufacturers, and service providers who navigate this rapidly shifting landscape.

Liability in Drone Operations

Drone operations introduce complex legal challenges regarding liability. Some of the unique legal questions include, How do the various aspects of liability associated with drone operations compare to traditional liability frameworks? What are the specific legal principles and concepts that underpin liability in the context of drone operations? How do existing liability frameworks address the unique characteristics and operational contexts of drones?

In addressing these questions and delve into the established legal doctrines and theories that help determine liability in this context. One such doctrine is negligence, which forms the foundation for many liability claims. How does the doctrine of negligence apply to drone operations, and what are the elements required to establish a breach of duty of care by drone operators? Negligence requires demonstrating that a drone operator breached their duty of care towards others, resulting in harm or damage. We will explore the elements of negligence later on and how they apply to drone-related violations, helping readers understand the standard of care expected from drone operators and the potential consequences of failing to meet that standard.

Another important concept in drone liability is strict liability, which holds manufacturers and suppliers responsible for any defects or hazards associated with their products. What are the principles of strict liability in relation to drone manufacturing and supply, and how do these principles govern the responsibilities of manufacturers and suppliers? Strict liability focuses on the inherent risks and dangers that may arise from drone operations, irrespective of negligence or fault. By exploring the principles of strict liability, we can gain insights into the legal framework that governs manufacturers' and suppliers' responsibilities in providing safe and reliable drone products.

Foreseeability is another key aspect of liability in drone operations. How is the principle of foreseeability interpreted and applied in assessing the liability of drone operators and other stakeholders in drone-related violations? It pertains to the ability to reasonably anticipate the potential harm or damage that may result from certain actions or omissions. Understanding the concept of foreseeability is crucial in assessing the liability of drone operators and other stakeholders. By examining past cases and legal precedents, we can explore how the courts have interpreted and applied the principle of foreseeability in the context of drone-related violations.

By exploring these legal concepts in more detail, I will provide a comprehensive understanding of the standards against which drone operators and other stakeholders may be held accountable. It is important to note that liability in drone operations is a complex, unique and evolving area of law, influenced by technological advancements and regulatory developments. Therefore, staying informed about the legal principles and concepts governing drone liability is crucial for both drone operators and the wider industry.

Liability of Drone Operators

Drone operators play a pivotal role in ensuring safe and responsible operations. The complex legal questions that comes to mind in relation to liability of drone operators include; What are the specific duties and responsibilities imposed on drone operators under traditional legal frameworks, and how do these obligations impact their liability in case of violation? How does the duty of care towards other airspace users and the public influence the legal consequences faced by drone operators in the event of violations or accidents? In what ways do operational limitations and compliance requirements shape the liability framework for drone operators, and what potential legal repercussions can arise from breaches of these conditions?

One crucial obligation for drone operators is compliance with applicable regulations. The aviation industry is subject to a comprehensive set of rules and regulations that govern the operation of drones in the EU. These regulations encompass various aspects, such as registration and licensing requirements, airspace restrictions, and operational limitations. Later on, we will examine the specific regulations that drone operators must adhere to and highlight the potential legal consequences of non-compliance.

Primarily, drone operators have a duty of care towards other airspace users and the public. As drones share the airspace with manned aircraft and interact with individuals on the ground, operators must exercise caution and take appropriate measures to avoid harm or damage. This duty of care extends to ensuring the safe operation of drones, maintaining proper control, and adhering to established safety protocols. Later on, I will delve into the legal obligations imposed on operators in terms of their duty of care and the potential liability they may face for breaches of this duty.

Furthermore, I will critically assess the potential consequences drone operators may face in the event of violations or accidents. Violations of regulations, negligence in operation, or failure to exercise the required duty of care can lead to a range of legal repercussions, including civil liability for property damage, personal injury, or privacy infringement. Understanding these potential consequences is crucial for drone operators to comprehend the gravity of their legal obligations and the need to prioritise safety and compliance.

Similarly, complex questions that comes to mind includes: what are the potential legal repercussions for drone operators in the event of regulatory violations, negligence in operation, or failure to exercise the required duty of care? How do instances of civil liability for property damage, personal injury, or privacy infringement impact the legal obligations and operational practices of drone operators? In what ways does understanding the potential consequences of non-compliance emphasise the importance of prioritising safety and adherence to regulations for drone operators?

By exploring the liability of drone operators and related legal questions, I aim to provide a comprehensive understanding of the legal landscape surrounding their responsibilities and highlight the significance of compliance with regulations, adherence to operational limitations, and the duty of care towards other airspace users and the public.

Liability of Manufacturers and Suppliers

Drone manufacturers and suppliers hold a significant share of responsibility in ensuring the safety and reliability of drone technology. In context, it is important to ask, what are the specific legal obligations and potential liabilities that manufacturers and suppliers of drones must adhere to under current product liability principles? How does the duty to warn about a drone's capabilities and limitations influence the liability of manufacturers and suppliers in cases where inadequate information leads to harm or damage? In what ways do legal standards governing the responsibility of manufacturers and suppliers shape the overall accountability structure within the drone.

The most significant area of liability for manufacturers and suppliers is product liability. Manufacturers have a responsibility to design and produce drones that meet industry standards and are safe for their intended use. They must ensure that the drones are free from defects that could pose risks to users or others. Similarly, suppliers have a duty to provide drones that meet the required quality standards and are fit for their intended purpose. Therefore, it is critical to ask, How do the principles of product liability apply within the legal framework to hold manufacturers and suppliers of drones accountable for defects or failures that result in harm or damage? What specific legal standards and criteria must be met to establish liability for manufacturers and suppliers in cases where drone defects cause injury or property damage? In what ways does the existing legal framework for product liability ensure that manufacturers and suppliers take responsibility for the safety and reliability of their drones? Later on, I will delve into these questions in detail, highlighting the role of manufactures and suppliers in ensuring that product safety and reliability.

Potential Vicarious Liability

When assessing liability in drone operations, the concept of vicarious liability becomes particularly relevant. Some complex legal questions come to mind; under what circumstances can an employer or principal be held vicariously liable for the actions or omissions of their employees or agents operating drones? What legal principles govern the determination of vicarious liability in drone-related violations, and how do these principles extend the scope of responsibility within the drone ecosystem? How does the concept of vicarious liability clarify the distribution of liability among various stakeholders in the drone industry, particularly in cases involving violations or accidents?

Vicarious liability is a legal principle that determines when an employer or principal may be held liable for the actions or omissions of their employees or agents. This principle is particularly relevant in the drone industry, where operators often act on behalf of their employers or principals. Vicarious liability serves as a valuable tool for clarifying the scope of responsibility and potential liability extensions within the drone ecosystem. It allows for the allocation of liability to those who have control or influence over the actions of drone operators. By holding employers or principals accountable for the actions of their employees or agents, vicarious liability promotes a sense of shared responsibility and encourages employers and principals to ensure proper training, supervision, and compliance with regulations.

The concept of vicarious liability serves to distribute liability among stakeholders in the drone industry by assigning responsibility to those who have a degree of control or oversight over the actions of others. In cases involving violations or accidents, vicarious liability clarifies that employers or principals may be held accountable for the wrongful acts of their employees or agents, even if they did not directly engage in the misconduct. This legal principle incentives companies and organizations to implement stringent operational protocols and training programs to mitigate risks associated with drone usage. By holding employers accountable, vicarious liability ensures that there is a broader net of accountability within the drone ecosystem, encouraging all stakeholders to prioritise safety and compliance.

Vicarious liability in the context of drone operations arises when an employer or principal is held accountable for the actions or omissions of their employees or agents. The traditional legal principle of *respondeat superior* applies here, where liability is imposed on an employer if the wrongful act was committed within the scope of employment. In the case of drones, this could mean that if an employee operates a drone negligently while performing their duties, the employer may be liable for any resulting harm. The key issue revolves around whether the drone operation was within the scope of the employee's role, and whether the employer exercised sufficient control over the operational environment. As drone technology becomes more integrated into business operations, the scope of what is considered "within employment" may expand, making vicarious liability a significant risk for employers.

The determination of vicarious liability in drone-related violations hinges on several legal principles, including the respondeat superior doctrine and the notion of "*control*" over the actions of the employee or agent. Courts often assess whether the wrongful act was committed in the course of employment and whether the employer had control over the manner in which the drone was operated. This extends the scope of responsibility within the drone ecosystem, particularly when companies employ drones for commercial purposes. Employers must be aware that their liability may extend to acts of negligence or misconduct by employees even if the employer did not directly authorise or foresee the specific action. The broader the control an employer exerts over the drone operations, the more likely they are to be held vicariously liable.

Furthermore, in assessing the potential for vicarious liability in the drone industry, it is crucial to consider the legal principles that govern its application. One key principle is the concept of *"course of employment"* or *"scope of agency."* This principle examines whether the actions of the employee or agent were within the scope of their employment or agency relationship. If the employee or agent was acting within the authorised scope of their employment or agency, the employer or principal may be held vicariously liable for their actions or omissions.

Another factor to consider is the level of control exercised by the employer or principal over the drone operations. The greater the control and direction exerted by the employer or principal, the stronger the argument for vicarious liability. However, it is important to note that the mere existence of an employment or agency relationship does not automatically establish vicarious liability. Courts consider various factors, such as the nature of the employment or agency, the degree of control, and the connection between the employee's or agent's actions and their authorised responsibilities.

The nuances of vicarious liability in the drone industry require careful analysis and consideration. As the industry continues to evolve and new business models emerge, it is essential to assess the extent to which vicarious liability may apply in different scenarios. This analysis helps determine the appropriate allocation of liability and ensures that those with control or influence over drone operations bear their fair share of responsibility.

Exploring the potential for vicarious liability in the context of drone-related violations brings forth the nuances of this legal principle and its value for clarifying the scope of responsibility and potential liability extensions within the drone ecosystem. By examining the legal principles that determine when employers or principals may be held liable for the actions or omissions of their employees or agents, we gain a deeper understanding of the dynamics of liability in the drone industry. This analysis contributes to a comprehensive assessment of liability within the drone ecosystem and can facilitate the establishment of robust frameworks that promotes accountability and protects the interests of all stakeholders involved.

Fault Liability in Drone Operations

Fault-based liability principles such as negligence imposes a responsibility on anyone along the causal chain to exercise a duty of care. Drone manufacturers, operators and service providers can potentially be held liable in negligence for failure to exercise care. This could have implications across the value chain. The law of negligence requires that reasonable care must be taken to avoid or reduce the likelihood of foreseeable harm arising from drone operations.

Some of the complex legal questions that may arise include; How can liability be attributed to multiple parties (tortfeasors) in the context of drone-related harm, given the complexities of human operation, software errors, and unexpected environmental conditions? What challenges arise in proving fault and attributing liability when both software and hardware issues contribute to a drone malfunction, particularly when the fault lies in complex systems that involve multiple contributors? Under what circumstances might a court impose liability on a software programmer for an unforeseen conflict in a code library that leads to a drone malfunction, especially when the programmer's code is intended for multiple applications?

The attribution of liability to multiple parties in drone-related incidents is complex due to the interplay between human operators, software developers, and environmental factors. The law of negligence allows for joint and several liabilities, where multiple tortfeasors can be held responsible for contributing to the harm. In the context of drones, if a malfunction or accident occurs, liability could potentially be distributed among the operator, who may have failed to respond appropriately to weather conditions, the software developer, whose coding error contributed to the malfunction, and even the manufacturer, if a design flaw is identified. This approach ensures that victims are compensated, even if multiple parties are involved. However, the complexity of drone systems raises challenges in identifying the precise contributions of each party, necessitating a careful examination of causation and the apportionment of responsibility.

Proving fault in cases where both software and hardware issues contribute to a drone malfunction is particularly challenging due to the technical complexities involved. The burden of proof typically lies with the plaintiff, who must establish a direct causal link between the defect and the harm suffered. In scenarios involving complex systems, pinpointing the exact source of the fault—whether in the software, hardware, or human operation—can be difficult. The challenges are exacerbated when the fault arises from an interaction between different components, such as a software bug triggering a hardware malfunction. The defence may argue that the fault was unforeseeable or that it was a minor flaw not directly causing the harm. The complexity of these systems often requires expert testimony, and courts may struggle to attribute liability fairly among the various contributors. This raises significant questions about whether current legal standards are equipped to handle the intricacies of modern technology.

Imposing liability on a software programmer for an unforeseen conflict in a code library that causes a drone malfunction presents significant legal challenges. Traditionally, courts have been reluctant to impose liability on programmers for code errors, especially when the code is used in multiple applications and the error was not reasonably foreseeable. However, the situation may differ if the programmer failed to follow industry standards or best practices, which could be construed as negligence. Furthermore, if the programmer was aware of the potential for conflicts and did not take steps to mitigate these risks, liability could be more easily established. Courts may also consider whether the software was subject to adequate testing and whether proper warnings were issued regarding its limitations. Ultimately, the imposition of liability would depend on the specific facts of the case, including the foreseeability of the error and the programmer's adherence to professional standards.

Tort Law and Drone-Related Violations

Tort law plays a crucial role in addressing civil wrongs and providing remedies for harms caused by drone-related incidents. Some of the crucial legal question that may arise in drone related violations includes: How do the principles and doctrines of tort law apply specifically to drone incidents, and in what ways do they assess liability in such cases? What legal theories are most effective in addressing drone-related violations, and how are they used to determine compensation for harm caused by drones? How does the tort law framework provide a comprehensive understanding of the liability landscape in the context of modern drone operations, especially when balancing the interests of various stakeholders?

The principles of tort law apply to drone incidents in much the same way they apply to other harmful activities, but with certain specificities due to the nature of drone technology. Tort law, primarily negligence and strict liability, forms the basis for determining liability in drone-related incidents. In negligence cases, the plaintiff must prove that the drone operator owed a duty of care, breached that duty, and caused harm as a result. For instance, if a drone operator flies a drone irresponsibly over a crowded area and it crashes, injuring a bystander, tort law would assess whether the operator breached a standard of care expected in such situations. Strict liability, on the other hand, may be applicable where drones are considered inherently dangerous. If a drone malfunctions due to a defect and causes harm, the manufacturer could be held liable regardless of fault, provided the defect was the direct cause of the damage. Tort law also considers contributory factors such as the operator's adherence to safety protocols, the foreseeability of the harm, and whether proper precautions were taken.

As discussed earlier, the most effective legal theories for addressing drone-related violations include negligence, strict liability, and product liability. These doctrines allow courts to analyze drone-related accidents and allocate responsibility. Negligence theory would focus on whether the operator or manufacturer failed to act as a reasonable person would under similar circumstances. For instance, if an operator did not follow required guidelines, such as maintaining a visual line of sight, this could be construed as negligence.

Product liability is especially relevant for drone manufacturers. If a drone was manufactured with a defect that caused injury, the legal doctrine of strict liability could apply. In such cases, the plaintiff does not need to prove negligence but merely show that the product was defective and caused harm. Compensation under these legal theories generally includes covering medical expenses, lost wages, property damage, and even non-economic damages like pain and suffering, depending on the severity of the harm caused.

Tort law provides a comprehensive framework for addressing liability in drone operations by balancing the interests of all parties involved—operators, manufacturers, the public, and even regulators. This balance is achieved by imposing responsibilities on drone operators to act with reasonable care and on manufacturers to ensure their products are free from defects that could cause harm. For example, tort law holds operators accountable for operating drones safely, adhering to regulations such as maintaining a safe distance from people, and avoiding restricted airspace.

On the manufacturer's side, tort law ensures that drones are safe for public use by imposing liability for defects in design or manufacturing that could lead to harm. The framework also allows victims of drone-related harm to seek compensation and holds stakeholders accountable for their actions, ensuring a level of safety in the industry. It serves as a deterrent for negligent behaviour while providing a mechanism for victims to recover damages.

Except for strict liability, defective product liability and liability under traffic laws, tortious liability in most EU legal systems are fault-based or subject to exculpation in cases where the tort has not been committed directly by the tortfeasor, but a third person for whom the tortfeasor is responsible and who has been assigned and supervised with due care. In such situations, a drone operator (employer) could be held vicariously liable for action of a drone controller (employee) working in the command centre or a remote pilot (employee) working on onsite where their action inflicts damage on a third party within the scope of employment. However, it might prove difficult to determine the level of assignment and supervision required to determine fault in vicarious liability cases.

Application of Tort Law to Drone Incidents

When drone-related incidents occur, tort law provides a legal framework for addressing the resulting harms and assigning liability. Some of the critical questions that may be asked includes: How do the traditional tort principles of negligence, strict liability, and product liability adapt to drone-related incidents, and what are the specific challenges in proving liability in cases involving personal injury, property damage, or privacy violations caused by drones? In what ways does the application of tort law account for the unique operational risks associated with drones, such as autonomous decision-making or software malfunctions, and how are these risks factored into liability determinations? What potential legal remedies are available to victims of drone-related harm, and how do courts assess the adequacy of these remedies within the framework of tort law given the evolving nature of drone technology and its uses?

In the context of drone-related incidents, traditional tort principles adapt to the evolving nature of drone technology, but their application faces challenges. Negligence requires proving that the operator failed to act with the standard of care expected, which may involve whether the operator adhered to aviation safety regulations or flew the drone recklessly. With strict liability, manufacturers may be held liable if a defect in the drone caused harm, regardless of whether the manufacturer acted negligently. Product liability is particularly relevant when drones malfunction due to design flaws, software errors, or inadequate safety warnings. The complexity of proving liability in these cases arises from the multiple factors that could contribute to an accident, including operator error, environmental conditions, and software glitches. Furthermore, the autonomy of drones complicates the assignment of responsibility, particularly when the drone acts independently of human intervention.

Tort law's application to drone operations must account for unique risks like autonomous decision-making, which separates the operator from direct control over the drone's actions. In negligence cases, courts must consider whether the operator reasonably relied on the drone's autonomous systems and whether they took appropriate precautions. Software malfunctions add another layer of complexity, as liability may extend to software developers if a flaw in the code contributed to the accident. The issue of foreseeability also plays a role in determining liability: were the risks of autonomous behaviour or software bugs foreseeable, and did the responsible parties (manufacturers, operators, developers) take reasonable steps to mitigate those risks? Courts are likely to assess whether operators have maintained their drones according to manufacturers' specifications and whether manufacturers have implemented sufficient safeguards to prevent software malfunctions.

Victims of drone-related harm may pursue remedies under tort law, including compensatory damages for personal injury, property damage, and economic loss. Additionally, courts may award punitive damages in cases of egregious misconduct, such as reckless operation of a drone or wilful disregard for safety protocols. The adequacy of these remedies is assessed by the courts based on the severity of the harm and the level of culpability of the defendant. However, given the evolving nature of drone technology, courts are increasingly confronted with new types of harm—such as privacy violations or drone crashes caused by software defects—that may require a re-evaluation of traditional compensation models. For example, privacy breaches caused by drones may not fit neatly into existing categories of harm, necessitating creative legal arguments and novel approaches to remedy these violations. Additionally, the rapid advancement of technology may push courts to consider future risks and the long-term impacts of drone incidents when awarding damages.

Negligence and Duty of Care

Negligence is a central concept in tort law that often arises in the context of drone-related incidents. What specific duty of care do drone operators and other stakeholders owe to the public, and how is this duty defined in the context of drone operations? How are the elements of negligence—duty of care, breach of duty, causation, and damages—applied to assess the conduct of drone operators, and what factors influence the determination of liability? What standards of conduct are expected from drone operators to avoid negligence, and how is liability established when these standards are not met in drone-related incidents?

The duty of care owed by drone operators to the public is inherently tied to the nature of drone operations and their potential to cause harm. Courts and regulators define this duty based on the operator's obligation to exercise reasonable care to avoid causing injury or damage to others. Drone operators must ensure that their drones are operated in accordance with safety protocols, aviation regulations, and technological standards specific to unmanned aircraft. This duty extends beyond the mere operation of the drone to include regular maintenance, adequate training, and awareness of no-fly zones.

The duty of care is heightened when drones are used in populated areas or near critical infrastructure. For instance, in some jurisdictions, drone operators are expected to avoid flying over people or property unless there is explicit permission. The duty is also impacted by the nature of the drone's use—commercial operators are held to a higher standard of care than hobbyists due to the increased risk and scale of their operations.

Additionally, stakeholders such as drone manufacturers, software developers, and service providers owe a duty of care to ensure that their products are safe for use and do not pose foreseeable risks to users or third parties. This means that all stakeholders in the drone ecosystem, from design to deployment, are bound by a duty to prevent harm through reasonable precautions and compliance with regulatory standards.

In assessing negligence in drone operations, courts first establish whether a duty of care exists between the operator and the harmed party. Once this duty is established, the next step is determining whether the operator breached that duty by failing to adhere to the standard of care expected under the circumstances. This could involve reckless flying, failing to maintain control of the drone, or ignoring safety protocols. Breach of duty is often established by comparing the operator's conduct to that of a reasonable person with similar experience and in similar circumstances.

Causation requires proof that the operator's breach of duty directly caused the harm. This can be complicated in drone cases due to factors like environmental conditions, software malfunctions, or third-party interference. Courts may rely on expert testimony to dissect the chain of events leading to the incident and to determine whether the operator's breach was the proximate cause of the damage.

Damages are the final element, where the court assesses the extent of the harm caused by the drone operator's actions. This could include compensation for property damage, medical expenses, loss of income, and non-economic damages such as pain and suffering. In some cases, punitive damages may also be awarded if the operator's conduct is found to be egregious.

To avoid negligence, drone operators are expected to adhere to a high standard of conduct that aligns with both regulatory requirements and the operational environment in which the drone is used. This includes conducting pre-flight safety checks, complying with airspace regulations, maintaining a visual line of sight with the drone (unless authorised for Beyond Visual Line of Sight (BVLOS) operations), and avoiding restricted zones such as airports, military bases, or crowded public spaces. The Remote Pilot Certificate in various jurisdictions establishes a baseline for competency, ensuring that operators possess the necessary knowledge of airspace rules and drone operation.

Liability is established when an operator deviates from these standards of conduct and causes harm. For example, if an operator flies a drone into restricted airspace or loses control of the drone due to poor maintenance, and this results in an accident, liability would likely be imposed. The court would analyze whether the operator's failure to comply with standard practices and regulations was a significant factor in causing the harm. Courts also consider the foreseeability of the harm—if the risk was foreseeable and could have been mitigated by adhering to safety protocols, the operator will likely be found liable.

As drone technology evolves, courts may also expect operators to stay updated on best practices and advancements in drone safety features, such as geofencing, collision avoidance systems, and fail-safe protocols. Failure to integrate these technologies where reasonably accessible could further expose operators to liability in the event of an accident.

Strict Liability in Drone Operations

In certain circumstances, strict liability may apply to drone operations, shifting the burden of proof and imposing liability regardless of fault or negligence. Under what specific conditions can strict liability be imposed on drone operators or manufacturers, and how does the application of this doctrine differ from traditional fault-based liability models? What are the potential legal implications and consequences for drone operators and manufacturers when strict liability is applied in the context of drone-related violations? How does the framework of strict liability extend the scope of responsibility for drone-related harm, particularly when addressing risks inherent in drone technology, such as autonomous systems and product defects?

Strict liability in the context of drone operations is primarily imposed when harm arises from inherently dangerous activities or defective products, without the need to prove fault or negligence. This doctrine is triggered under specific conditions, such as when drones are considered ultra-hazardous—flying in densely populated areas, transporting hazardous materials, or operating in environments where potential harm to people or property is foreseeable. For manufacturers, strict liability is imposed when drones have design defects, manufacturing flaws, or inadequate warnings that lead to harm.

The key difference between strict liability and fault-based liability is that under strict liability, there is no need to establish that the operator or manufacturer was careless. The focus is instead on the inherent risk posed by the activity or the defect in the product. For example, a drone delivery company transporting hazardous materials would face strict liability if a drone crashes and causes a toxic spill, regardless of whether the company exercised the utmost care. Similarly, a manufacturer could be held liable for a drone malfunction due to a latent design defect, even if the manufacturing process met industry standards. Strict liability thus broadens the scope of accountability in drone-related incidents, ensuring that victims have recourse without the high burden of proving negligence.

The imposition of strict liability carries significant legal and financial implications for both drone operators and manufacturers. For operators, strict liability raises the stakes in terms of operational risk management. Companies must anticipate potential harm not only from operator error but also from the inherent risks of the drone activity itself. This may lead to increased insurance premiums, as insurers assess the heightened risk of claims even in the absence of fault. Additionally, operators may need to invest in advanced safety features, such as geofencing or collision avoidance systems, to mitigate the risks associated with strict liability.

For manufacturers, strict liability forces greater attention to product safety and compliance. The possibility of being held accountable for defects—regardless of fault—pushes manufacturers to conduct rigorous testing, implement stringent quality controls, and provide clear and adequate warnings about the drone's limitations. Failure to meet these standards could result in costly lawsuits, product recalls, and reputational damage. Moreover, the legal consequences of strict liability may encourage manufacturers to innovate toward safer designs and preventive technologies, such as self-diagnostic systems that alert users to potential malfunctions before they cause harm.

The broad application of strict liability in the drone industry also raises concerns about innovation. While strict liability promotes safety, it may stifle innovation by increasing the cost and risk associated with developing new drone technologies. Manufacturers may become more conservative in their approach, slowing the pace of innovation to ensure compliance with safety standards.

The framework of strict liability extends the scope of responsibility by holding operators and manufacturers accountable for harm arising from risks that are inherent to drone technology. This includes not only physical defects in the drone itself but also systemic risks associated with autonomous systems. For example, if a drone equipped with autonomous navigation software fails to avoid an obstacle due to a programming error, the manufacturer of the software may be held strictly liable for any resulting damage, even if the drone operator followed all safety protocols.

Autonomous systems present unique challenges because they often involve complex algorithms and machine learning, which may behave unpredictably in certain environments. The strict liability framework effectively extends responsibility to cover these technological risks by focusing on the outcome—harm to third parties—rather than the process. This legal approach pushes manufacturers and developers to prioritise fail-safe mechanisms and real-time monitoring to prevent accidents.

Furthermore, the scope of strict liability can encompass third-party service providers, such as those responsible for maintaining or updating drone software. If a service provider's failure to properly maintain the software leads to a malfunction, they too could be brought under the strict liability umbrella. This creates a broader ecosystem of accountability within the drone industry, ensuring that all actors involved in the lifecycle of a drone are incentivised to prioritise safety.

When strict liability is applied to drone cases, the burden of proof is shifted from the victim to the drone operator or manufacturer. This means that the operator or manufacturer is held liable regardless of fault or negligence, as long as certain conditions are met. In the context of drone operations, strict liability may be imposed when significant harm occurs, even in the absence of fault, defects, malperformance, or non-compliance with the law. This approach recognises that establishing fault or proving specific elements can be challenging for the victim and may result in inadequate compensation or inefficiency.

One of the advantages of strict liability is the simplicity it brings to the compensation process for victims of drone accidents. In strict liability regimes, victims generally only need to prove that they have suffered a loss as a result of the drone operation, rather than delving into the specific technical details or investigating faults. This streamlined process can save time and resources, ensuring that victims are adequately compensated without unnecessary delays or burdensome procedures.

While strict liability assigns responsibility to the drone operator, it does not preclude the possibility of making claims against other parties involved in the drone ecosystem. For example, if the harm was caused by a defect in the drone itself, the victim may also seek compensation from the manufacturer under product liability principles. This allows for a comprehensive approach to liability, ensuring that multiple parties can be held accountable when their actions or products contribute to the harm suffered by the victim.

Adopting strict liability in drone operations provides a higher level of legal certainty, as it clearly identifies the party responsible for compensating the victim. When national laws define liability as strict, the obligation to compensate the victim automatically falls on the drone operator, regardless of the contributing factors that led to the damage or injury. However, a crucial aspect is the clear identification of the drone operator, which can be facilitated by implementing drone registration systems or requiring permission for drone operations. These measures help ensure that the responsible party can be readily identified and held accountable for any harm caused by the drone operation.

Understanding the concept of strict liability and its application in the context of drone-related violations is vital for comprehending the scope of responsibility and liability extensions in drone operations. Strict liability provides a simpler compensation process for victims, removes the need to establish fault or defects, and offers more legal certainty. By exploring the principles of strict liability, we can better navigate the complex landscape of liability in the drone industry and ensure that victims are appropriately compensated for the harm they have suffered.

Product Liability and Defective Drones

Product liability principles become particularly relevant when addressing issues related to defective drones. What constitutes a defective product under EU law, particularly in the context of drones, and how is product safety assessed? How does the Product Liability Directive in EU law regulate the liability of manufacturers for defective drones, and what are the legal consequences for manufacturers when their products cause harm? What is the scope of liability for drone manufacturers under EU law, and how does the principle of liability without fault affect their responsibilities?

A drone manufacturer whose products causes damage or personal injury to customers may be held liable for defective product in most European jurisdictions. Product Liability Directive regulates liability for defective products in EU law. Product in EU law means all movables, with the exception of primary agricultural products and game, even though incorporated into another movable or into an immovable, defective products are products that do not provide the safety which a person is entitled to expect, considering, all of the circumstances, including, the presentation of the product, such as adequacy of the warning, the use to which it could reasonably be expected. The directive establishes the principle of liability without fault to products sold in the internal market. Therefore, where a defective products causes harm or damage to an EU consumer, the manufacturer is automatically liable.

The application of product liability principles to drone manufacturers and suppliers is rooted in the idea that these entities have a fundamental responsibility to ensure that their products are safe for consumer use. When defects or failures occur, leading to harm or damage, the legal framework typically imposes strict liability on manufacturers and suppliers. This means that they can be held accountable without the need to prove negligence, as long as the injured party can demonstrate that the product was defective and that the defect caused the harm. The question then becomes whether the existing legal standards are sufficiently robust to address the complexities of drone technology, which often involves sophisticated autonomous systems that may malfunction in unforeseen ways.

To establish liability in cases where drone defects cause injury or property damage, plaintiffs must typically prove that the product was defective, the defect existed when the product left the manufacturer's control, and that the defect was the direct cause of the injury or damage. Legal standards often categorise defects into design defects, manufacturing defects, and failures to warn. Each category presents its own challenges, particularly in the context of drones, where the interplay between software and hardware can complicate the identification of defects. Furthermore, the rapid pace of technological advancement raises questions about whether traditional product liability doctrines can adequately address issues such as software updates, autonomous decision-making, and cybersecurity vulnerabilities in drones.

The manufacturer may avoid liability by proving the absence of a causation between the defective product and the damage suffered or that the product complies with the product requirements by the regulator. Strict liability could be excluded if the manufacturer is able to prove that the damage was caused by an unforeseen defect given the technical and scientific knowledge available at the time the product was put into circulation. Defective product liability caused by design defect, fabrication defects, user instruction defects and product supervision defects under the complete control of the manufacturer may be limited by product specification, user instruction, and supervision by the manufacturer. Drone manufacturers may also spread their liability that may arise from defective products through product insurance and pricing mechanisms aimed at reducing the impact of losses that may be incurred.

The existing legal framework for product liability serves as a critical mechanism for ensuring that manufacturers and suppliers prioritise the safety and reliability of their drones. By imposing strict liability, the law encourages these entities to adopt rigorous quality control measures, conduct thorough testing, and provide comprehensive warnings and instructions to consumers. However, the effectiveness of this framework in the drone industry depends on how well it can adapt to the unique challenges posed by advanced technologies. The question here is whether the current legal framework sufficiently incentives manufacturers and suppliers to address the complex risks associated with drones, including potential software glitches, cybersecurity threats, and the need for continuous updates and maintenance.

Traffic Liability in Drone Operations

Can drones be classified as vehicles under existing traffic laws, and if so, could traffic laws be applied to impose strict liability on drone owners or operators in cases of accidents? How might courts interpret drones' participation in public traffic, and is there a basis for equating drone regulation with the legal framework applied to motorised wheelchairs or self-driving cars? Could traffic laws serve as an alternative or supplement to product liability in addressing accidents involving drones, particularly in the absence of specific drone traffic regulations?

The question of whether drones could be classified as vehicles under existing traffic laws is critical to determining if traffic law principles, such as strict liability for vehicle owners, could apply to drone accidents. Traditional traffic laws were designed with ground-based vehicles in mind, defining a vehicle as any device by which a person or property may be transported upon a highway or public road. The extension of these definitions to airspace is complex, but there is a growing body of regulatory work dealing with autonomous and remotely operated vehicles, such as self-driving cars. These developments provide a precedent for adapting traffic laws to new forms of mobility.

Courts might interpret drones as vehicles participating in public traffic if they are used in urban environments for tasks such as deliveries or surveillance. The primary argument for classifying drones as vehicles under traffic law lies in their ability to transport goods and, indirectly, to facilitate human-related activities. The low speed and weight of certain drones may make them analogous to motorised wheelchairs or bicycles, which are also subject to specific traffic regulations despite not fitting the traditional mode of motor vehicles.

However, unlike motorised wheelchairs or bicycles, drones operate in the airspace rather than on roads or walkways, raising the question of whether traditional traffic laws, which are designed to govern the flow of land-based vehicles, can extend to cover airspace regulation. The airspace is already subject to its own regulatory framework through aviation law, raising potential conflicts or overlaps between these legal regimes. For instance, would a drone flying at low altitudes be regulated under traffic law, aviation law, or both?

Traffic laws often impose strict liability on the owner of a vehicle for any damage caused by that vehicle, regardless of whether the owner was operating it at the time. This concept could theoretically be extended to drones, with the owner being held liable for any damage caused by their drone, even if a third party or autonomous system was controlling the drone. In the case of self-driving cars, which share characteristics with drones (such as autonomy and reliance on software), strict liability frameworks have already been proposed or implemented in some jurisdictions.

This shift could see traffic laws evolve to accommodate drones in much the same way they are being adapted for self-driving cars. The key legal question here would be how courts and legislators define participation in public traffic. If drones are considered to be engaging in public traffic when flying in low-altitude airspace over cities, owners may face similar liability obligations as car owners do on the ground. This could also bring insurance requirements similar to those imposed on vehicle owners, which would be an extension of current aviation insurance requirements.

Traffic law could serve as an alternative or complement to product liability in addressing accidents involving drones. Under product liability, a manufacturer may be held responsible for defects that cause harm, while traffic laws extend liability to operators or owners regardless of defect. Thus, while product liability focuses on the safety and functionality of the drone itself, traffic law could impose liability based on how the drone is operated within public spaces.

The convergence of these two areas of law could be beneficial in situations where proving a product defect is difficult or where accidents result from operator error rather than a mechanical issue. For example, if a drone crashes into a person while delivering a package, the victim might pursue a claim under both product liability (if the crash was caused by a defect) and traffic law (if the operator failed to follow proper flight paths or safety protocols). This dual liability approach could ensure broader protection for victims while holding both manufacturers and operators accountable for safe drone use.

Drones present unique legal challenges, particularly in balancing innovation with public safety. Given their autonomous capabilities, drones are often seen as analogous to self-driving cars, both of which require new legal frameworks to manage liability effectively. As drones become increasingly integrated into public and commercial use, the legal landscape will likely continue to evolve. Legislators and courts may need to create hybrid legal frameworks that borrow from both traffic law and aviation law to comprehensively address the risks posed by drones in public spaces.

IX

9. Private International Law in Drone Violations

Jurisdictional Issues in Cross-Border Drone Operations

In the realm of cross-border drone operations, determining the appropriate jurisdiction for addressing legal issues and enforcing rights can be complex. What are the specific jurisdictional challenges that arise in drone-related violations, and how do legal principles such as territoriality and nationality affect the determination of jurisdiction in cross-border drone operations? How does the legal framework governing cross-border drone operations influence the ability of affected parties to seek legal remedies, and what factors determine which jurisdiction's laws apply in such cases? In what ways do international treaties, conventions, and bilateral agreements shape the jurisdictional rules for drone-related violations, and how do they impact enforcement and accountability in a globalised drone market?

Determining Applicable Jurisdiction

The determination of applicable jurisdiction is a fundamental aspect of resolving legal disputes in cross-border drone operations. Jurisdictional challenges in drone-related violations are multifaceted due to the cross-border nature of drone operations, which can involve activities that span multiple legal systems. The principle of territoriality, which grants jurisdiction to the country where the incident occurs, is the foundational doctrine in most cases. For example, if a drone causes property damage while flying over Country A, the courts in Country A would typically have jurisdiction to hear the case.

However, drones complicate the territoriality principle because their operations can span multiple countries within a short timeframe. Consider a scenario where a drone operated by a company based in Country B flies into Country A's airspace and causes harm. Here, both Country A and Country B may claim jurisdiction based on the nationality principle (Country B's jurisdiction over its citizens or companies) and territoriality principle (Country A's jurisdiction over incidents within its borders). Courts will need to analyze the drone's location, the operator's domicile, and the applicable international treaties to determine jurisdiction.

An additional complication arises with transnational drone operations,   where drones are used for services such as international deliveries or cross-border surveillance. In such cases, jurisdictional claims may intersect with laws regulating airspace under aviation law treaties, such as the Chicago Convention of 1944, which governs international civil aviation. These complexities highlight the need for clear bilateral or multilateral agreements between states to pre-empt jurisdictional conflicts in drone operations.

The legal framework for cross-border drone operations significantly influences an injured party's ability to seek legal remedies. The Brussels I Regulation (recast) in the European Union, for example, allows plaintiffs to sue a defendant in the courts of the country where the harmful event occurred, even if the drone operator or manufacturer is based in another EU country. This regulation simplifies jurisdictional determinations within the EU but leaves questions open for incidents that cross into non-EU jurisdictions, especially when dealing with multinational drone operations.

One of the major factors in determining which jurisdiction's laws apply in cross-border drone cases is the choice of law rules. These rules typically dictate that the applicable law is the law of the country most closely connected to the incident, which often coincides with where the damage occurred or where the drone was operated. Courts also consider contractual agreements between parties involved in the drone operation, such as user agreements that specify the jurisdiction for any disputes. This is common in commercial drone services, where terms of service might include a jurisdiction clause favouring the country of the service provider.

Private international law plays a crucial role in cross-border disputes, particularly where multiple jurisdictions have legitimate claims. In cases involving non-EU jurisdictions, international treaties such as the Hague Convention on the Recognition and Enforcement of Foreign Judgments may determine how judgments are enforced across borders. Additionally, conflicts of law rules may require courts to balance the interests of the involved jurisdictions, considering factors such as the location of harm, the domicile of the parties, and the nature of the incident.

International treaties and conventions provide an essential framework for determining jurisdiction and enforcing judgments in cross-border drone cases. The Chicago Convention of 1944 and its annexes establish the rules for the use of airspace by civil aviation, including drones. The convention grants each state sovereignty over its airspace, implying that drones crossing into another state's airspace must comply with that state's laws and regulations. This principle of airspace sovereignty can lead to jurisdictional claims by the state over which the drone is flying, particularly when incidents occur within its borders.

Bilateral agreements between countries are becoming increasingly important as the drone market globalises. For instance, the US-EU Open Skies Agreement allows for greater regulatory cooperation in civil aviation matters. Similar agreements could be developed to manage drone operations across borders, reducing jurisdictional conflicts and clarifying the applicable legal standards. These agreements can also streamline processes for cross-border enforcement of judgments, which is critical for holding drone operators and manufacturers accountable across multiple jurisdictions.

The enforcement of judgments in cross-border drone disputes hinges on international cooperation. Treaties such as the New York Convention on the Recognition and Enforcement of Arbitral Awards and the Hague Convention facilitate the recognition of foreign judgments and arbitral decisions, which is essential for ensuring that parties can enforce liability claims and obtain remedies even when the defendant is located in a different jurisdiction.

These frameworks not only determine jurisdiction but also shape the legal obligations for drone operators in the international arena. They ensure that operators are subject to consistent rules when conducting cross-border operations and that victims have access to remedies regardless of where the harm occurs. However, the global nature of the drone industry continues to challenge existing frameworks, necessitating further legal evolution to ensure comprehensive accountability in a globalised drone market.

It can also be argued that the Rome Convention of 1952 equally applies to non-contractual liability caused by drones. The Rome Convention prescribes a strict liability regime for aircrafts operators and attributes liability for the damage caused to third parties by an aircraft without proof of the operator's intent or negligence. While the Rome Convention does not provide a definition for the term aircraft. It can be argued that the definition provided by the Chicago Convention is applicable. This definition states that 'an aircraft is any machine that can derive support in the atmosphere from the reactions of the air'. The definition is wide, and could include drones, which are generally defined as aircrafts designed to be operated without an on-board pilot. Unfortunately, the Rome Convention has been ratified by only by four EU Member States and eight countries in total. Thus, the impact of this convention on the EU framework for drone liability is limited. Also, the scope of Rome Convention is limited to cross-border flights. Drone the operation are mostly conducted locally therefore, the Rome convention will be mostly inapplicable.

National Aviation Laws of the EU Member States often make provisions use of drones below 150 kg concerning certification, licensing, third-party liability and insurance. National regulations mostly refer to the EC Regulation 785/2004 on insurance for air carriers and aircraft operators, which defines requirements for third-party liability insurance for manned aircraft operators and, in connection therewith, has set forth a regime of strict liability. Studies on third-party liability and insurance requirements of drones found that "in the most of EU Member States, national law defines that the liability regime for drones as strict".

Despite the fact that the Rome Convention was only ratified by eight countries, its legal principles are similar to those applied in different jurisdictions, as they are rooted in international conventions on non-contractual liability adopted by most countries. Principles of strict liability, regardless of the negligence of the liable party, which implies that the liable party be identified and compensation for damages be capped with compulsory insurance requirements by the liable party are recognisable in European jurisdictions. It could be argued that principles of strict liability would be designated in future harmonised EU framework for a third-party liability regime for drones.

Jurisdictional challenges in drone-related incidents require an intersection of multiple legal principles, ranging from territoriality and nationality to the evolving international agreements that govern airspace and cross-border operations. In answering these complex legal questions, it's clear that both national and international frameworks must adapt to the rapid advancement of drone technology to ensure effective legal remedies and enforcement across borders.

Conflict of Laws and Choice of Law Principles

Cross-border drone operations often give rise to conflicts between different legal systems. How do the conflict of laws and choice of law principles apply in the context of cross-border drone-related violations, particularly when jurisdictions have divergent regulations governing drones? What methodologies and legal doctrines are used to resolve conflicts of laws in drone operations, and how do courts determine the applicable law when incidents span multiple jurisdictions? In what ways can the application of conflict of laws principles ensure fairness and consistency in legal outcomes, especially in cases involving international drone operations with varying degrees of regulatory oversight? How do courts balance the interests of different jurisdictions in cross-border drone disputes, and what factors influence the determination of the governing law, such as the location of the harm, the domicile of the parties, and the intent of the operation?

The principles of conflict of laws and choice of law become particularly complex in cross-border drone violations due to the intersection of aviation law, commercial law, tort law, and privacy regulations. Typically, the courts apply conflict of laws principles to determine which jurisdiction's laws should govern the dispute. These principles are rooted in public international law and are meant to resolve inconsistencies that arise when parties from different legal systems are involved in a single dispute.

In the drone context, conflict of laws issues may arise when a drone is manufactured in one country, operated in another, and causes harm in a third. Here, the *"closest connection"* doctrine is often applied, which evaluates which jurisdiction has the most significant relationship to the incident. Courts would consider factors such as the location of the harm, the place where the drone was registered, the residence of the operator, and the country where the drone was designed and manufactured.

For example, if a drone manufactured in China and operated by a company based in the United States causes an accident in the European Union, conflict of laws principles will guide which jurisdiction's laws apply—be it Chinese product liability law, U.S. aviation law, or EU consumer protection law. This requires courts to balance multiple interests and legal principles, particularly when considering international treaties such as the Rome I Regulation (applicable in the EU) that govern cross-border contractual obligations and non-contractual obligations (Rome II).

The methodologies and legal doctrines employed to resolve conflicts of laws in drone operations typically center on connecting factors and choice of law rules. The main methodologies include:

1. **Lex loci delicti** (the law of the place where the harm occurred): This doctrine applies when the harm occurs in a specific location, such as where a drone crashes. The courts in the jurisdiction where the incident took place will typically apply their domestic law to the dispute.

2. **Party autonomy in contracts:** In commercial drone operations, contracts often contain a choice of law clause, where the parties agree in advance on the jurisdiction that will govern any disputes. This clause can streamline conflict resolution by predetermining the applicable legal framework, although courts will still ensure that the chosen law does not infringe on public policy.

3. **Most significant relationship or closest connection doctrine:** When the conflict arises in a tort context, courts often look to the jurisdiction with the most significant relationship to the event or the parties involved. For instance, in a case where a drone malfunction occurs during an international delivery, courts would examine where the contract for delivery was formed, where the malfunction occurred, and where the delivery was destined to decide which jurisdiction's laws apply.

An example would be drone delivery companies that operate internationally. When an incident occurs, the courts might examine not only the location of the drone operator but also the nature of the drone operation, the airspace regulations in each jurisdiction, and the laws governing air traffic management.

The fairness and consistency of legal outcomes in cross-border drone operations depend on the correct and nuanced application of conflict of laws principles. Courts strive to ensure fairness by considering a range of factors that balance the rights of the parties involved and the public interest. The principle of comity—the legal respect for the laws and judicial decisions of another jurisdiction—often plays a role in ensuring fairness. Courts may defer to the laws of a foreign jurisdiction if that jurisdiction is deemed to have a stronger connection to the case or if applying the foreign law would lead to a more equitable outcome.

Consistency in legal outcomes is supported by international treaties and harmonization efforts. The Hague Convention on the Recognition and Enforcement of Foreign Judgments, for instance, helps ensure that a judgment in one country will be recognised and enforceable in another, provided certain criteria are met. This reduces the risk of conflicting judgments and encourages more predictable outcomes.

Moreover, courts may invoke public policy exceptions to avoid the application of foreign law that may result in unfair or inequitable outcomes. For example, if the laws of one country are too permissive or fail to protect individual rights adequately (e.g., lack of privacy protections for drone surveillance), courts in another jurisdiction might refuse to apply those laws on the grounds of public policy, ensuring that legal proceedings align with broader societal values.

Courts balance the interests of different jurisdictions by employing a multi-factor analysis that considers the connections of each jurisdiction to the incident. Factors influencing the determination of the governing law include:

1. **Location of the harm**: Courts generally prioritise the jurisdiction where the harm occurred, as that location has the strongest interest in regulating behaviour within its borders and providing remedies for victims.

2. **Domicile of the parties:** The residence or domicile of the drone operator, the manufacturer, and the victim are key considerations. If the operator and the victim reside in different jurisdictions, courts will evaluate which jurisdiction has the greater interest in the case.

3. **Intent of the operation:** Courts may also consider the intent and purpose behind the drone's operation. For example, if a drone was explicitly intended for use in a particular jurisdiction, the courts in that jurisdiction might claim authority over the case, particularly if the drone's operation was subject to local regulatory oversight.

4. **Nature of the drone operation:** Commercial or state-related drone operations, such as drone delivery services or military uses, are treated differently than recreational drone activities. The more structured and regulated the operation, the more likely it is that international regulatory frameworks, such as aviation treaties, will apply, guiding the court's determination of jurisdiction and governing law.

In balancing these factors, courts aim to respect the sovereignty of each involved jurisdiction while seeking a fair resolution for all parties. They may also look to international regulatory bodies, such as the International Civil Aviation Organization (ICAO), for guidance on best practices in managing cross-border drone incidents. The increasing globalization of drone technology underscores the need for harmonized legal frameworks to reduce uncertainty and facilitate legal recourse for affected parties across different jurisdictions.

Forum Shopping in Drone-Related Litigation

Forum shopping, the practice of selecting a favourable jurisdiction to pursue legal action, can present challenges in drone-related litigation. What are the motivations behind forum shopping in the context of drone-related violations, and how do parties strategically select favourable jurisdictions to gain procedural or legal advantages?How does the phenomenon of forum shopping in cross-border drone litigation affect the balance of power between plaintiffs and defendants, and what implications does this practice have for the fairness and consistency of legal outcomes? What factors influence the choice of forum in drone-related litigation, and how do legal principles, procedural rules, and jurisdictional strategies play a role in the selection of a favourable court for pursuing claims? How do courts and legal systems address the challenges posed by forum shopping in cross-border drone disputes, and what mechanisms exist to prevent abuse of jurisdictional selection in complex international cases?

Forum shopping arises in drone litigation when parties seek to exploit the differences between jurisdictions to secure a more favourable legal outcome. The primary motivations behind this practice often include choosing jurisdictions with favourable substantive laws, lenient procedural rules, or less stringent standards for proving liability. In drone-related violations, plaintiffs may opt for jurisdictions where strict liability or low burdens of proof exist, making it easier to hold drone operators, manufacturers, or service providers accountable without having to prove intent or negligence. Defendants, conversely, may seek out jurisdictions that impose more restrictive standards on liability or offer favourable defences such as contributory negligence or caps on damages.

For example, a manufacturer might prefer to litigate in the U.S. due to product liability defences that may limit exposure, while a plaintiff might seek to bring the case in the EU, where the Product Liability Directive establishes strict liability for defective products. The strategic selection of a forum can also hinge on procedural aspects, such as discovery rules—which may be broader in the U.S.—or jurisdictional thresholds, such as minimum contacts tests that allow access to courts.

Forum shopping can be viewed as an inevitable result of globalised litigation, but courts often scrutinise such behaviour under the doctrine of forum non convenient, which allows a court to dismiss a case if it believes another jurisdiction is significantly more appropriate for hearing the case. Forum shopping can significantly affect the balance of power between plaintiffs and defendants in drone litigation. Plaintiffs often gain a tactical advantage by choosing a jurisdiction that might offer a faster trial process, greater likelihood of success, or higher potential damages. Conversely, defendants might be disadvantaged by being drawn into a court with unfavourable procedural rules or onerous discovery obligations. This imbalance can pressure defendants into settling cases early or incurring high litigation costs in unfamiliar legal environments.

The implications for fairness are profound. On one hand, forum shopping allows plaintiffs to avoid jurisdictions that may have onerous burdens of proof or limited access to justice. On the other hand, it can lead to inconsistent outcomes where identical cases result in vastly different rulings simply because they were adjudicated in different legal systems. For example, a drone incident causing injury in the U.S. may lead to significantly higher damages than a similar incident adjudicated in Japan due to differing tort standards.

Courts may intervene to mitigate these imbalances by invoking doctrines such as anti-suit injunctions, which prevent parties from initiating parallel or abusive litigation in other jurisdictions. Moreover, certain treaties, such as the Brussels I Regulation in the EU, seek to streamline jurisdictional rules across member states, ensuring more consistent legal outcomes.

The choice of forum in drone-related litigation is influenced by a combination of substantive legal factors, procedural rules, and strategic considerations. Key factors include:

1. **Jurisdictional reach:** Courts must establish jurisdiction over the defendant, which often relies on the minimum contacts doctrine or the forum's connection to the drone incident. For example, the U.S. Supreme Court case Daimler AG v. Bauman (2014) limited the application of jurisdiction in U.S. courts based solely on a defendant's business activities in the state, potentially affecting international drone manufacturers.

2. **Applicable law:** The substantive law of the chosen forum is a key consideration. Plaintiffs may choose a jurisdiction with strict liability laws for defective products or pro-plaintiff tort rules. For example, German product liability law, though stringent, may favour manufacturers compared to U.S. tort law, which may allow for punitive damages.

3. **Procedural rules:** Procedural advantages such as jury trials, broad discovery, class action mechanisms, and the availability of punitive damages are critical factors. A plaintiff might choose the U.S. as a forum due to the possibility of a jury award for emotional distress, which may not be available in civil law jurisdictions like France.

4. **Enforcement of judgments:** Plaintiffs also consider the ease of enforcing judgments across borders. Jurisdictions that are signatories to treaties like the Hague Convention on the Recognition and Enforcement of Foreign Judgments offer greater certainty that any awarded damages will be recoverable.

Choice of forum is a calculated move, often based on where the balance of legal risks and rewards tilts in a party's favour. Courts are increasingly aware of strategic forum shopping and may respond with mechanisms like consolidation of cases or dismissal of duplicative claims under res judicata principles to prevent fragmentation of litigation.

Courts and legal systems address forum shopping primarily through jurisdictional doctrines and procedural safeguards designed to ensure fairness and prevent abuse. One primary tool is forum non convenient, where courts may decline to hear a case if there is a more appropriate forum available. In cases of clear forum shopping, courts will analyze whether the plaintiff's choice of forum has significant connections to the dispute or whether the choice is purely strategic.

In cross-border drone disputes, another common mechanism is the doctrine of comity, which promotes respect for the laws and judgments of foreign courts. For instance, if a party has initiated litigation in one jurisdiction, courts in another may defer to that jurisdiction to prevent conflicting judgments or duplicative litigation. Additionally, anti-suit injunctions can be employed to restrain parties from pursuing claims in multiple jurisdictions simultaneously.

Furthermore, international agreements, such as the Brussels I Regulation within the European Union, establish clear rules for jurisdictional choice and the recognition of judgments across member states. These regulations seek to reduce forum shopping by ensuring that cases are brought in the most appropriate and connected jurisdiction. In the U.S., multi-district litigation (MDL) serves to consolidate related cases to ensure consistency in rulings and limit the strategic abuse of forum selection.

Courts play an active role in preventing forum shopping from undermining the integrity of legal proceedings. By invoking doctrines like forum non convenient and relying on international agreements for cross-border disputes, courts strive to balance litigants' interests with procedural fairness while maintaining judicial efficiency.

Protecting Fundamental Rights in the Context of Drones

The widespread use of drones raises important considerations regarding the protection of fundamental rights. How do drone operations intersect with the protection of fundamental rights, and what legal challenges arise in balancing the rights of individuals with the operational freedom of drone operators? Which specific fundamental rights are most at risk due to widespread drone usage, and how do legal frameworks address the potential violations of these rights in both national and international contexts? What legal and ethical frameworks are in place to safeguard fundamental rights such as privacy, freedom of expression, and security in the context of drone operations, and are these frameworks sufficient to address the evolving technological landscape? How should courts and lawmakers reconcile the potential conflict between the right to privacy and the expanding use of drones in commercial, governmental, and recreational activities?

The intersection between drone operations and the protection of fundamental rights presents a complex legal challenge, primarily because drone technologies operate in ways that can easily infringe upon fundamental rights, particularly the right to privacy. The widespread use of drones for surveillance, data collection, and photography often encroaches upon individuals' rights to personal privacy, family life, and even property rights.

Internationally, Article 8 of the European Convention on Human Rights (ECHR) provides for the right to respect for private and family life, home, and correspondence. Drones, especially those equipped with cameras and sensors, have the potential to violate this right by recording images or collecting data without consent. Similarly, the International Covenant on Civil and Political Rights (ICCPR), under Article 17, protects individuals from arbitrary or unlawful interference with their privacy.

Balancing these rights with the operational freedom of drone operators—whether commercial, governmental, or recreational—requires carefully crafted legislation. For instance, while drones can facilitate crucial activities such as law enforcement, infrastructure monitoring, and delivery services, these activities must be weighed against individuals' fundamental rights. National data protection laws, such as the General Data Protection Regulation (GDPR) in the European Union, impose stringent obligations on drone operators to ensure that personal data collected by drones is processed lawfully, fairly, and transparently. This creates a balancing act between operational efficiency and the protection of fundamental rights, requiring courts and regulators to consistently evaluate proportionality and necessity in the context of drone use.

Courts often face the challenge of reconciling technology's expanding capabilities with the need to preserve rights enshrined in international conventions. The legal framework governing drone operations must evolve to address these issues, ensuring that drone usage is compatible with fundamental rights through clear regulatory limitations and judicial oversight.

The specific fundamental rights most at risk due to drone usage include the right to privacy, freedom of expression, and, in some instances, the right to life and security. Privacy remains the most immediately threatened, particularly in cases where drones are equipped with high-resolution cameras, thermal imaging, or data-gathering sensors. These capabilities make it easier to engage in covert surveillance, raising concerns about the infringement of personal space and the unauthorized collection of sensitive personal data.

Internationally, the European Court of Human Rights (ECtHR) has consistently emphasised the importance of the right to privacy, particularly under Article 8 of the ECHR, which has been invoked in cases involving surveillance technologies. Similar protections are found in the ICCPR under Article 17, which prohibits arbitrary interference with privacy. Additionally, the United Nations Guiding Principles on Business and Human Rights encourage businesses involved in drone operations to ensure that their activities do not infringe upon the rights of individuals and that adequate mechanisms are in place for the protection of those rights.

To address the potential violations of these rights, national laws have been developed to provide safeguards. For instance, GDPR sets out clear guidelines for the processing of personal data, including data collected by drones, ensuring that individuals retain control over their data. In the United States, privacy concerns surrounding drones have led to the development of state-specific laws that restrict drone surveillance and regulate the use of drones near private property. At the federal level, the Fourth Amendment of the U.S. Constitution, which protects against unreasonable searches and seizures, may apply when drones are used for surveillance by government entities.

The protection of fundamental rights in the context of drone operations requires a careful balancing of interests through both international conventions and domestic legislation. National courts and international tribunals will need to increasingly address these conflicts, ensuring that evolving drone technologies do not erode the rights of individuals, particularly in areas of privacy and security.

The current legal and ethical frameworks governing drone operations and fundamental rights primarily derive from a combination of international human rights law, national privacy legislation, and sector-specific regulations. For privacy, GDPR remains a leading standard, ensuring that drone operators collecting personal data adhere to the principles of lawful and transparent processing, data minimisation, and consent. The United States' Fourth Amendment jurisprudence, while not specifically geared towards drones, provides some degree of protection against government surveillance through drones without a warrant.

The Chicago Convention on International Civil Aviation (1944) provides another layer of regulation, ensuring that the operation of drones in international airspace adheres to safety standards, while respecting the sovereignty of national airspace. Ethical frameworks, such as those recommended by the European Union Agency for Fundamental Rights (FRA), advocate for a "privacy by design" approach to drone technology, urging manufacturers to incorporate privacy-enhancing features such as geo-fencing or restricted access to certain data from the outset.

However, these frameworks are increasingly coming under pressure as drone technology evolves rapidly. Drones are becoming smaller, faster, and equipped with more advanced sensors, making traditional regulatory models potentially inadequate. For instance, facial recognition technology mounted on drones presents unique ethical and legal challenges, particularly concerning data protection, profiling, and mass surveillance.

Existing legal frameworks, while robust in certain jurisdictions, may not fully keep pace with rapid technological advancements. There is a pressing need for the development of international norms and harmonized regulations to address emerging threats to fundamental rights posed by the next generation of drone technology.

Courts and lawmakers are faced with the challenge of reconciling the expansion of drone usage with the fundamental right to privacy by creating nuanced and technology-specific regulations that strike a balance between innovation and rights protection. One approach that has gained traction is proportionality analysis, where courts assess whether the intrusion on privacy by drones is proportionate to the legitimate aim pursued—such as law enforcement surveillance or commercial efficiency.

In the Klass v. Germany (1978) case before the ECtHR, the court established that while national security may justify surveillance measures, such measures must be accompanied by safeguards against abuse. This principle can be extended to drone operations, particularly in governmental activities. National courts must ensure that drone surveillance is used in a manner that is narrowly tailored to serve its purpose without unnecessarily infringing on individual privacy rights.

In the regulatory domain, lawmakers are increasingly looking to technology-specific solutions such as geo-fencing and no-fly zones over sensitive areas (e.g., private homes, schools, hospitals) to preemptively mitigate privacy violations. Additionally, enhanced transparency requirements can obligate drone operators—particularly in the commercial sector—to disclose the purposes of their data collection activities and provide individuals with rights to access, correct, or erase data.

Lawmakers need to draft clear and adaptable legislation that accommodates the rapid evolution of drone technology while safeguarding privacy rights. Courts, on the other hand, play a critical role in ensuring that such legislation is interpreted in ways that uphold fundamental rights, requiring a careful balancing act between competing interests.

Privacy Rights and Drone Surveillance

Privacy rights are at the forefront of concerns surrounding drone operations, particularly when it comes to drone surveillance. To what extent do privacy rights serve as a limiting factor in drone surveillance, and how should legal frameworks balance the rights of individuals with the legitimate interests of drone operators engaged in activities such as aerial surveillance and data collection? How do existing legal frameworks—both domestic and international—define and regulate the boundaries of privacy rights in the context of drone operations, particularly regarding the collection and processing of personal data? What limitations must be imposed to ensure that drone surveillance does not infringe upon these rights? What ethical considerations must be taken into account when determining the permissible scope of drone surveillance, and how should courts and regulators assess the proportionality of drone activities that impact individual privacy in both public and private spaces? In cases where drone surveillance conflicts with privacy rights, what legal remedies or safeguards should be implemented to protect individuals from potential abuse, and how should these safeguards be enforced to ensure compliance with privacy laws?

Privacy rights serve as a significant limiting factor in drone surveillance, especially in jurisdictions with robust privacy protection laws like the European Union (EU) and California under the California Consumer Privacy Act (CCPA). In the EU, the General Data Protection Regulation (GDPR) governs the collection and processing of personal data, including data captured by drones. Article 5 of the GDPR requires that data collection be lawful, fair, and transparent, meaning drone operators must establish a legal basis for their surveillance activities and inform individuals of their rights to access, correct, or delete data.

The European Court of Human Rights (ECtHR) has consistently reinforced the right to privacy under Article 8 of the European Convention on Human Rights (ECHR), particularly in cases involving surveillance. In Peck v. United Kingdom (2003), the court emphasised that public surveillance must balance individual privacy rights with state interests. This precedent is crucial when assessing drone surveillance conducted by governmental agencies or private actors.

In the U.S., the Fourth Amendment provides protection against unreasonable searches, which may extend to drone surveillance by government entities. However, case law is still evolving. Kyllo v. United States (2001) held that using technology to gain information from within a home that could not otherwise be obtained without physical intrusion violates the Fourth Amendment. This principle could apply to drones equipped with advanced imaging technology used in private spaces.

Courts and lawmakers must balance privacy rights with the interests of drone operators by ensuring that drone surveillance complies with proportionality and necessity principles. The legal framework should mandate that any intrusion into personal privacy must serve a legitimate interest, be limited in scope, and involve the least invasive means possible.

Legal frameworks such as the GDPR in the EU, the CCPA in California, and state-specific laws in the U.S. (such as Texas' Privacy Protection Act) impose strict boundaries on the collection and processing of personal data by drones. The GDPR mandates that personal data be collected only for specified, explicit, and legitimate purposes (Article 5). This means that drone operators must disclose the purpose of the data collection, obtain consent when necessary, and implement data minimisation techniques to ensure that only relevant data is gathered.

At the international level, privacy by design is emerging as a key principle in technology law, advocating that privacy features be built into drone systems at the development stage. This approach aligns with guidelines issued by regulatory bodies such as the International Association of Privacy Professionals (IAPP) and the Council of Europe in its recommendation on the protection of personal data in the context of video surveillance.

Limitations imposed by privacy frameworks should include strict data retention policies, transparency in data usage, and robust security measures to prevent unauthorized access to drone-captured data. These safeguards must also provide individuals with avenues to challenge improper surveillance and seek redress.

Ethical considerations in drone surveillance revolve around the principle of proportionality, which requires that the scope of surveillance be commensurate with the risks posed to individual privacy. Courts and regulators must weigh the legitimate interests of surveillance—such as crime prevention, public safety, or commercial purposes—against the potential harms to privacy.

In both public and private spaces, the location and manner of surveillance play a crucial role in assessing proportionality. For example, drones monitoring public spaces for traffic management may be justified under public interest exceptions, whereas drones flying near private residences or in densely populated areas raise concerns about covert surveillance and intrusive data collection. The right to be let alone, which forms a core component of privacy law in jurisdictions like the U.S. (derived from Warren and Brandeis' 1890 article on privacy), reinforces the need for such proportionality.

Ethical frameworks like PCAST (President's Council of Advisors on Science and Technology) Privacy Principles and EU Code of Conduct on Privacy for Drones encourage drone operators to adopt geo-fencing, anonymization techniques, and no-fly zones over sensitive areas to mitigate privacy risks.

Courts should require drone operators to justify their surveillance activities based on the specific context of the operation, ensuring that the least intrusive measures are used. Regulatory bodies must create industry standards that limit the use of drones in sensitive areas and impose strict conditions on the use of advanced surveillance technologies such as facial recognition.

Legal remedies for conflicts between drone surveillance and privacy rights include both civil actions for damages and injunctive relief. Under the GDPR, individuals have the right to file complaints with data protection authorities (DPAs) or directly bring claims before courts for unlawful data processing. In Greece v. United Kingdom (2008), the ECtHR ruled that individuals must have access to effective remedies in cases of privacy violations, establishing a precedent for holding public authorities accountable for invasive surveillance.

In the U.S., the Electronic Communications Privacy Act (ECPA) and state-level legislation allow individuals to seek statutory damages and injunctive relief if their privacy rights are violated by drone surveillance. These statutes are complemented by trespass and nuisance laws, which provide additional remedies for drone overflights that interfere with property rights.

To enforce compliance, regulatory bodies like the EU's DPAs and the Federal Trade Commission (FTC) in the U.S. have the authority to investigate breaches of privacy laws and impose significant penalties. For instance, under the GDPR, penalties can reach up to €20 million or 4% of global annual turnover, ensuring strong deterrence against non-compliance. Additionally, privacy audits and impact assessments are critical tools for regulators to assess compliance proactively.

Robust enforcement mechanisms, including penalty provisions, compliance monitoring, and class-action lawsuits, are essential to ensuring that drone operators adhere to privacy laws. Legal remedies must also allow for swift action, particularly in cases of ongoing or imminent privacy violations, to protect individuals from further harm.

Data Protection and Drone Operations

The use of drones often involves the collection and processing of personal data, raising significant concerns regarding data protection. What specific obligations do drone operators have under data protection laws concerning the collection and processing of personal data, and how do these obligations vary depending on the nature of the data collected (e.g., sensitive data vs. general personal data)? How does the legal framework governing data protection address the unique challenges posed by drone operations, particularly concerning the real-time collection of data in public and private spaces, and what limitations or safeguards must be imposed to protect against unlawful data processing?

What responsibilities do drone operators and stakeholders bear in relation to the storage and retention of personal data collected by drones, and how do data protection principles such as data minimisation and purpose limitation apply to drone activities? How do data protection authorities monitor and enforce compliance with data protection laws in the context of drone operations, and what penalties or sanctions can be imposed on operators who fail to meet their legal obligations? How do data protection laws reconcile the tension between the operational needs of drone technologies and the fundamental rights to privacy, particularly when drones are used in areas where individuals have a reasonable expectation of privacy?

Under data protection laws, such as the General Data Protection Regulation (GDPR) in the European Union, drone operators are regarded as data controllers if they collect and process personal data using their drones. The GDPR requires that drone operators adhere to several obligations when handling personal data, including lawfulness, fairness, and transparency (Article 5 GDPR). Operators must ensure that data processing has a legal basis, such as consent, legitimate interest, or contractual necessity. Moreover, they are required to inform individuals about the purposes of data collection and how their data will be processed (Articles 12-14 GDPR).

When drones collect sensitive data (e.g., biometric data, health data), the obligations increase under Article 9 GDPR, which restricts the processing of such data unless specific conditions are met, such as explicit consent or public interest justifications. For example, drones collecting facial recognition data during public events or for law enforcement purposes would fall under this higher threshold of protection.

Operators must be diligent in assessing the nature of the data collected, implementing stronger safeguards and legal justifications when processing sensitive data. They also need to conduct Data Protection Impact Assessments (DPIAs) under Article 35 GDPR when there is a high risk to individuals' rights and freedoms, particularly in the context of surveillance by drones.

The GDPR and similar data protection laws globally address the challenges posed by drones by placing significant emphasis on transparency, consent, and minimisation of data processing. One key challenge with drones is their ability to collect real-time data, often without individuals being aware. To mitigate this, regulators require that drone operators clearly disclose their data processing activities and, where necessary, obtain consent from individuals whose data is being collected.

A potential safeguard for real-time data collection is the implementation of privacy by design and default (Article 25 GDPR), where drones must be configured to minimise data collection to what is strictly necessary. Additionally, the deployment of geo-fencing technology can help ensure drones avoid sensitive areas such as private residences or locations where individuals have a reasonable expectation of privacy.

The challenge is ensuring that these obligations are not only theoretical but actively enforced. Privacy regulators must work closely with drone operators to ensure compliance, including through audits and ensuring drones are equipped with mechanisms that prevent overreach in data collection, such as blurring or anonymising personal data captured in public areas. Under Article 5 GDPR, drone operators are required to adhere to the principles of data minimisation and purpose limitation. Data minimisation mandates that operators should only collect data that is adequate, relevant, and necessary for the stated purposes, while purpose limitation restricts the use of the data to the specific purposes communicated at the time of collection.

For example, if a drone is used for infrastructure inspection, the operator should ensure that personal data (e.g., images of people or vehicles) is not collected unless absolutely necessary for the inspection purposes. If such data is collected, operators must securely store and manage it to prevent unauthorized access or breaches. Data retention policies must also be put in place, ensuring that personal data is stored only for as long as necessary and deleted afterward (Article 17 GDPR, right to erasure).

Operators must implement robust security measures under Article 32 GDPR, such as encryption and access controls, to protect the stored data. Furthermore, they should establish retention schedules that automatically delete or anonymise data after its intended use is completed. Data protection authorities (DPAs) such as the Information Commissioner's Office (ICO) in the UK or the European Data Protection Board (EDPB) in the EU play a critical role in monitoring and enforcing compliance with data protection laws. DPAs are responsible for conducting audits, investigating complaints, and ensuring that drone operators comply with relevant laws, including the GDPR.

Non-compliance with data protection obligations can lead to significant penalties under GDPR. For serious violations, such as failing to obtain consent for data collection or failing to implement adequate security measures, drone operators can face fines of up to €20 million or 4% of global annual turnover, whichever is higher (Article 83 GDPR). Beyond financial penalties, DPAs may also issue compliance orders, forcing operators to stop processing data until corrective measures are in place.

DPAs often take a proactive approach to drone-related data protection issues, releasing guidance notes and best practice documents for drone operators. Operators should closely follow these developments to ensure they are not only compliant with current laws but are also prepared for stricter future regulations, particularly as drone use expands.

Data protection laws like the GDPR seek to strike a balance between the innovation potential of drone technologies and the protection of individuals' privacy rights. The principle of proportionality is central to this balance—drone operations that significantly intrude upon privacy (such as drones with high-resolution cameras operating in private residential areas) must be justified by a legitimate interest and comply with privacy safeguards, including obtaining consent where appropriate.

For instance, in cases involving surveillance of private spaces, courts and regulators may consider whether the drone operator has taken reasonable steps to minimise the privacy impact, such as restricting flight paths or limiting the data captured to what is strictly necessary. The right to privacy, recognised under Article 8 ECHR and Article 17 ICCPR, requires that any interference must be lawful, necessary, and proportionate to the purpose it serves.

Courts are likely to scrutinise drone operations in contexts where individuals have an expectation of privacy, such as in their homes or private gardens. Drone operators must be prepared to justify such operations rigorously, balancing the economic and operational needs of their activities with the privacy rights of individuals.

Freedom of Expression and Drone Journalism

Drones have become powerful tools for journalism and the exercise of freedom of expression, but their use also raises legal and ethical questions. How do current legal frameworks governing freedom of expression and media freedom accommodate the use of drones for journalistic purposes, and what specific limitations or restrictions apply to drone journalism to balance the public's right to information with privacy and security concerns? What legal and ethical challenges arise from the use of drones in news gathering, particularly in situations where the line between public interest journalism and intrusion into private lives or sensitive areas is blurred?

To what extent does the protection of freedom of expression under constitutional and international human rights laws extend to drone journalism, and how should courts and regulatory bodies interpret these protections when other competing interests, such as national security, public safety, or individual privacy, are at stake? What obligations and responsibilities do drone journalists have in ensuring that their operations comply with both media laws and aviation regulations, and how do these intersect with ethical standards in journalism to promote responsible and lawful reporting practices? How should courts and legislators reconcile the tension between promoting innovation in journalism through the use of drones and enforcing necessary regulatory safeguards to protect the rights of third parties, including privacy and property rights?

Freedom of expression, including media freedom, is protected under Article 10 of the European Convention on Human Rights (ECHR) and the First Amendment of the U.S. Constitution. However, this protection is not absolute, particularly when the exercise of freedom conflicts with other rights, such as privacy (protected under Article 8 ECHR) or national security.

Drone journalism, while enhancing the ability to gather news in ways previously unimaginable, must comply with laws governing privacy, data protection, and airspace regulations. In the European Union, the General Data Protection Regulation (GDPR) imposes obligations on drone operators, including journalists, who collect personal data via drones. Journalists must ensure that they do not collect data unnecessarily and must obtain consent when required.

In the United States, the Federal Aviation Administration (FAA) regulates drone usage through the Part 107 Rules, which prohibit drones from flying over people, unless the operator has a waiver. This significantly limits drone usage in crowded public events, a common setting for journalistic activities. Additionally, state laws governing intrusion upon seclusion provide further limitations on drone journalism by protecting individuals' right to privacy within their homes or private spaces.

Drone journalists must navigate between their right to inform the public and existing laws that impose restrictions to protect privacy, security, and public safety. Courts and regulatory bodies, in interpreting these laws, often prioritise privacy rights, especially in cases where drone usage could be seen as intrusive or disproportionate to the public interest served by the news gathering.

The key challenge in drone journalism is balancing the public interest in news gathering with the risk of intruding into private lives or restricted areas. Legally, the challenge revolves around how far journalists can go in capturing information that is in the public's interest without breaching privacy or trespassing laws.

For example, in the U.S., courts may apply the newsworthiness doctrine to justify certain invasions of privacy by the media if the information gathered is of legitimate public concern (see Shulman v. Group W Productions, Inc.). However, this does not give journalists carte blanche to invade private spaces. If a drone is used to capture footage inside a private residence, without the homeowner's consent, this could result in liability under intrusion torts. Similarly, capturing footage of private moments in sensitive locations (e.g., hospitals, prisons) could constitute invasion of privacy, even if done under the guise of public interest journalism.

Journalists must ensure they are not crossing ethical or legal boundaries, particularly when reporting on private individuals. Media outlets should adopt robust editorial policies that balance public interest with respect for personal privacy and adhere to aviation regulations regarding restricted airspace.

The protection of freedom of expression under Article 10 ECHR and First Amendment jurisprudence extends to drone journalism insofar as it serves the public interest, but it is subject to limitations that protect national security, public safety, and individual privacy. Courts have consistently upheld that the right to free expression must be balanced against other protected rights. For example, in Handyside v. United Kingdom, the European Court of Human Rights held that freedom of expression can be subject to restrictions when necessary in a democratic society to protect the rights of others.

Regulatory bodies, like the FAA in the U.S. and the European Union Aviation Safety Agency (EASA) in the EU, place restrictions on drone flights in sensitive areas, such as near military installations or airports, to prevent security breaches and ensure public safety. National security concerns could also justify limiting drone use during high-risk events, such as protests or demonstrations, where there is a threat of violence or terrorism.

Courts and regulators must interpret freedom of expression protections with a view to ensuring that the exercise of media rights does not unduly infringe on national security or public safety. Drone journalists must be prepared to justify their use of drones in sensitive situations by demonstrating that their actions are proportionate and serve a legitimate public interest.

Drone journalists must adhere to a dual set of obligations: compliance with aviation regulations (e.g., FAA's Part 107 in the U.S. or EASA's regulations in the EU) and compliance with media laws, including those concerning defamation, privacy, and data protection. Aviation regulations often impose strict operational limits, such as flying within the visual line of sight, maintaining altitude restrictions, and avoiding no-fly zones.

Simultaneously, journalists must follow ethical standards, such as those outlined by the Society of Professional Journalists (SPJ) or the International Federation of Journalists (IFJ). These standards emphasise the need for accuracy, minimising harm, and maintaining independence. Journalists using drones should take care not to intrude on private moments without a compelling public interest, avoid sensationalism, and be mindful of the safety risks posed by their operations to the public.

Journalists must navigate the intersection of these regulatory frameworks by ensuring that their drone operations are lawful under aviation laws while simultaneously upholding journalistic integrity and respecting the rights of individuals. Failure to comply with these dual obligations can result in both legal consequences and damage to journalistic credibility.

Courts and legislators face a significant challenge in reconciling innovation with regulation. On one hand, drones represent a powerful tool for journalists, enabling new forms of reporting, particularly in hard-to-reach or dangerous locations. On the other hand, there is a need to safeguard the rights of third parties, including protecting privacy, avoiding trespass, and preventing accidents caused by reckless drone operations.

Courts may consider adopting a balancing approach, similar to that seen in privacy vs. free speech cases, where they assess the public interest served by drone journalism against the harm caused to individuals. Legislators, meanwhile, could promote innovation-friendly regulations by creating specific exemptions or fast-track licensing processes for accredited journalists, enabling them to operate drones in restricted airspace under certain conditions, such as during natural disasters or political protests.

Legislators must craft nuanced regulations that promote journalistic innovation while preserving fundamental rights. This could include special licenses for journalists, clear guidelines on what constitutes the public interest, and stricter penalties for drone misuse that infringes on personal privacy or property rights. Courts, too, must ensure they provide adequate recourse for those harmed by unlawful drone journalism, balancing the need for a free press with the protection of individual rights.

Drone journalism operates within a complex legal and ethical landscape where the exercise of freedom of expression must constantly be weighed against privacy, safety, and regulatory compliance. Drone journalists, courts, and legislators must navigate this dynamic space carefully to ensure that the benefits of drone technology are realised while safeguarding the rights of individuals and the public at large.

Balancing Fundamental Rights and Public Interests

The protection of fundamental rights must be balanced with the public interests and societal concerns associated with drone operations. How can legal frameworks be designed to strike an appropriate balance between the protection of fundamental rights, such as privacy and freedom of movement, and public interests, including national security and public safety, in the context of drone operations? What legal principles and doctrines should be applied to reconcile the competing interests of drone innovation with fundamental rights, and how do courts and regulators determine when public interests justify restrictions on these rights?

How should national security and public safety concerns be weighed against fundamental rights in regulatory frameworks governing drone operations, and what limitations are necessary to ensure that regulatory measures are both proportionate and effective? What role does proportionality play in determining the legality of restrictions on fundamental rights in the interest of public safety and security, and how should courts assess the necessity and legitimacy of such measures in the context of drone activities? How can regulators craft policies that both protect fundamental rights and promote societal concerns like innovation, national security, and public safety without overstepping the bounds of individual freedoms and human rights?

Legal frameworks must integrate principles of proportionality, necessity, and legitimate aim to balance fundamental rights with public interests. Under European Union law, the Charter of Fundamental Rights and the European Convention on Human Rights (ECHR) guide this balancing act. For example, under Article 8 ECHR, the right to privacy can be interfered with if the interference is lawful, necessary in a democratic society, and pursues legitimate objectives such as national security or public safety.

In the context of drone operations, frameworks like the General Data Protection Regulation (GDPR) provide guidance on protecting privacy, particularly when drones collect data. Simultaneously, drone-specific regulations, such as EASA's drone regulations or the FAA's Part 107 in the United States, permit certain intrusions into fundamental rights when justified by public safety or security needs.

The challenge for legislators is to ensure these frameworks do not disproportionately limit fundamental rights. Judicial oversight and impact assessments, such as Data Protection Impact Assessments

(DPIAs) under the GDPR, are essential in ensuring that restrictions on rights are continually evaluated and justified by evolving public interest needs.

Legislators must employ dynamic, adaptive legal frameworks that accommodate rapidly changing technology and societal needs, while still preserving fundamental rights as a core principle. Flexibility in these frameworks, paired with robust judicial review mechanisms, allows for constant recalibration as drone technology and its applications evolve.

The primary legal doctrine applied to reconcile competing interests in drone innovation with fundamental rights is proportionality. This principle requires that any interference with rights must be suitable, necessary, and proportionate to the legitimate aim pursued, such as public safety or national security. Proportionality ensures that measures affecting fundamental rights are no more restrictive than necessary. Courts often rely on a three-part test to assess whether restrictions on fundamental rights are justified:

- First, is the interference lawful and based on a legitimate legal framework?

- Second, is the interference pursuing a legitimate aim, such as public safety or national security?

- Third, is the interference proportionate to the aim, meaning there are no less restrictive alternatives?

For example, in Delfi AS v. Estonia, the European Court of Human Rights (ECHR) used this test to assess whether restrictions on freedom of expression online were proportionate to the legitimate aim of protecting individual reputation. Similar reasoning applies to drone technology, where courts evaluate if restrictions on drone use (e.g., limiting flight over private properties) are proportionate to protecting privacy. Courts are likely to require drone regulations that strike a balance between encouraging innovation and protecting rights. Innovations that enhance public safety (e.g., drones for search and rescue) are often justified, but courts will scrutinise innovations that disproportionately infringe upon individual privacy or freedom without sufficient safeguards.

National security and public safety concerns often justify restrictions on fundamental rights, but these concerns must be carefully balanced against the individual's rights. For instance, Article 15 of the ECHR allows states to derogate from certain rights during national emergencies, but such derogations must be strictly necessary and proportional.

In the context of drone regulations, national security concerns (e.g., preventing drones from entering restricted airspace around military installations or airports) are often weighted heavily. Public safety laws, such as FAA regulations on no-fly zones and EASA prohibitions on drone flights near sensitive infrastructure, prioritise the prevention of accidents, sabotage, or surveillance threats.

However, the limitations imposed must still respect minimum standards of legality and proportionality. For instance, geo-fencing technology can be employed to prevent drones from flying into sensitive areas without completely banning drones from broader usage.

Courts generally uphold restrictions that are grounded in legitimate security concerns, but they require evidence that such restrictions are necessary and proportionate. Drone operators must be aware that regulations limiting drone use in certain areas are legally defensible, but only if the government can justify them through specific threats and clear public safety benefits.

Proportionality plays a central role in determining the legality of restrictions on fundamental rights. Courts assess proportionality by asking whether the interference with a right is appropriate to achieve the desired objective, whether less restrictive means could have been used, and whether the interference is justified by the benefits gained.

For example, in S. and Marper v. United Kingdom, the European Court of Human Rights held that retaining DNA data indefinitely was a disproportionate violation of privacy, even if it was justified by public safety concerns. Similarly, in drone cases, courts are likely to scrutinise whether broad restrictions (e.g., total bans on drone use in certain regions) are proportionate to the security risks involved or whether more targeted measures (e.g., limiting specific drone functionalities) would suffice.

Courts will likely view granular, specific regulations as more proportionate than sweeping prohibitions on drone use. Proportionality tests ensure that regulatory bodies do not excessively restrict rights for minor security gains, and operators may challenge overly restrictive measures that do not pass this rigorous assessment.

Regulators must strike a balance between promoting innovation in drone technology and preserving fundamental rights by crafting tailored policies that address specific risks without imposing blanket restrictions. For instance, the use of differentiated regulations based on drone size, purpose, and operational area can ensure that fundamental rights are only restricted where necessary.

Policies should be evidence-based and take into account the specific context in which drones are used. For example, emergency services may require looser regulations on drone use for public safety purposes, whereas private companies using drones for surveillance might face stricter scrutiny under privacy laws.

Regulators can also promote innovation by introducing exemptions or waivers for certain drone activities, provided that operators implement privacy safeguards and security measures. For instance, a company developing drones for environmental monitoring could be granted expanded operational permissions if they adhere to stringent data protection protocols.

Regulators should focus on nuanced, flexible frameworks that promote innovation while embedding necessary safeguards. Impact assessments, regular reviews, and stakeholder engagement can help ensure that regulations evolve alongside drone technology, preventing overreach while addressing emerging societal concerns.

The legal landscape surrounding drone operations requires a careful balance between innovation, fundamental rights, and public safety. Courts and regulators must use principles of proportionality, legality, and necessity to navigate this balance, ensuring that rights are protected without stifling technological advancements.

X

10. Non-Normative Rules and Drone Regulation

Non-normative rules play a significant role in shaping the regulatory landscape for disruptive technologies such as drones. What are the defining characteristics of non-normative rules, and how do these rules differ from traditional normative regulations when applied to the governance of disruptive technologies such as drones? How can non-normative rules effectively complement traditional normative approaches in regulating emerging technologies, and what are the potential benefits and limitations of using non-normative frameworks in the context of drone regulation? In what ways can non-normative rules be utilised to create more flexible and adaptive regulatory environments for disruptive technologies, and what challenges might arise in ensuring compliance with these less formalised rules? How do courts and regulatory bodies interpret and enforce non-normative rules within the drone industry, and what impact do these rules have on the evolution of technology law and innovation?

Non-normative rules, as opposed to normative rules, do not rely on explicit prohibitions or requirements backed by legal sanctions. Instead, they encompass guidelines, best practices, industry standards, and technological measures that indirectly influence behaviour and promote compliance. These rules often emerge organically within industries and are shaped by the collective actions and agreements of stakeholders, including manufacturers, operators, regulatory bodies, and users. In the rapidly evolving field of drone technology, non-normative rules provide the flexibility and adaptability that traditional normative regulations may lack.

One of the key characteristics of non-normative rules is their dynamic and responsive nature. Unlike rigid legal frameworks that can take years to develop and amend, non-normative rules can evolve quickly in response to technological advancements and emerging risks. For instance, industry-led standards for drone safety protocols can be updated promptly to address new threats or incorporate innovative safety features. This agility is crucial in the context of drone regulation, where technology and its applications are advancing at a pace that often outstrips the ability of traditional regulatory processes to keep up.

Non-normative rules, unlike traditional normative regulations, are not legally binding but provide guidance, best practices, and industry standards that influence behaviour without direct enforcement. These rules are often adaptive and evolve rapidly in response to technological advancements, offering flexibility that traditional regulations may lack. For example, soft law instruments like guidelines issued by industry bodies (e.g., the International Civil Aviation Organization (ICAO)) or recommendations from data protection authorities help shape the practices of drone operators without creating strict legal obligations.

In contrast, normative rules, such as EASA's drone regulations in Europe or the FAA's Part 107 Rules in the U.S., are codified into law and have clear consequences for non-compliance. These rules establish legally binding obligations, such as registration requirements, altitude restrictions, and licensing, which drone operators must adhere to.

Non-normative rules complement traditional regulations by offering flexibility and responsiveness to new challenges, especially in rapidly evolving industries like drone technology. These rules can be easily updated and tailored to specific technological developments, allowing industry players to adapt quickly without waiting for the legislative process to catch up. For instance, industry guidelines on geo-fencing technology or privacy-by-design principles can be implemented faster than statutory regulations.

One major benefit of non-normative rules is that they foster innovation by providing guidance without stifling new technologies with rigid laws. For example, privacy-focused guidelines for drone data collection under the GDPR are often framed as best practices rather than fixed rules, allowing operators to innovate while still protecting privacy.

However, the limitations of non-normative rules lie in their lack of enforceability. Compliance is often voluntary, and there may be insufficient mechanisms to hold parties accountable for breaches. Additionally, there is a risk of fragmentation where different jurisdictions or industry bodies issue conflicting guidelines, leading to uncertainty for operators who work across borders.

Non-normative rules also encourage a collaborative approach to regulation. By involving a broad spectrum of stakeholders in the development and implementation of these rules, a more comprehensive and practical set of guidelines can be achieved. This collaboration can lead to higher levels of buy-in and adherence, as stakeholders are more likely to follow rules they had a hand in creating. For example, the development of geo-fencing standards, which prevent drones from entering restricted areas, typically involves input from drone manufacturers, aviation authorities, and security agencies. This collaborative process ensures that the resulting standards are both technically feasible and effective in enhancing safety and security.

Non-normative rules can be utilised to encourage innovation and flexibility by setting forth best practices and guidelines that can quickly evolve alongside technology. For drones, non-normative rules might address data protection, cybersecurity, and environmental impacts, allowing the industry to implement cutting-edge technologies like autonomous navigation and AI-powered surveillance while adhering to ethical standards.

For instance, self-regulation through industry codes of conduct can promote a culture of compliance without the need for heavy-handed state intervention. The Partnership on AI, an industry body that promotes ethical AI use, serves as a model for how non-normative rules could apply to drone AI technologies, setting standards for responsible development.

However, the challenges of relying on non-normative rules include inconsistent adherence, as companies may not be legally compelled to follow them. Additionally, without legal backing, there is a lack of enforcement mechanisms, which can lead to varying degrees of compliance across different entities. To mitigate these challenges, regulators may need to incentives adherence, such as through certification programs or by offering legal protections (e.g., reduced liability) for companies that follow industry best practices.

Furthermore, non-normative rules can serve as a bridge between innovation and regulation. They allow for experimentation and the adoption of new technologies without the immediate imposition of restrictive legal constraints. This is particularly beneficial in the early stages of technological adoption, where overly stringent regulations could stifle innovation and delay the deployment of beneficial technologies. In the case of drones, non-normative guidelines on testing and pilot programs enable the safe exploration of new uses and capabilities, facilitating the gradual integration of drones into various sectors while managing risks responsibly.

Courts and regulatory bodies generally view non-normative rules as persuasive rather than binding. They may look to these rules to inform decisions, particularly in cases where the law is silent or ambiguous. For example, in tort claims involving drone accidents, courts might consider whether the operator followed industry best practices, even if those practices are not legally required. This could influence the court's assessment of negligence or due diligence.

Regulatory bodies, like EASA or the FAA, may use non-normative rules to guide policy development. For instance, early iterations of drone regulations often started as guidelines before being codified into law. The impact of non-normative rules on technology law is significant because they allow regulators to test new ideas and frameworks in a low-risk environment before committing to legal mandates.

In essence, non-normative rules offer a versatile and adaptive approach to regulating disruptive technologies like drones. Their ability to evolve rapidly, foster collaboration, and balance innovation with risk management makes them an invaluable complement to traditional normative regulations. As we delve deeper into the specific non-normative rules applicable to drone regulation, it will become evident how these rules can shape a more effective and responsive regulatory environment, ensuring the safe and innovative use of drone technology in the age of artificial intelligence.

Non-normative rules are essential in contexts where innovation outpaces legislation. They provide a framework for responsible behaviour while allowing regulatory bodies to assess the impact of disruptive technologies before enacting binding laws. Non-normative rules also facilitate global cooperation, as seen in the aviation industry, where cross-border operations require common guidelines even when jurisdictions have different laws. The most effective regulatory frameworks for drones will likely involve a hybrid approach where non-normative guidelines fill the gaps in normative laws, ensuring that operators are guided by industry best practices while still being held to enforceable standards.

Non-normative rules are most effective when paired with incentives for compliance, such as certification schemes or market rewards for responsible behaviour. Regulatory bodies should also provide a clear pathway for non-normative rules to transition into enforceable law once their efficacy is proven. Non-normative rules act as a testing ground for new legal concepts and regulatory frameworks. Courts often look to these rules to determine industry standards of care, and regulatory bodies may adopt non-normative standards into law once they have been shown to work effectively. This fluid relationship between non-normative and normative rules facilitates the evolution of law in tandem with technological innovation.

Characteristics of Non-Normative Rules

Non-normative rules encompass a set of principles, guidelines, and best practices that are not legally binding but hold significant influence over the behaviour and governance of technologies. What are the legal implications of relying on non-normative rules, such as principles and guidelines, in regulating disruptive technologies like drones, given their voluntary nature and lack of legal enforceability?

To what extent can non-normative rules, grounded in industry practices, technological standards, and self-regulation, be considered legally persuasive or influential in litigation or regulatory decision-making related to emerging technologies? How do the distinctive features of non-normative rules, such as their flexibility and adaptability, affect the governance of rapidly evolving technologies like drones, and what challenges arise in ensuring compliance and accountability when such rules are not legally binding?

Non-normative rules, while not legally binding, hold significant influence in shaping industry behaviour, particularly in sectors where technology is evolving faster than law can keep pace. Their voluntary nature makes them persuasive rather than enforceable. For example, in litigation or regulatory actions, courts and regulatory bodies may refer to non-normative rules as a benchmark for industry standards or best practices. A drone operator who adheres to non-normative guidelines from recognised bodies, such as the International Civil Aviation Organization (ICAO) or European Aviation Safety Agency (EASA), may be able to demonstrate a good faith effort to comply with expected norms, even if these rules do not have the force of law. This can influence legal outcomes, particularly in negligence or liability cases, where a court may assess whether the operator acted reasonably in relation to accepted standards.

However, relying on non-normative rules also creates legal uncertainty. For example, ISO standards or industry codes of conduct might be viewed differently in various jurisdictions. In some cases, these guidelines could be used as a shield against liability by arguing that the operator followed widely accepted practices. Yet, their voluntary nature also means that they lack a clear enforcement mechanism, which can undermine consistency across the industry. In this sense, non-normative rules may create a soft legal environment that lacks the rigour and predictability of statutory law, making it difficult to ensure universal compliance.

Courts are increasingly willing to blend normative and non-normative frameworks. The concept of lex mercatoria in international commercial law, for example, shows how non-normative practices have historically influenced legally enforceable outcomes. In the drone industry, this blending may evolve, leading to the development of hybrid legal frameworks where voluntary compliance with best practices begins to carry quasi-legal weight.

Non-normative rules, particularly those grounded in industry self-regulation, can play a crucial role in influencing litigation and regulatory decisions by establishing customary practices within the industry. Courts often look to these practices to assess what a reasonable operator would do under similar circumstances. For example, if a drone-related accident occurs and the operator can show adherence to widely accepted non-normative standards (such as those issued by Drone Industry Associations or EASA's guidelines on privacy and safety), courts may view this compliance as evidence that the operator exercised due diligence.

Furthermore, in areas where formal regulation lags behind technological development, non-normative rules can fill the regulatory gap. For instance, drone data collection raises significant privacy concerns, but formal regulations may not fully address these issues. In such cases, adherence to GDPR-compliant best practices, even if voluntary, can be influential in demonstrating responsible data management in court proceedings. Regulatory agencies may also view companies that adhere to non-normative rules more favourably, which could affect the regulatory enforcement actions they face.

However, this influence is context-dependent. In highly regulated environments, such as aviation safety, adherence to non-normative standards may not absolve operators from liability if statutory regulations are breached. The voluntary nature of these rules means they do not create a legal shield, but they offer mitigating factors that courts and regulators might consider when determining culpability or penalty severity.

There is a growing trend toward the codification of non-normative rules into binding regulations, particularly as regulators seek to formalise industry practices that have proven effective. This shift highlights the legal community's recognition of the value that industry-driven standards bring to emerging technology governance. The challenge lies in striking a balance between flexibility and enforceability, as over-regulation can stifle innovation, but under-regulation may lead to gaps in accountability.

The flexibility and adaptability of non-normative rules allow for rapid adjustments in response to technological advancements. In the drone industry, where innovation often outpaces regulatory development, non-normative rules can serve as dynamic guidelines that evolve as technology progresses. For example, the development of new sense-and-avoid technologies in drones could be quickly incorporated into industry best practices long before formal regulations are updated. This adaptability is crucial in maintaining a balanced regulatory environment that encourages innovation while mitigating risks.

However, the voluntary nature of these rules presents challenges in ensuring consistent compliance. Without a formal enforcement mechanism, adherence to non-normative rules depends on industry cooperation and corporate ethics. This can lead to fragmentation, where some operators follow industry best practices, while others do not, creating an uneven playing field. Accountability becomes more difficult to achieve, especially when non-compliant actors are not legally required to adhere to the same standards.

Moreover, non-normative rules often lack transparency and public participation in their development, which raises questions about their legitimacy and inclusiveness. For instance, guidelines developed by a small group of industry leaders may not fully represent the interests of all stakeholders, including consumers and civil society. This can lead to regulatory capture, where the rules serve the interests of the industry rather than the public good.

The integration of multi-stakeholder processes into the development of non-normative rules could enhance their legitimacy and effectiveness. By involving regulators, industry, civil society, and legal experts, non-normative rules could evolve into consensus-driven standards that have broader acceptance and voluntary compliance across the sector. Additionally, as these rules gain traction, regulatory bodies may increasingly rely on them as precursors to formal regulation, thus creating a soft enforcement mechanism through industry pressure and reputational risk.

Non-normative rules serve as valuable tools in the governance of emerging technologies like drones, providing flexibility and adaptability in a fast-paced environment. While they cannot replace formal regulation, they complement normative laws by filling gaps and offering best practices that help ensure responsible innovation. However, their lack of enforceability presents challenges, and careful attention must be paid to incentivising compliance and preventing fragmentation within the industry.

Voluntary Compliance

How do non-normative rules function as regulatory tools within industries, and what mechanisms drive voluntary compliance in the absence of legal enforcement? The essence of non-normative rules lies in their voluntary nature. Compliance is not enforced through legal sanctions but through the incentives and pressures inherent within industries. Companies and individuals choose to adhere to these rules to gain competitive advantages, such as enhanced safety records, customer trust, and operational efficiency. Voluntary compliance fosters a sense of ownership and commitment among stakeholders, as they see the direct benefits of following these guidelines.

Industry-Driven Standards

How do non-normative rules, developed by industry bodies and consortiums, gain widespread adoption and influence across industries, despite their lack of legal binding force? Non-normative rules are often developed and propagated by industry bodies and consortiums that bring together various stakeholders. These entities create standards that reflect the collective expertise and practical experiences of the industry. For example, the International Organization for Standardisation (ISO) develops numerous non-normative standards that, while not legally binding, are widely adopted across industries to ensure quality and safety.

Flexibility and Adaptability

How do non-normative rules provide industries with the agility to respond to technological advancements and emerging challenges more effectively than traditional normative regulations, and what are the implications of this adaptability for regulatory frameworks? One of the most significant advantages of non-normative rules is their ability to adapt quickly to technological advancements and emerging challenges. Unlike normative regulations, which can be slow to change due to the legislative process, non-normative rules can be revised and updated promptly. This flexibility allows industries to stay ahead of potential risks and integrate the latest innovations more seamlessly. For instance, as drone technology evolves, non-normative guidelines can be swiftly adjusted to incorporate new safety protocols or operational practices.

Self-Regulation and Ethical Governance

How do non-normative rules facilitate self-regulation within industries, and what are the legal and ethical implications of relying on self-regulation to address issues of governance, transparency, and accountability, particularly in emerging technologies like AI-driven drones? Non-normative rules promote self-regulation within industries, encouraging companies to govern themselves according to agreed-upon standards. This approach not only reduces the regulatory burden on government agencies but also fosters a culture of ethical governance. Companies that adhere to these rules often demonstrate a commitment to responsible practices, which can enhance their reputation and build public trust. Ethical guidelines for AI, for example, help ensure that autonomous drones operate transparently and fairly, addressing concerns about bias and accountability.

Collaborative Development

How does the collaborative process in the development of non-normative rules ensure that these guidelines are comprehensive and reflect diverse perspectives, and what are the legal and practical implications of such inclusivity in setting industry standards, particularly in emerging fields like drone technology? The development of non-normative rules is typically a collaborative process involving multiple stakeholders. This inclusivity ensures that the rules are comprehensive and reflect a wide range of perspectives and expertise. Collaboration can occur at various levels, from industry-wide initiatives to cross-sector partnerships. For example, the development of geofencing standards for drones might involve input from aviation authorities, drone manufacturers, and security experts to ensure that the resulting guidelines are both technically feasible and effective in enhancing safety.

Market Incentives and Competitive Advantages

Adherence to non-normative rules can provide significant market incentives and competitive advantages. How do non-normative rules, such as industry standards and best practices, provide competitive advantages and market incentives for companies within highly regulated industries like the drone sector? What challenges and benefits do non-normative approaches bring to the regulatory landscape of the drone industry, particularly in addressing the limitations of traditional regulatory frameworks? How do non-normative rules, through collaboration and stakeholder engagement, influence the development of a regulatory environment that fosters innovation while ensuring safety and compliance in the drone industry? In what ways do non-normative rules contribute to the proactive management of risks in drone operations, and how do they complement traditional normative regulations? How can the integration of non-normative approaches with existing drone regulations promote the harmonization of international standards, particularly in facilitating cross-border drone operations?

Companies that comply with industry standards often find it easier to enter new markets, attract customers, and build partnerships. Certifications and endorsements based on adherence to non-normative guidelines can serve as quality marks, differentiating companies from their competitors. In the drone industry, for example, manufacturers that adhere to recognised safety and operational standards may be more appealing to commercial operators and end-users.

Understanding the advantages and challenges of adopting non-normative approaches is essential for a sophisticated regulatory approach for commercial drones. As the drone industry rapidly evolves, traditional regulatory frameworks often struggle to keep pace with technological advancements and emerging risks. Non-normative approaches, which include self-regulation, industry standards, and technological solutions, offer a flexible and adaptive alternative that can complement normative regulations. This subsection examines the potential benefits of non-normative approaches, such as flexibility, industry expertise, and faster adaptation to technological advancements, while also considering the challenges, including potential conflicts of interest, lack of legal enforceability, and the need for industry cooperation. By exploring these aspects, this subsection aims to provide a balanced perspective on the incorporation of non-normative approaches in drone regulation.

Non-normative rules can address many of the limitations identified in current drone regulations by providing flexible mechanisms, adaptive guidelines, and industry-led standards. Enhancing flexibility and responsiveness is one way non-normative rules can complement normative regulations. Industry standards and best practices, which can be updated more frequently and with greater ease than formal regulations, allow the regulatory framework to remain relevant and responsive to technological advancements and emerging risks. For example, industry consortia can develop and revise standards for AI-driven drones, ensuring that safety protocols keep pace with innovation.

Promoting collaboration and stakeholder engagement is another benefit of non-normative rules. The development of these rules typically involves extensive collaboration among stakeholders, including manufacturers, operators, regulators, and end-users. This collaborative approach ensures that the rules are practical, comprehensive, and aligned with industry needs. By fostering a sense of ownership and commitment among stakeholders, non-normative rules can lead to higher levels of compliance and cooperation.

Non-normative rules can also provide a regulatory environment that encourages experimentation and innovation. Pilot programmes and sandbox initiatives, guided by industry standards and best practices, allow companies to test new technologies and operational models in a controlled setting. This fosters innovation while ensuring that risks are managed responsibly.

Addressing the nuances of autonomous drone operations is another area where non-normative rules can be beneficial. These rules can offer detailed guidelines for the safe and ethical operation of autonomous drones, addressing issues such as algorithm transparency, bias mitigation, and accountability for autonomous actions. By providing a clear framework for autonomous operations, non-normative rules can enhance safety and public trust in AI-driven drones.

Industry-led standards and best practices can help bridge gaps between different national regulations, promoting greater harmonisation and consistency. Organisations such as ISO and ASTM can play a crucial role in developing international standards that facilitate cross-border drone operations and ensure a common level of safety and compliance.

Non-normative rules can promote proactive risk management by providing continuous monitoring and feedback mechanisms. These mechanisms can detect emerging risks and prompt timely updates to guidelines and best practices, reducing the likelihood of accidents and incidents. For example, continuous data collection and analysis can identify patterns and trends that inform the development of more effective safety protocols.

By integrating non-normative approaches, the EU can address the limitations of current drone regulations, fostering a regulatory environment that is flexible, responsive, and conducive to innovation. This will ensure the safe, sustainable, and innovative use of drones across various sectors, supporting the continued growth and development of the drone industry.

Evaluating the sophistication of current drone regulations reveals both strengths and areas for improvement. While existing frameworks provide a solid foundation, they often struggle to keep pace with technological advancements and emerging risks. Integrating non-normative approaches can enhance the flexibility, responsiveness, and effectiveness of drone regulations. By leveraging industry-led standards, collaborative development, and adaptive guidelines, regulatory authorities can create a more dynamic and comprehensive regulatory environment. This approach ensures the safe and innovative use of drone technology in the age of artificial intelligence, balancing the need for robust oversight with the imperative to foster innovation and growth in the industry.

Advantages of Non-Normative Approaches

One of the most significant advantages of non-normative approaches is their inherent flexibility. How does the flexibility of non-normative approaches provide an advantage over traditional normative regulations in rapidly evolving industries like drone technology, and what are the legal implications of this adaptability? What role does industry expertise play in the development of non-normative rules, and how does this influence the technical soundness and practical applicability of the standards in the drone industry?In what ways do non-normative approaches enable faster adaptation to technological advancements in the drone industry, and what are the legal and operational consequences of this accelerated regulatory process?

Unlike traditional normative regulations, which are often rigid and slow to adapt, non-normative rules can be updated quickly in response to technological advancements and changing industry needs. This flexibility is crucial in the drone industry, where innovations such as AI-driven autonomous drones and new applications are constantly emerging. Industry standards and best practices can be revised more frequently, ensuring that they remain relevant and effective. This adaptability allows the regulatory framework to keep pace with innovation, fostering an environment that supports technological advancement while maintaining safety and compliance.

Another advantage of non-normative approaches is the leverage of industry expertise. Industry-led standards and best practices are developed by experts who have a deep understanding of the technology and its operational contexts. This expertise ensures that the standards are technically sound and practically applicable. For instance, organizations like the International Organization for Standardisation (ISO) and ASTM International develop comprehensive technical standards that reflect the latest advancements and insights from the industry. By involving those who are most familiar with the technology, non-normative approaches can create more effective and implementable regulations.

Non-normative approaches also enable faster adaptation to technological advancements. The traditional regulatory process can be lengthy, involving multiple stages of drafting, consultation, and approval. In contrast, industry standards and best practices can be developed and disseminated more rapidly. This speed is particularly important in a fast-moving industry like drones, where delays in updating regulations can hinder innovation and create uncertainty for operators and manufacturers. The ability to quickly integrate new technologies and address emerging risks ensures that the regulatory framework remains robust and relevant.

Challenges of Non-Normative Approaches

Despite their advantages, non-normative approaches also present several challenges. One significant challenge is the potential for conflicts of interest. How do conflicts of interest impact the credibility and effectiveness of non-normative approaches in industry-led standards, and what legal mechanisms can be implemented to mitigate these risks while ensuring that safety and public interest are prioritised over profitability? What are the legal implications of the lack of enforceability in non-normative rules, and how can a hybrid regulatory model that integrates non-normative and normative approaches enhance compliance and ensure more consistent adherence to industry standards? What challenges arise from the need for industry cooperation in the development and implementation of non-normative rules, and what legal or policy strategies can foster collaboration and consensus among stakeholders in competitive and emerging industries like commercial drones? How does the maturity and capability of an industry influence the effectiveness of non-normative approaches, and what legal or regulatory support is necessary to build capacity for effective self-regulation in emerging sectors like the commercial drone industry?

Industry-led standards and self-regulation may be influenced by commercial interests, potentially prioritising profitability over safety and public interest. This conflict can undermine the credibility and effectiveness of non-normative rules. To mitigate this risk, it is essential to establish transparent processes for developing standards and to involve independent experts and stakeholders in the process. Ensuring that the development of non-normative rules is driven by a commitment to safety and ethical practices is crucial for maintaining public trust.

Another challenge is the lack of legal enforceability. Unlike normative regulations, which are backed by legal sanctions and enforcement mechanisms, non-normative rules are typically voluntary. This lack of legal enforceability can lead to inconsistent adherence and gaps in compliance. Operators and manufacturers may choose to ignore voluntary standards if they perceive that there are no significant consequences for non-compliance. To address this issue, it is important to integrate non-normative approaches with normative regulations, creating a hybrid model where voluntary standards are supported by legal requirements and enforcement mechanisms. This integration can enhance compliance and ensure that non-normative rules are taken seriously.

The need for industry cooperation is another challenge associated with non-normative approaches. Effective self-regulation and the development of industry standards require a high level of collaboration and consensus among stakeholders. Achieving this cooperation can be difficult, particularly in a competitive industry where different players may have divergent interests and priorities. Building a culture of collaboration and shared responsibility is essential for the success of non-normative approaches. This can be facilitated through industry associations, collaborative research initiatives, and regular stakeholder consultations. Encouraging a cooperative mindset among industry players can lead to more effective and widely accepted standards.

Additionally, the effectiveness of non-normative approaches depends on the maturity and capability of the industry. In well-established industries with strong professional organizations and a history of self-regulation, non-normative rules may be highly effective. However, in emerging industries like commercial drones, where stakeholders are still developing best practices and standards, the reliance on non-normative approaches alone may be premature. It is important to recognize the developmental stage of the industry and to provide the necessary support and guidance to build capacity for effective self-regulation.

Application in Drone Regulation
In the context of drone regulation, non-normative rules play a crucial role in addressing the rapidly evolving technological landscape. How do non-normative rules effectively complement traditional normative regulations in the context of drone regulation, and what are the legal implications of relying on voluntary industry standards for safety and compliance? In what ways can non-normative rules facilitate innovation in drone technology, and what legal frameworks or safeguards are necessary to balance the benefits of a flexible regulatory environment with the need for public safety and accountability? What are the potential legal challenges and advantages of adopting a regulatory approach that integrates both normative and non-normative rules in the governance of emerging technologies like drones, and how can this integration be managed effectively?

The voluntary adoption of safety protocols, operational guidelines, and ethical standards helps ensure that drones are integrated into the airspace safely and responsibly. For instance, industry standards for drone design and manufacturing can promote the development of safer and more reliable drones. Similarly, best practice guidelines for drone operators can enhance operational safety and compliance with airspace regulations. Moreover, non-normative rules can facilitate innovation by providing a flexible regulatory environment that encourages experimentation and the adoption of new technologies. Pilot programs and sandbox initiatives, guided by non-normative principles, allow companies to test new drone applications and business models in a controlled setting. These initiatives can generate valuable insights and inform the development of future regulations.

Essentially, non-normative rules offer a versatile and adaptive approach to regulating disruptive technologies like drones. Their voluntary nature, industry-driven standards, flexibility, self-regulation, collaborative development, and market incentives make them well-suited to address the challenges posed by rapidly evolving technologies. By complementing traditional normative regulations, non-normative rules can create a more effective and responsive regulatory environment, ensuring the safe and innovative use of drone technology in the age of AI.

Role and Function in Regulating Disruptive Technologies

Non-normative rules play a vital role in regulating disruptive technologies, including drones, by providing complementary approaches to traditional normative regulations. How do non-normative rules interact with traditional normative regulations in the context of drone technology, and what are the legal implications of relying on these complementary approaches to effectively regulate disruptive technologies? In what ways do non-normative rules contribute to fostering innovation and encouraging responsible practices in drone operations, and how can these rules address emerging risks that traditional normative regulations may not adequately cover? What is the role of non-normative rules in designing a comprehensive and effective regulatory framework for drone operations, and how can the integration of normative and non-normative rules create a more holistic approach to regulating the rapidly evolving drone industry?

Non-normative rules and traditional normative regulations interact in a symbiotic manner, especially in the context of drone technology, where rapid advancements often outpace the legislative process. Non-normative rules, such as industry guidelines, best practices, and voluntary standards, provide an agile response to emerging technologies, filling the gaps left by slower-moving normative regulations. This interaction allows for a more comprehensive regulatory framework that can adapt quickly to technological changes while maintaining a foundation of legal enforceability.

The legal implications of relying on non-normative rules involve the balance between flexibility and enforceability. On the one hand, non-normative rules offer a mechanism to address new risks and innovations that traditional regulations may not cover. On the other hand, their voluntary nature means that compliance is not legally mandated, which can lead to inconsistencies in enforcement and potential gaps in safety and accountability. For instance, while the International Organization for Standardisation (ISO) may develop industry standards that enhance drone safety, these standards lack the binding force of law unless they are integrated into normative regulations or adopted by regulatory bodies.

This dynamic raises questions about the effectiveness of a regulatory framework that heavily relies on non-normative rules. It suggests the need for a hybrid approach where non-normative rules inform and complement normative regulations, ensuring that the legal framework remains both responsive and robust. Regulatory bodies could incorporate industry standards into legal requirements, thus giving them a normative force while retaining their flexibility to adapt to technological advancements.

Non-normative rules play a crucial role in fostering innovation by providing a flexible framework that encourages experimentation and the development of new technologies. In the rapidly evolving drone industry, where innovations such as AI-driven autonomous drones are emerging, non-normative rules allow companies to explore these new frontiers without the constraints of rigid regulatory processes. This flexibility is particularly important in pilot programs and sandbox initiatives, where companies can test new applications and business models in a controlled environment, gaining insights that can inform future regulations.

These rules also encourage responsible practices by setting industry-led standards that reflect the collective expertise of stakeholders. For instance, guidelines developed by industry bodies may emphasise ethical considerations, such as data privacy and security, in drone operations. By adhering to these standards, companies can demonstrate their commitment to responsible practices, which in turn can enhance their reputation and build public trust.

However, the effectiveness of non-normative rules in addressing emerging risks depends on the industry's willingness to self-regulate and the comprehensiveness of the standards developed. While traditional normative regulations are often slow to adapt to new risks, non-normative rules can be updated more frequently to reflect the latest technological advancements and emerging threats. For example, as drone technology evolves, non-normative guidelines can be revised to incorporate new safety protocols or address issues such as algorithmic bias in AI-driven drones. This adaptability ensures that the regulatory framework remains relevant and capable of addressing new challenges. Non-normative rules are essential in designing a comprehensive and effective regulatory framework for drone operations because they provide the flexibility needed to address the unique challenges posed by disruptive technologies. By complementing normative regulations, non-normative rules ensure that the regulatory framework remains adaptable and responsive to the rapid pace of innovation in the drone industry.

The integration of normative and non-normative rules creates a holistic approach to regulation by combining the enforceability of traditional legal norms with the agility of industry-led standards. This hybrid model allows regulators to establish a baseline of legal requirements while encouraging industry stakeholders to go beyond mere compliance through the adoption of best practices and voluntary guidelines. For example, while normative regulations may set mandatory safety standards for drone operations, non-normative rules can provide more detailed guidance on specific operational practices, such as geo-fencing or the ethical use of AI in drone surveillance.

Moreover, this integrated approach can help harmonise regulatory frameworks across different jurisdictions, promoting greater consistency and reducing the regulatory burden on companies operating in multiple markets. By aligning non-normative rules with international standards, such as those developed by ISO, regulators can facilitate cross-border drone operations while ensuring a common level of safety and compliance.

However, the success of this approach depends on the effective collaboration between regulators and industry stakeholders. Regulatory bodies must ensure that non-normative rules are developed through transparent and inclusive processes, with input from a wide range of stakeholders, including manufacturers, operators, and end-users. This collaboration is key to creating standards that are not only technically sound but also aligned with public interests and ethical considerations. Additionally, regulators should consider mechanisms to incentives compliance with non-normative rules, such as through certification schemes or market incentives, to enhance their effectiveness and integration into the broader regulatory framework.

Industry Organizations that facilitate non-normative dimensions.

The development of non-normative rules and standards for drones is a collaborative process that involves various industry organizations, each playing a crucial role in establishing guidelines to ensure safety, reliability, and ethical use of drone technology. How do industry organizations ensure that the guidelines they establish for drone safety, reliability, and ethical use are comprehensive and equitable across diverse stakeholders, including manufacturers, operators, regulators, and end-users? What legal obligations, if any, do these industry organizations have when developing and disseminating non-normative standards? How do these non-normative standards interact with existing legal frameworks, and what legal challenges could emerge from the reliance on industry-developed guidelines in the absence of formal regulatory oversight? Moreover, what are the implications of stakeholder collaboration in setting these standards, particularly in terms of potential conflicts of interest or biases, and how might these issues be addressed within the legal and regulatory landscape?

Industry organizations ensure that guidelines for drone safety, reliability, and ethical use are comprehensive and equitable by adopting an inclusive and collaborative approach. This typically involves forming committees and working groups that include representatives from various stakeholder groups—such as manufacturers, operators, regulators, and end-users—who bring diverse perspectives and expertise to the table. These organizations often conduct extensive consultations and public comment periods to gather input from a broad range of stakeholders, ensuring that the guidelines are balanced and reflect the interests of all parties involved.

Moreover, these organizations may employ a consensus-based decision-making process, where the final standards are agreed upon by all participants, ensuring that no single group dominates the process. By fostering open dialogue and transparency throughout the development process, industry organizations strive to create guidelines that are not only technically robust but also socially and ethically sound.

While industry organizations are typically not bound by legal obligations in the same way as governmental regulatory bodies, they do have a responsibility to ensure that the standards they develop are accurate, reliable, and do not cause harm. In some jurisdictions, these organizations might be subject to oversight by regulatory agencies, especially when their standards are widely adopted and integrated into legal frameworks.

There may also be a duty of care owed to the stakeholders who rely on these standards, particularly if the standards are marketed or presented as being essential for safety or compliance. If an organization knowingly disseminates a flawed or dangerous standard, it could potentially face legal action, particularly if harm arises from adherence to those standards.

Non-normative standards often function as supplements to existing legal frameworks, filling gaps where formal regulations may not yet exist or providing detailed technical guidance that regulations may not cover. In some cases, these standards may be incorporated by reference into legal frameworks, thereby gaining a quasi-legal status. However, their voluntary nature means that compliance is not mandated by law, which can lead to inconsistencies in adherence and enforcement.

Legal challenges can arise if there is a significant reliance on these standards in the absence of formal regulatory oversight. For example, if a drone operator strictly follows industry guidelines but still causes harm or violates rights, questions may arise regarding the sufficiency of those guidelines. Courts may need to determine whether adherence to non-normative standards constitutes a reasonable defence in cases of liability or negligence. Furthermore, if these guidelines are seen as favouring certain stakeholders over others, there could be legal disputes over their fairness and equity.

Stakeholder collaboration is essential for developing comprehensive and practical standards, but it can also lead to potential conflicts of interest or biases, particularly if dominant industry players exert undue influence. These issues can be addressed through several mechanisms within the legal and regulatory landscape:

1. **Transparency Requirements**: Industry organizations can be required to disclose the composition of their working groups, the nature of their deliberations, and any potential conflicts of interest among participants. This transparency helps ensure that the process is open to scrutiny and that any undue influence is identified and mitigated.

2. **Independent Oversight:** Regulatory bodies or independent third parties can oversee the development of non-normative standards to ensure that they are free from bias and serve the public interest. This oversight can take the form of reviews, audits, or the inclusion of independent experts in the standard-setting process.

3. **Public Participation:** Encouraging or mandating public participation in the standard-setting process can help balance the influence of powerful stakeholders and ensure that a wider range of interests is represented.

4. **Legal Recourse:** Affected parties should have legal recourse if they believe that a standard has been developed in a way that unfairly disadvantages them or fails to meet safety or ethical standards. This could include the ability to challenge the adoption of a standard or to seek remedies if adherence to a biased standard causes harm.

By addressing these potential issues through a combination of transparency, oversight, and legal mechanisms, the risks associated with conflicts of interest and biases in the development of non-normative rules can be mitigated, ensuring that these standards serve the broader public good.

A. International Organization for Standardisation (ISO)

The International Organization for Standardisation (ISO) is one of the most prominent global organizations responsible for developing and publishing international standards across a wide range of industries, including drone technology. ISO's work in the drone sector focuses on creating comprehensive standards that address various aspects of drone operations, safety, and manufacturing. For instance, ISO 21384-1:2019 sets out general specifications and requirements for unmanned aircraft systems (UAS), ensuring that drones are designed and operated safely and efficiently. ISO standards are widely recognised and adopted, providing a common framework that facilitates international trade and regulatory harmonization.

B. ASTM International

ASTM International, formerly known as the American Society for Testing and Materials, is another leading organization that develops technical standards for a broad spectrum of materials, products, systems, and services. In the drone industry, ASTM has established several standards that cover critical aspects such as drone design, performance, maintenance, and operator training. For example, ASTM F3266-18 provides guidelines for the design and construction of small unmanned aircraft systems (sUAS) to ensure safety and reliability. These standards help manufacturers produce high-quality drones and assist regulators in establishing safety protocols.

C. European Union Aviation Safety Agency (EASA)

The European Union Aviation Safety Agency (EASA) is the primary regulatory body for aviation safety in the European Union, including the regulation of unmanned aircraft systems (UAS). While EASA primarily issues legally binding regulations, it also develops non-normative standards and guidelines that support the safe integration of drones into European airspace. EASA's guidelines cover a wide range of topics, from risk assessment frameworks to operational best practices. These guidelines are developed in consultation with industry stakeholders and serve as a reference for both regulators and drone operators across Europe.

D. Federal Aviation Administration (FAA)

The Federal Aviation Administration (FAA) is the United States government agency responsible for regulating all aspects of civil aviation, including drones. In addition to its regulatory functions, the FAA collaborates with industry organizations to develop standards and guidelines that enhance the safety and efficiency of drone operations. The FAA's Unmanned Aircraft Systems Integration Office works closely with stakeholders to establish standards for drone manufacturing, operation, and maintenance. The FAA also promotes initiatives like the UAS Traffic Management (UTM) system, which aims to safely integrate drones into the national airspace through collaborative industry efforts.

E. Joint Authorities for Rule-making on Unmanned Systems (JARUS)

The Joint Authorities for Rule-making on Unmanned Systems (JARUS) is a global group of experts from national aviation authorities and regional aviation safety organizations. JARUS's mission is to recommend a single set of technical, safety, and operational requirements for the certification and safe integration of UAS into airspace and at aerodromes. By developing harmonized standards, JARUS helps to streamline regulatory processes and facilitate international drone operations. JARUS's guidelines are used by regulators worldwide to develop consistent and effective drone regulations.

F. RTCA, Inc.

RTCA, Inc. (formerly known as the Radio Technical Commission for Aeronautics) is a non-profit organization that develops consensus-based standards for aviation, including unmanned aircraft systems. RTCA collaborates with industry experts and regulatory authorities to create standards that address the technical and operational challenges of integrating drones into the airspace. For example, RTCA's DO-362 standard provides guidelines for UAS command and control (C2) data links, ensuring reliable communication between drones and their operators.

G. International Civil Aviation Organization (ICAO)

The International Civil Aviation Organization (ICAO) is a specialised agency of the United Nations responsible for establishing international civil aviation standards and regulations. ICAO's work in the drone sector includes developing standards and recommended practices (SARPs) for the safe and efficient integration of UAS into global airspace. ICAO collaborates with member states and industry stakeholders to address issues such as airworthiness, operator certification, and air traffic management for drones. ICAO's global standards facilitate international cooperation and ensure a harmonized approach to drone regulation.

H. Global UTM Association (GUTMA)

The Global UTM Association (GUTMA) is an international organization that focuses on the development and implementation of unmanned traffic management (UTM) systems. GUTMA brings together stakeholders from the drone industry, air navigation service providers, and regulatory bodies to create standards and protocols for the safe and efficient integration of drones into airspace. By promoting interoperability and collaboration, GUTMA aims to establish a seamless UTM ecosystem that supports the growth of the drone industry.

Industry organizations play a crucial role in developing non-normative standards for drones, providing a framework that ensures safety, reliability, and ethical operation. These standards, while not legally binding, carry significant weight and influence, helping to shape the regulatory landscape and drive industry best practices. Through collaboration and consensus-building, these organizations facilitate the safe integration of drones into the airspace, promoting innovation while addressing the challenges posed by this rapidly evolving technology. By adhering to these standards, stakeholders in the drone industry can enhance their operations, improve safety outcomes, and build public trust in drone technology.

Insights from Non-Normative Technology Management.

Incorporating insights from non-normative technology management can enhance the effectiveness and adaptability of drone regulations. How can the integration of non-normative approaches within the regulatory framework for drones provide regulatory authorities with enhanced insights into technological advancements, risk mitigation strategies, and operational best practices? What are the potential challenges and limitations of incorporating non-normative rules into drone regulation, particularly in terms of achieving industry-wide adoption and maintaining a balance between regulatory flexibility and oversight? In what ways can non-normative technology management contribute to a more nuanced understanding of the complexities associated with drone operations, and how might this influence the development of a collaborative and dynamic regulatory environment? What are the legal implications of relying on non-normative rules in drone regulation, and how might regulatory authorities address the need for consistency and predictability while allowing for innovation and flexibility within the industry?

The integration of non-normative approaches within the regulatory framework for drones can provide regulatory authorities with enhanced insights into technological advancements, risk mitigation strategies, and operational best practices by offering a flexible mechanism for adapting to the rapid evolution of drone technology. Unlike rigid prescriptive regulations, non-normative rules—those that are not legally binding but serve as guidelines—allow regulators to stay informed about industry developments without the need to constantly amend formal regulations. This flexibility is crucial in a field like drone technology, where innovation is constant and swift. By relying on non-normative rules, regulators can encourage creativity and experimentation within the industry, fostering an environment that supports technological growth while maintaining a degree of oversight. Non-normative guidelines also serve as a sandbox for testing risk mitigation strategies before they are codified into law, enabling authorities to assess their effectiveness in real-world applications. Furthermore, by observing industry-led best practices, regulators can gain a deeper understanding of the operational realities of drone technology, leading to more informed and balanced regulatory decisions that align with practical needs.

However, the incorporation of non-normative rules into drone regulation is not without its challenges and limitations. One significant challenge is achieving industry-wide adoption of these voluntary guidelines. Since non-normative rules are not legally enforceable, there is a risk that different operators will adhere to them inconsistently, potentially leading to disparities in safety and operational standards. This inconsistency can undermine the overall safety framework and create an uneven playing field within the industry. Additionally, the voluntary nature of these rules can introduce uncertainty among operators regarding compliance expectations. Without the force of law behind them, companies may struggle to determine the extent to which they should invest in aligning with these guidelines, especially if there is no clear incentive or penalty structure in place. Balancing the flexibility offered by non-normative rules with the need for regulatory oversight is another critical concern. While flexibility is essential for fostering innovation, it can also lead to inadequate oversight and enforcement, increasing the risk of accidents or misuse of drone technology. Regulators must carefully delineate the boundaries between non-normative guidelines and mandatory regulations, ensuring that essential safety and privacy protections are not compromised. Moreover, the lack of legal enforceability of non-normative rules poses a challenge in holding operators accountable in cases of non-compliance, particularly when such non-compliance leads to safety incidents or privacy breaches.

Non-normative technology management can contribute significantly to a more nuanced understanding of the complexities associated with drone operations, ultimately influencing the development of a collaborative and dynamic regulatory environment. By fostering a collaborative learning environment, non-normative approaches encourage continuous dialogue between industry stakeholders and regulators, allowing for the sharing of insights and experiences that can inform regulatory policies. This ongoing exchange of knowledge helps regulators stay attuned to the practical challenges and opportunities within the industry, leading to more responsive and adaptive regulatory measures. Additionally, the use of non-normative rules as a testing ground for emerging technologies and risk management strategies enables regulators to adopt an incremental approach to regulation. This approach allows for the gradual introduction of more formal requirements as technology matures and more data becomes available, preventing the imposition of overly stringent rules that might stifle innovation. The dynamic nature of non-normative rules also allows regulators to quickly adapt to new developments, such as advancements in autonomous flight or AI integration, ensuring that the regulatory framework remains relevant and supportive of technological progress. Furthermore, the engagement of industry stakeholders in the development of non-normative guidelines fosters a sense of ownership and acceptance, leading to a more robust and widely supported regulatory framework that reflects the operational realities of the industry.

The reliance on non-normative rules in drone regulation introduces several legal implications, particularly concerning consistency, predictability, and enforceability. One of the primary legal challenges is ensuring consistency and predictability across the industry. The voluntary nature of non-normative rules can lead to varying interpretations and applications, resulting in discrepancies in safety and operational practices.

To address this, regulators could issue official interpretations or guidance documents that clarify how non-normative rules should be applied, providing a baseline for consistency. Another critical issue is the enforceability of non-normative rules. While these rules are not legally binding, regulatory authorities can incorporate them by reference into formal regulations or licensing requirements. For example, adherence to specific non-normative guidelines could be made a condition for obtaining or renewing a drone operator's license, thus preserving the flexibility of non-normative rules while adding a layer of enforceability. Legal safeguards must also be maintained to ensure that reliance on non-normative rules does not compromise due process or equal protection under the law. This requires transparency in the application of guidelines and a clear distinction between non-normative and legally binding regulations. Finally, maintaining a balance between fostering innovation and ensuring legal compliance is crucial. A tiered regulatory approach could be employed, where non-normative rules apply to lower-risk operations, while higher-risk activities are subject to more stringent, legally binding regulations. This approach allows for innovation in emerging areas of drone technology while ensuring that key safety and privacy concerns are adequately addressed.

The integration of non-normative approaches into the regulatory framework for drones offers numerous benefits, including flexibility, the promotion of innovation, and a deeper understanding of operational realities. However, these benefits must be weighed against potential challenges such as inconsistent adoption, regulatory uncertainty, and

294 / Drone Law 3.0

enforceability issues. By carefully managing these challenges, regulatory authorities can create a dynamic and effective regulatory environment that supports the growth and safe operation of drone technology.

Benefits of Integrating Non-Normative Rules

A. Enhanced Responsiveness to Technological Change: Non-normative rules can adapt swiftly to technological advancements, providing a more agile regulatory environment. As drone technology evolves, non-normative standards and guidelines can be updated regularly to incorporate new safety features, operational capabilities, and risk management practices. This adaptability is crucial in an industry where innovation occurs rapidly, often outpacing the development of formal regulations.

B. Encouraging Industry Innovation: By offering a flexible regulatory framework, non-normative rules encourage innovation within the drone industry. Companies can experiment with new technologies and operational models without the immediate burden of strict legal compliance. Pilot programs and sandbox initiatives, guided by non-normative principles, allow for the safe exploration of new applications and business models. This fosters an environment where innovation can thrive, driving the industry forward.

A. Promoting Best Practices and Standardisation: Non-normative rules often emerge from industry consensus, reflecting the collective expertise and experiences of stakeholders. These rules promote best practices and standardisation, ensuring that drones are designed, manufactured, and operated safely and efficiently. For example, industry-led initiatives on drone safety protocols and operational guidelines can set benchmarks that all stakeholders strive to meet, enhancing overall safety and reliability.

B. Facilitating Collaboration and Stakeholder Engagement: The development and implementation of non-normative rules typically involve collaboration among various stakeholders, including manufacturers, operators, regulators, and users. This collaborative approach ensures that the rules are practical and comprehensive, addressing the needs and concerns of all parties involved. By fostering a sense of ownership and commitment, this approach can lead to higher levels of compliance and cooperation.

C. Risk Mitigation and Proactive Management: Non-normative rules can provide proactive risk mitigation strategies, helping to identify and address potential issues before they become significant problems. For instance, continuous monitoring and feedback mechanisms can detect emerging risks and prompt timely updates to guidelines and best practices. This proactive approach reduces the likelihood of accidents and incidents, enhancing the overall safety of drone operations.

Considerations and Challenges

A. Need for Industry-Wide Adoption: For non-normative rules to be effective, there must be widespread adoption across the industry. Achieving this requires strong leadership and coordination among industry bodies, as well as buy-in from all stakeholders. Without broad acceptance, the effectiveness of these rules may be limited, leading to inconsistencies and gaps in safety and compliance.

B. Balancing Flexibility and Oversight: While flexibility is a key advantage of non-normative rules, it must be balanced with sufficient regulatory oversight to ensure that safety and ethical standards are maintained. Regulators need to establish clear criteria and monitoring mechanisms to evaluate the effectiveness of non-normative rules and ensure they are being followed. This balance is crucial to prevent potential abuses and ensure that the primary objectives of safety and public interest are upheld.

C. Addressing Legal and Ethical Concerns: Non-normative rules must align with existing legal and ethical frameworks to ensure they do not inadvertently undermine fundamental rights and principles. For example, guidelines on data privacy and ethical AI usage in drones must be consistent with broader regulations such as the GDPR and ethical standards for AI. Ensuring this alignment requires ongoing dialogue and cooperation between regulators, industry bodies, and legal experts.

D. Ensuring Transparency and Accountability: To build trust and credibility, the development and implementation of non-normative rules must be transparent and accountable. Stakeholders need to understand how these rules are created, who is responsible for their enforcement, and how compliance is monitored. Clear communication and reporting mechanisms can enhance transparency and ensure that all parties are held accountable for their roles in maintaining safety and compliance.

E. Integrating with Normative Regulations: Non-normative rules should complement, not replace, traditional normative regulations. Integrating these two approaches requires careful coordination to ensure they work together effectively. For example, non-normative guidelines on drone operations can be supported by normative regulations that provide a legal framework for enforcement. This integration ensures a comprehensive regulatory environment that leverages the strengths of both approaches.

Incorporating insights from non-normative technology management into the regulatory framework for drones offers significant benefits, including enhanced responsiveness to technological change, encouragement of innovation, promotion of best practices, facilitation of collaboration, and proactive risk mitigation. However, achieving these benefits requires careful consideration of challenges such as the need for industry-wide adoption, balancing flexibility with oversight, addressing legal and ethical concerns, ensuring transparency and accountability, and integrating with normative regulations. By thoughtfully incorporating non-normative rules, regulatory authorities can strengthen their understanding of drones' complexities and foster a collaborative and dynamic regulatory environment, ensuring the safe and innovative use of drone technology in the age of artificial intelligence.

Evaluating the Sophistication of Current Drone Regulations

Evaluating the sophistication of current drone regulations is a crucial step in identifying areas where non-normative approaches can enhance existing frameworks. This subsection examines the current state of drone regulations and assesses their ability to address the evolving challenges posed by drone technology. It explores the gaps, limitations, and potential areas of improvement in current regulations. Additionally, it discusses how non-normative rules can complement normative regulations by providing flexible mechanisms, adaptive guidelines, and industry-led standards. By evaluating the sophistication of current drone regulations, stakeholders can determine the opportunities for integrating non-normative approaches effectively.

Current State of Drone Regulations

Looking ahead, several future perspectives can guide the development of EU drone regulations to ensure they remain relevant and effective in a rapidly evolving technological landscape. These perspectives include continuous adaptation and flexibility, international cooperation and standardisation, innovation-friendly regulations, enhanced enforcement and compliance, and stakeholder collaboration.

Continuous adaptation and flexibility are essential for drone regulations to keep pace with technological advancements and emerging use cases. Regulations need to be regularly updated and revised, taking into account stakeholder feedback, technological developments, and societal needs. This dynamic approach ensures that the regulatory framework is future-proof and responsive to the evolving challenges of the drone industry.

International cooperation and standardisation are also crucial. By collaborating with international partners and standardisation bodies, the EU can establish common frameworks that promote interoperability of drone operations across borders. Aligning EU regulations with international best practices and fostering global cooperation facilitates the harmonisation of rules, creating a conducive environment for cross-border drone operations. This alignment is essential for the seamless integration of drones into global airspace, supporting international commerce and enhancing safety standards worldwide.

Balancing safety with innovation through innovation-friendly regulations is another key perspective. Encouraging responsible innovation within the regulatory framework is vital for the growth of the drone industry. Implementing regulatory sandboxes and experimentation programmes can provide controlled environments where new technologies and applications can be tested and validated without compromising safety. These initiatives help foster technological advancements while ensuring that safety standards are rigorously maintained.

Enhanced enforcement and compliance mechanisms are paramount to address safety risks and prevent unauthorised operations. Strengthening these mechanisms involves effective monitoring, surveillance, and imposing penalties for non-compliance. Such measures can deter unsafe practices and promote responsible drone operations. Ensuring that drone operators adhere to regulations is critical for maintaining public trust and safety in the skies.

Stakeholder collaboration plays a vital role in developing inclusive and effective regulations. Engaging all relevant stakeholders, including drone operators, manufacturers, industry associations, regulatory bodies, and the public, is essential. Collaborative approaches that foster dialogue, knowledge sharing, and multi-stakeholder participation can lead to well-informed and balanced regulations. These regulations should reflect the diverse perspectives and interests involved, ensuring that the needs and concerns of all parties are addressed.

By addressing these challenges and adopting future-oriented perspectives, the EU can build an effective regulatory framework that enables the safe, sustainable, and innovative use of drones across various sectors. The next chapters of this book will delve further into specific aspects of EU drone regulations, examining their effectiveness and exploring potential strategies for improvement and advancement.

Drone Regulations in the European Union

Current drone regulations within the European Union have established foundational frameworks aimed at ensuring safety, security, and privacy. Key regulations include the EU's Implementing Regulation (EU) 2019/947 on the rules and procedures for the operation of unmanned aircraft and the Delegated Regulation (EU) 2019/945 on unmanned aircraft systems and third-country operators. These regulations provide comprehensive guidelines on various aspects of drone operations, such as certification, operational restrictions, and safety protocols.

The EU regulations mandate that drones meet specific safety standards before they can be operated. This includes requirements for design and manufacturing, as well as operational limitations based on the risk level of the intended use. For instance, drones used in high-risk environments, such as urban areas, must adhere to stricter standards compared to those used in isolated rural areas.

The regulations categorise drone operations into three main risk categories: Open, Specific, and Certified. Each category has distinct requirements and operational limitations, ensuring that higher-risk activities undergo more rigorous scrutiny and safety measures. Current regulations also address privacy concerns, particularly regarding data collection and surveillance. Drones equipped with cameras and sensors must comply with the General Data Protection Regulation (GDPR), ensuring that data collected is processed lawfully, transparently, and for specific purposes. While these regulations provide a robust framework, they are not without limitations. As drone technology and its applications rapidly evolve, several gaps and challenges have become evident.

Gaps and Limitations in Current Regulations

One of the primary limitations of current drone regulations is their relatively slow adaptation to technological advances. The legislative process can be lengthy and inflexible, often lagging behind the pace of innovation in the drone industry. This delay can lead to outdated regulations that fail to address new risks and opportunities presented by emerging technologies, such as AI-driven autonomous drones.

Current regulations are often insufficiently detailed when it comes to the nuances of autonomous drone operations. Autonomous drones, which operate with minimal human intervention, introduce complexities that traditional regulations may not fully anticipate. Issues such as decision-making algorithms, machine learning biases, and the delegation of accountability for autonomous actions require more sophisticated regulatory approaches. Traditional normative regulations are often rigid, with fixed rules that do not easily adapt to changing circumstances or technological innovations. This rigidity can stifle innovation and limit the ability of companies to experiment with new applications and operational models. The lack of flexibility also means that regulatory responses to emerging risks can be slow and reactive rather than proactive.

While efforts have been made to harmonise drone regulations across different jurisdictions, significant differences still exist. This fragmentation can create challenges for international drone operations, where operators must navigate a complex landscape of varying rules and standards. Inconsistent regulations can lead to confusion, increased compliance costs, and potential safety risks.

The development of drone regulations has traditionally been a top-down process, with limited involvement from industry stakeholders. This can result in regulations that are not fully aligned with industry needs or practical realities. Greater collaboration with manufacturers, operators, and other stakeholders could lead to more effective and relevant regulatory frameworks.

Incorporating Non-Normative Technology Management

Incorporating insights from non-normative technology management can offer fresh perspectives on regulating disruptive technologies like commercial drones. Non-normative approaches, which include self-regulation, industry standards, and technological solutions, can provide flexible and adaptive mechanisms that complement traditional normative regulations. This section explores the advantages and challenges of adopting non-normative approaches in the regulatory context, and examines how these concepts can enhance the regulatory framework for commercial drones. By considering the insights from non-normative technology management, this chapter aims to contribute to a more comprehensive and adaptable regulatory approach for commercial drones.

One of the primary advantages of non-normative technology management is its inherent flexibility. Unlike rigid normative regulations, non-normative approaches can quickly adapt to technological advancements and emerging challenges. In the rapidly evolving drone industry, where new technologies and applications are constantly being developed, this flexibility is crucial. For instance, industry standards and best practices can be updated more frequently than formal regulations, ensuring that they remain relevant and effective. This adaptability allows the regulatory framework to keep pace with innovation, fostering a conducive environment for technological advancement.

Self-regulation is a key component of non-normative technology management. It involves industries and companies setting their own standards and practices to ensure safety, compliance, and ethical behaviour. In the context of commercial drones, self-regulation can empower manufacturers and operators to take proactive measures in addressing risks and maintaining high standards. For example, industry-led initiatives like the Dronecode Project, which develops open-source software standards for drones, demonstrate how self-regulation can lead to the creation of robust and widely adopted standards. By taking ownership of regulatory compliance, industry stakeholders can ensure that their practices are aligned with the latest technological developments and safety protocols.

Industry standards also play a vital role in non-normative regulation.

Organizations such as the International Organization for Standardisation (ISO) and ASTM International develop technical standards that provide detailed guidelines for drone design, manufacturing, and operation. These standards, while not legally binding, carry significant weight in the industry and are often adopted voluntarily by manufacturers and operators. The widespread adoption of such standards can lead to a higher baseline of safety and reliability across the industry. Moreover, industry standards can serve as a foundation for formal regulations, providing regulators with a tested and validated framework to build upon.

Technological solutions are another critical aspect of non-normative technology management. By embedding regulatory compliance directly into the technology, it is possible to preclude the practical option of non-compliance. For example, geofencing technology, which prevents drones from entering restricted areas, is a technological solution that enhances compliance with airspace regulations. Similarly, remote identification systems allow authorities to track and identify drones in real time, facilitating enforcement and enhancing security. These technological measures can complement traditional regulatory approaches by providing real-time, automated compliance mechanisms that reduce the burden on both regulators and operators.

However, incorporating non-normative approaches into the regulatory framework also presents several challenges. One significant challenge is ensuring accountability. While self-regulation and industry standards can drive high levels of compliance, there must be mechanisms in place to hold stakeholders accountable for their actions. Without formal oversight, there is a risk that some operators may not adhere to voluntary standards, leading to safety and security concerns. To mitigate this risk, it is essential to establish clear accountability frameworks that delineate the responsibilities of manufacturers, operators, and regulators.

Another challenge is achieving consistency and harmonization across the industry. While non-normative approaches offer flexibility, they can also lead to a fragmented regulatory landscape if different stakeholders adopt varying standards and practices. Ensuring that industry standards are widely adopted and harmonized with formal regulations is crucial for maintaining a coherent regulatory framework. This requires collaboration and coordination among industry bodies, regulatory authorities, and international organizations to develop and promote unified standards that are recognised globally.

Balancing innovation with safety is also a critical consideration when incorporating non-normative approaches. While flexibility and adaptability are essential for fostering innovation, it is important to ensure that safety and security are not compromised. Regulatory sandboxes, which allow for the controlled testing of new technologies and operational models, can provide a balanced approach. These sandboxes enable regulators to monitor and evaluate innovative practices in a controlled environment, ensuring that they meet safety standards before being widely adopted.

Engaging stakeholders is vital for the success of non-normative regulatory approaches. Involving a diverse range of stakeholders, including industry experts, technology developers, regulatory bodies, and end-users, ensures that different perspectives and expertise are considered. This collaborative approach can lead to the development of more effective and balanced standards and practices. Public consultations, industry forums, and collaborative research initiatives are valuable tools for gathering input and building consensus among stakeholders.

Incorporating insights from non-normative technology management can significantly enhance the regulatory framework for commercial drones. The flexibility and adaptability of non-normative approaches, such as self-regulation, industry standards, and technological solutions, provide valuable complements to traditional normative regulations. By addressing the challenges of accountability, consistency, and stakeholder engagement, it is possible to develop a more comprehensive and adaptable regulatory approach that supports the safe and innovative use of drone technology. This balanced regulatory framework can ensure that the commercial drone industry continues to thrive while maintaining high standards of safety, compliance, and ethical behaviour in the age of artificial intelligence.

Balancing Non-Normative and Normative Approaches

To maximize the benefits and mitigate the challenges of non-normative approaches, a balanced regulatory framework that integrates both non-normative and normative elements is essential. This hybrid model can leverage the flexibility and expertise of non-normative rules while ensuring accountability and legal enforceability through normative regulations. For example, non-normative industry standards can inform the development of normative regulations, providing a tested and validated foundation. Conversely, normative regulations can mandate adherence to certain non-normative standards, enhancing compliance and consistency.

Stakeholder engagement and public consultation play a crucial role in developing this balanced approach. Involving a diverse range of stakeholders, including industry experts, regulatory bodies, independent researchers, and the public, ensures that different perspectives are considered and that the regulatory framework is comprehensive and inclusive. Transparent and collaborative processes for developing both normative and non-normative rules can build trust and ensure that the regulations are well-informed and widely accepted.

In essence, adopting non-normative approaches in the regulatory framework for commercial drones offers significant advantages in terms of flexibility, industry expertise, and rapid adaptation to technological advancements. However, it also presents challenges related to conflicts of interest, lack of legal enforceability, and the need for industry cooperation. By balancing non-normative and normative elements and fostering a collaborative regulatory environment, it is possible to develop a sophisticated and effective regulatory approach that supports the safe and innovative use of drone technology in the age of artificial intelligence.

Case Studies on Non-Normative Approaches in Drone Regulation

Case studies provide valuable insights into the practical implementation and effectiveness of non-normative approaches in drone regulation. This section presents case studies that showcase real-world examples of non-normative rules being applied to the regulation of drones. It examines successful initiatives, industry collaborations, and self-regulatory frameworks that have effectively addressed drone-related challenges through non-normative means. By analysing these case studies, regulators and stakeholders can draw lessons and best practices for incorporating non-normative approaches in their own regulatory frameworks.

Case Study 1: The Dronecode Project – Collaborative Open-Source Development

The Dronecode Project, an initiative under the Linux Foundation, exemplifies the transformative potential of collaborative, open-source development in establishing industry standards and best practices for drone technology. By uniting developers, manufacturers, and regulatory bodies, the project focuses on creating and maintaining open-source software for drones, including the widely used PX4 autopilot system.

The primary aim of the Dronecode Project is to accelerate the development of safe and reliable drone technology through community-driven efforts. By pooling resources and expertise, the project develops common software platforms that can be utilised by various drone manufacturers and operators. This collaborative approach ensures that safety features and best practices are inherently integrated into the core technology. The initiative's implementation strategy revolves around fostering an environment where developers from different backgrounds can contribute to and refine the software, leading to robust and reliable outcomes.

The effectiveness of the Dronecode Project is evident in its rapid development and dissemination of high-quality software that adheres to industry standards. The PX4 autopilot system, a notable outcome of the project, includes advanced features such as real-time obstacle detection and avoidance, significantly enhancing operational safety. The open-source nature of the project allows for continuous improvement and adaptation to emerging risks and technological advancements, ensuring that the software remains at the forefront of innovation.

Several key lessons have emerged from the Dronecode Project. First, collaborative development has proven to be highly effective in creating robust and widely accepted standards and best practices. The pooling of diverse expertise and perspectives leads to more comprehensive and resilient solutions. Second, the open-source model offers significant benefits, enabling continuous innovation and a rapid response to new challenges. This approach fosters an environment where improvements and updates can be swiftly implemented, keeping pace with technological advancements and evolving threats. Lastly, engaging a broad community of stakeholders enhances the quality and relevance of regulatory solutions. By involving developers, manufacturers, and regulatory bodies, the project ensures that the software developed is practical, effective, and aligned with industry needs.

The Dronecode Project highlights the power of collaborative, open-source development in advancing drone technology. Through its community-driven approach, the project not only accelerates innovation but also ensures that safety and best practices are deeply embedded in the technology. The lessons learned from this initiative underscore the importance of collaboration, open-source development, and broad stakeholder engagement in creating effective and adaptive regulatory frameworks for emerging technologies.

Case Study 2: FAA's Low Altitude Authorisation and Notification Capability (LAANC) – Integrating Non-Normative Rules

The Federal Aviation Administration (FAA) in the United States has revolutionised the process for commercial drone operations in controlled airspace through the Low Altitude Authorisation and Notification Capability (LAANC). This initiative leverages non-normative rules to enhance regulatory efficiency and safety, marking a significant step forward in the integration of drones into national airspace.

LAANC is the result of a collaborative effort between the FAA and private sector UAS Service Suppliers (USS). It provides drone operators with real-time access to controlled airspace, allowing them to request and receive authorisation almost instantly via automated systems. This innovative system replaces the traditionally lengthy manual approval process, facilitating more efficient and safer drone operations. By automating the authorisation process, LAANC not only speeds up approvals but also ensures that drone operations are better coordinated with air traffic control.

Since its implementation, LAANC has dramatically reduced the time required for airspace authorisation, thereby supporting more efficient commercial drone operations. The system's automation ensures that drone flights are integrated smoothly into the existing air traffic management framework, significantly reducing the risk of airspace conflicts. The use of automated, data-driven processes enhances compliance with airspace regulations and supports the continued growth of the drone industry.

Several key lessons have emerged from the LAANC initiative. First, the use of automated systems can significantly streamline regulatory processes, enhancing both efficiency and compliance. The ability to process authorisations in real-time allows for more dynamic and responsive airspace management. Second, the success of LAANC underscores the value of public-private collaboration. Partnerships between regulatory bodies and private sector companies can yield innovative regulatory solutions that are both effective and efficient.

Finally, the integration of real-time data and technology has proven critical for improving safety and operational coordination. By providing up-to-date information, LAANC helps ensure that drone operations are conducted safely and in harmony with other airspace users.

In conclusion, the LAANC initiative by the FAA demonstrates how leveraging non-normative rules and automated systems can enhance regulatory efficiency and safety in drone operations. The lessons learned from this initiative highlight the importance of automation, public-private collaboration, and real-time data integration in developing innovative regulatory solutions. LAANC serves as a model for how regulatory frameworks can adapt to support the growth and safe integration of emerging technologies like drones.

Case Study 3: DJI's GEO System – Industry-Led Geofencing Standards

DJI, a leading drone manufacturer, has pioneered the Geospatial Environment Online (GEO) system, an industry-led initiative designed to enhance the safety and compliance of drone operations through non-normative rules. The GEO system utilises geofencing technology to prevent drones from entering restricted areas, thereby ensuring safer skies.

The implementation of the GEO system involves using real-time geospatial data to establish virtual boundaries around sensitive locations such as airports, prisons, and critical infrastructure. When a DJI drone approaches one of these geofenced areas, the system automatically restricts its flight capabilities, effectively preventing unauthorized access. This system is continually updated to reflect new no-fly zones and temporary restrictions, ensuring that it remains current and effective.

The effectiveness of the GEO system is evident in its ability to prevent drones from entering restricted airspace, thereby reducing the risk of accidents and security breaches. By proactively managing airspace compliance, the system significantly enhances overall safety and fosters public trust in drone technology. The regular updates ensure that the system adapts to changing security needs, maintaining its relevance and responsiveness.

Several key lessons have emerged from the development and implementation of the GEO system. First, proactive compliance can be effectively enforced through technological solutions, significantly reducing the risk of violations. The GEO system exemplifies how technology can be used to enforce rules without requiring constant human oversight. Second, the importance of dynamic updates is clear. Regularly updating geofencing systems ensures they remain effective and relevant in response to new security challenges and regulatory changes. Finally, the role of industry leadership is crucial. As a leading manufacturer, DJI has driven the development and adoption of non-normative rules, setting standards that others in the industry can follow. DJI's GEO system showcases the potential of industry-led initiatives in enhancing drone safety and compliance. By leveraging real-time geospatial data and dynamic updates, the system ensures drones operate within safe parameters, preventing unauthorized access to restricted areas. The proactive and adaptive nature of the GEO system highlights the benefits of integrating non-normative rules into drone technology, setting a precedent for future innovations in the industry.

Case Study 4: European U-Space Concept – Integrating Non-Normative with Normative Approaches

The European Union's U-Space concept seeks to establish a comprehensive framework for managing drone traffic in low-altitude airspace. By integrating non-normative rules with traditional regulatory approaches, U-Space aims to ensure safe and efficient drone operations across Europe, fostering a harmonious and secure aerial environment.

U-Space is composed of a series of services and procedures designed to support drone operations from takeoff to landing. These services include e-registration, e-identification, geofencing, traffic management, and dynamic airspace management. Leveraging digital technologies and industry standards, U-Space provides real-time information and coordination between drones and manned aircraft. This integration of digital tools enables seamless communication and ensures that all aerial activities are conducted safely and efficiently.

Since its implementation, U-Space has significantly improved the safety and efficiency of drone operations in Europe. By blending non-normative rules with normative regulations, U-Space offers a flexible yet robust framework that can readily adapt to technological advancements and changing operational needs. This comprehensive system facilitates the seamless integration of drones into the airspace, supporting a wide array of commercial applications, from delivery services to agricultural monitoring.

Several key lessons have emerged from the U-Space initiative. First, an integrated approach that combines non-normative and normative regulations creates a comprehensive framework capable of adapting to evolving technological and operational needs. This blended approach ensures that regulatory measures remain relevant and effective. Second, the digital transformation facilitated by U-Space has proven essential. The use of digital technologies enables real-time coordination and management of airspace, significantly enhancing safety and efficiency. Finally, the scalability of the U-Space framework is a crucial advantage.

It supports a diverse range of drone operations and applications, promoting growth and innovation within the industry.

These case studies illustrate the practical implementation and effectiveness of non-normative approaches in regulating drone technology. Through collaborative efforts, automated systems, industry-led initiatives, and integrated frameworks, these examples demonstrate how safety, efficiency, and compliance in drone operations can be significantly enhanced. By analysing these real-world implementations, regulators and stakeholders can draw valuable insights and best practices for incorporating non-normative approaches into their own regulatory frameworks. This ensures that drone technology continues to advance safely and responsibly in the age of artificial intelligence.

Case Study 5: GoDrone in the Netherlands - A Non-Normative Approach in Drone Regulation

The GoDrone platform in the Netherlands represents a pioneering non-normative approach to drone regulation, leveraging technological solutions to enhance safety, compliance, and operational efficiency for both recreational and commercial drone pilots. This case study explores the functionalities and impact of GoDrone, highlighting how its suite of applications provides vital safety information, flight management tools, and streamlined regulatory processes. By examining the integration of GoDrone into the Dutch regulatory framework, this case study illustrates the potential of non-normative approaches in managing the complexities of drone operations.

GoDrone is a comprehensive suite of applications designed to provide flight management tools and critical safety information to drone pilots in the Netherlands. The platform consists of a mobile app (available for iOS and Android) and a companion website, both offering a range of services to enhance the safety and efficiency of drone operations. Key features of GoDrone include an interactive safety map, an operator portal for commercial flight planning and approvals, and a user-friendly mobile app for real-time operational management.

The interactive safety map is a cornerstone of the GoDrone platform, offering drone operators a registration-free tool to identify potential flight risks in the air and on the ground. The map integrates various data sources, including Notices to Airmen (NOTAMs), ground hazards, terrain, and current weather conditions. By providing this information in an easily accessible format, the safety map helps operators make informed decisions about their flight plans, reducing the risk of accidents and enhancing overall safety.

Designed specifically for commercial drone operators, the operator portal offers a sophisticated flight planning system that simplifies the process of submitting and receiving digital clearance to operate in controlled traffic regions (CTR). This feature streamlines regulatory compliance by allowing operators to create mission plans, submit them for approval, and receive digital authorisations through a secure web portal. The efficiency of the operator portal not only saves time but also ensures that all necessary regulatory steps are followed, promoting safe and lawful drone operations.

The GoDrone mobile app extends the platform's capabilities to the field, enabling operators to plan their operations, submit voluntary mission reports, and receive digital clearances on the go. The app provides real-time updates on airspace conditions, ensuring that operators are aware of any changes that might affect their flights. By facilitating seamless communication between operators and regulatory authorities, the mobile app enhances situational awareness and operational safety.

GoDrone's integration into the Dutch regulatory framework exemplifies the potential of non-normative approaches in drone regulation. By providing tools and information that empower operators to proactively comply with safety standards, GoDrone complements traditional regulatory measures and enhances overall compliance. The platform's user-friendly design and accessibility encourage widespread adoption, making it easier for operators to adhere to regulatory requirements.

One of the primary benefits of GoDrone is its ability to enhance safety and compliance through real-time information and streamlined processes. The interactive safety map and real-time updates ensure that operators are well-informed about potential risks, reducing the likelihood of accidents. The operator portal's efficient clearance process ensures that commercial operators can quickly obtain necessary authorisations, minimising delays and ensuring that all flights are conducted within legal parameters.

GoDrone also supports innovation by reducing the regulatory burden on operators and encouraging the adoption of new technologies. The platform's ease of use and comprehensive features enable operators to focus on developing innovative applications for drones without being bogged down by complex regulatory processes. By providing a supportive environment for drone operations, GoDrone fosters a culture of innovation and growth within the industry.

The development and implementation of GoDrone reflect a collaborative approach between regulatory authorities, industry stakeholders, and technology providers. This collaboration ensures that the platform meets the needs of operators while aligning with regulatory objectives. The inclusion of industry feedback in the platform's design and functionality demonstrates a commitment to creating a tool that is both practical and effective.

While GoDrone represents a successful integration of non-normative approaches in drone regulation, it also highlights some challenges and lessons learned. Ensuring widespread adoption of the platform requires ongoing education and outreach efforts to inform operators about its benefits and functionalities. Additionally, maintaining and updating the platform to keep pace with technological advancements and regulatory changes is essential to its continued success.

The GoDrone platform in the Netherlands serves as a model for how non-normative approaches can enhance drone regulation. By providing critical safety information, streamlining regulatory processes, and supporting innovation, GoDrone exemplifies the potential of technological solutions in managing the complexities of drone operations. This case study underscores the importance of a balanced approach that integrates non-normative tools with traditional regulatory measures to create a comprehensive and adaptive regulatory framework. As drone technology continues to evolve, platforms like GoDrone will play an increasingly vital role in ensuring safe, compliant, and innovative drone operations.

Case Study 6: Amazon Prime Air MK30 Drone Delivery - An Industry-Led Obstacle Detection and Avoidance Standard

Amazon Prime Air has revolutionised the logistics industry with its innovative drone delivery system. Central to this advancement is the MK30 drone, which exemplifies how industry-led standards can effectively address regulatory challenges through non-normative approaches. This case study explores the implementation, functionalities, and regulatory implications of the MK30 drone, highlighting its role in setting a benchmark for obstacle detection and avoidance in commercial drone operations.

Amazon's Prime Air initiative aims to deliver packages to customers in under an hour using fully electric drones. The MK30 drone, introduced in 2024, represents a significant leap forward from its predecessors. It is designed to be quieter, smaller, and lighter, with enhanced capabilities to navigate various weather conditions and environments. The MK30 drone employs sophisticated sense-and-avoid technology to autonomously navigate and avoid obstacles in its flight path. This technology allows the drone to detect objects such as other aircraft, buildings, and even dynamic obstacles like people and pets, ensuring safe and reliable operations. The AI-powered system analyses real-time data from multiple sensors, including cameras and radar, to make split-second decisions that prevent collisions.

Equipped with advanced machine learning algorithms, the MK30 can autonomously plan and adjust its flight paths based on real-time environmental data. This capability ensures optimal routing, efficient energy use, and adherence to safety protocols. The drone can adapt to unexpected changes in its environment, such as sudden weather shifts or new obstacles, without human intervention. Safety is a paramount concern for Amazon. The MK30 drone's design incorporates redundancies and fail-safes to mitigate risks. For instance, if the drone encounters an unforeseen obstacle or system error, it can automatically return to a safe landing zone. These features are designed following rigorous aerospace standards to ensure reliability and safety comparable to traditional ground transportation.

The development of the MK30 drone exemplifies a non-normative regulatory approach. Instead of relying solely on prescriptive rules, Amazon collaborates with regulatory bodies like the Federal Aviation Administration (FAA) to set industry standards that promote innovation while ensuring safety. This partnership enables a flexible, adaptive regulatory framework that evolves with technological advancements. Through the use of advanced technology and industry collaboration, the MK30 drone simplifies compliance with airspace regulations. The automated systems integrated into the drone and the broader Prime Air network ensure that all flights are pre-authorised and monitored in real-time, significantly reducing the administrative burden on regulatory agencies and operators. Amazon's proactive engagement with international regulators has facilitated the global deployment of the

MK30 drone. By aligning with international standards and best practices, Amazon ensures that its drone operations are compliant across different jurisdictions, promoting interoperability and smoother cross-border operations.

One of the primary challenges in deploying AI-driven drones is ensuring the transparency and fairness of the algorithms used. Amazon has addressed this by continuously refining its machine learning models and incorporating feedback from diverse data sets to minimise bias and enhance decision-making transparency. Building public trust in drone technology is crucial for widespread adoption. Amazon has invested in public awareness campaigns and community engagement initiatives to educate the public about the safety and benefits of drone delivery, thereby fostering greater acceptance. Scaling the drone delivery system to handle a large volume of deliveries while maintaining safety and reliability presents significant technical challenges. Amazon's iterative approach to design and testing, along with its robust infrastructure, has been key to addressing these challenges and ensuring operational scalability.

The Amazon Prime Air MK30 drone delivery system illustrates the potential of industry-led standards in shaping the future of drone regulation. By leveraging advanced AI technologies for obstacle detection and avoidance, Amazon sets a high benchmark for safety and efficiency in commercial drone operations. This case study underscores the importance of a collaborative, adaptive regulatory framework that embraces technological innovation while ensuring public safety and regulatory compliance. As the drone industry continues to evolve, the lessons learned from the MK30's implementation will be invaluable in guiding future regulatory and technological developments.

XI

11. Evolving the Legal Typology for Commercial Drones

In this chapter, several critical legal questions emerge as we navigate the evolving regulatory landscape for commercial drones across different jurisdictions. These questions set the tone for a deeper exploration of how existing legal frameworks are currently shaping, and ought to shape, the future of drone operations. How effective are current regulatory frameworks, at the national, regional, and international levels, in addressing the unique challenges posed by commercial drones? Are these frameworks adequately aligned to ensure safety, security, and accountability while encouraging innovation? What are the inherent limitations of these regulatory structures in terms of scalability, flexibility, and adaptability, particularly when dealing with rapidly advancing technologies such as AI-driven drones?

Another fundamental question arises concerning the liability structures in place—how are different jurisdictions handling liability in the context of commercial drones? What are the specific legal principles that determine responsibility in cases of accidents, privacy violations, or breaches of airspace regulations? As drones become more autonomous, how should liability be distributed between drone manufacturers, operators, service providers, and software developers? Should strict liability regimes be more widely adopted, or is a negligence-based approach more effective in governing the operation of commercial drones?

Given the cross-border nature of drone technology, how do international legal frameworks reconcile the differences between local, regional, and global regulations? What legal principles should guide the harmonisation of regulations to ensure both operational freedom and strict compliance with safety standards? Looking ahead, what role should non-normative rules—such as industry-led standards and best practices—play alongside traditional legal regulations in shaping the future regulatory environment for commercial drones? Should we evolve toward a hybrid regulatory model, where non-normative guidelines complement formal legal frameworks? Lastly, what direction should legal typologies take as drone technology continues to evolve, and how can future frameworks balance innovation, public safety, privacy, and the global expansion of drone-based services?

Mapping the International Regulatory Landscape

Mapping the regulatory landscape is a crucial step in comprehending the complex network of regulations that govern commercial drones, especially in the context of advancing AI technologies. This subsection delves into the intricate web of legal provisions, guidelines, and standards applicable to the operation and use of commercial drones. It explores the diverse sources of regulation, including aviation authorities, telecommunications agencies, data protection bodies, and other relevant entities. By mapping out the regulatory landscape, this section aims to provide a systematic overview of the legal framework that guides the activities of drone operators, manufacturers, and other stakeholders.

A. Aviation Authorities

European Union Aviation Safety Agency (EASA)
EASA plays a central role in regulating the safety of civil aviation, including the operation of unmanned aircraft systems (UAS) in the

European Union. EASA's regulations, particularly the Implementing Regulation (EU) 2019/947 and the Delegated Regulation (EU) 2019/945, establish comprehensive rules for the certification, operation, and maintenance of drones. These regulations categorise drone operations into Open, Specific, and Certified categories based on risk levels, providing a structured approach to safety and compliance.

Federal Aviation Administration (FAA)
In the United States, the FAA regulates all aspects of civil aviation, including drones. The FAA's Part 107 rules outline the requirements for commercial drone operations, including pilot certification, operational limitations, and safety protocols. The FAA's integration initiatives, such as the LAANC system, further enhance regulatory compliance by streamlining airspace authorisations.

International Civil Aviation Organization (ICAO)
ICAO provides international standards and recommended practices (SARPs) for the safe and efficient integration of UAS into global airspace. These guidelines help harmonise national regulations and facilitate international cooperation in drone operations.

B. Telecommunications Agencies

European Telecommunications Standards Institute (ETSI)
ETSI develops standards for telecommunications and broadcasting, including those applicable to drone communications and control systems. Standards such as EN 303 213 for UAS ground control stations and EN 302 617 for UAS data links ensure reliable and secure communication channels, which are critical for safe drone operations.

Federal Communications Commission (FCC)
In the United States, the FCC regulates the use of radio frequencies for drone communication. The FCC's rules ensure that drones operate within designated frequency bands, minimising interference with other critical communications and enhancing operational safety.

C. Data Protection Bodies

General Data Protection Regulation (GDPR)
The GDPR sets stringent rules for data protection and privacy within the EU. Drones equipped with cameras and sensors must comply with GDPR requirements, ensuring that data collection, processing, and storage are conducted lawfully and transparently. Compliance with GDPR is crucial for maintaining public trust and protecting individual privacy.

National Data Protection Authorities
Each EU member state has a national data protection authority responsible for enforcing GDPR and other data protection laws. These authorities provide guidelines and oversight for the use of drones in data collection activities, ensuring that privacy rights are upheld.

D. Other Relevant Entities

National Aviation Authorities (NAAs)
NAAs in each EU member state, such as the Civil Aviation Authority (CAA) in the UK and the Dirección General de Aviación Civil (DGAC) in Spain, implement and enforce EASA regulations at the national level. NAAs also provide additional guidelines and requirements tailored to specific national contexts, addressing unique operational environments and risks.

Industry Organizations
Organizations like ASTM International, the International Organization for Standardisation (ISO), and the Joint Authorities for Rulemaking on Unmanned Systems (JARUS) develop industry standards that complement regulatory frameworks. These standards provide detailed technical specifications and best practices that enhance safety, reliability, and interoperability of drone systems.

Integrated Regulatory Framework

The regulatory landscape for commercial drones is shaped by a multi-layered framework that seamlessly integrates both normative and non-normative rules. This comprehensive system involves a diverse array of stakeholders, including regulatory bodies, industry organizations, and standard-setting entities, all contributing to a dynamic and adaptive regulatory environment.

Normative regulations serve as the legal backbone for drone operations, establishing mandatory rules and requirements that ensure safety, security, and privacy. These regulations are enforced by governmental and international aviation authorities, providing a structured and standardised approach to drone activity oversight.

In contrast, non-normative rules, such as industry standards and best practices, offer a level of flexibility and adaptability that normative regulations cannot. These rules can evolve more rapidly, keeping pace with technological advancements and emerging risks. Collaborative initiatives and self-regulatory frameworks are pivotal in developing and disseminating these non-normative rules, which complement the more rigid normative regulations by addressing specific industry needs and innovations.

Effective regulation of commercial drones hinges on interagency collaboration among various regulatory and standard-setting bodies. Aviation authorities often work in conjunction with telecommunications agencies to ensure that drones operate safely within designated frequency bands. Additionally, data protection bodies collaborate with drone manufacturers and operators to ensure that privacy laws are adhered to, balancing technological capability with regulatory compliance.

One of the primary challenges in the regulatory landscape is the harmonization of regulations across different jurisdictions. Inconsistent rules and standards can create significant barriers to international drone operations and complicate compliance efforts. Initiatives such as the International Civil Aviation Organization (ICAO)'s Standards and Recommended Practices (SARPs) and the Joint Authorities for Rulemaking on Unmanned Systems (JARUS)'s recommendations aim to harmonise these regulations, facilitating smoother cross-border drone activities.

The rapid pace of technological innovation in the drone industry necessitates the continuous adaptation of regulatory frameworks. Normative regulations must evolve to address new capabilities and risks, particularly those associated with AI-driven autonomous drones. Non-normative rules, with their inherent flexibility, are crucial in this adaptation process, allowing for a more responsive approach to emerging technologies.

Ensuring compliance and effective enforcement mechanisms are essential for the success of any regulatory framework. Technological solutions, such as automated airspace management systems and real-time monitoring tools, can enhance compliance and reduce the burden on regulatory bodies. Collaborative efforts between regulators and industry stakeholders can also promote voluntary compliance through incentives and the adoption of best practices.

Balancing innovation and safety remains a fundamental objective of drone regulations. While strict regulations are necessary to ensure safety and security, overly restrictive rules can stifle innovation. A balanced approach that leverages both normative and non-normative rules can create an environment where innovation thrives without compromising safety. This dual strategy ensures that the regulatory framework is robust yet adaptable, capable of supporting the dynamic nature of the drone industry while maintaining high standards of safety and compliance.

Mapping the regulatory landscape for commercial drones reveals a complex network of regulations and standards governed by diverse entities. This multi-layered framework integrates both normative and non-normative rules, providing a comprehensive and adaptive regulatory environment. By understanding the intricate web of regulations and the roles of various stakeholders, regulators and industry participants can navigate the complexities of drone operations more effectively. Continuous evaluation and adaptation of this regulatory landscape are essential to address the evolving challenges and opportunities presented by drone technology in the age of artificial intelligence.

Identifying Regulatory Gaps and Overlaps

In the rapidly evolving field of commercial drones, identifying regulatory gaps and overlaps is vital to ensure an effective and coherent legal framework. This subsection examines the existing regulations and identifies areas where there may be gaps in coverage or inconsistencies in the legal requirements. It scrutinises the interplay between different regulatory bodies and the potential challenges that arise from overlapping jurisdictions or conflicting rules. By identifying regulatory gaps and overlaps, this section seeks to provide insights into areas where regulatory adjustments or harmonization may be necessary to enhance the clarity and effectiveness of the regulatory regime.

Regulatory Gaps

1. Autonomous Drone Operations

The rapid development of AI and autonomous technologies in drones has outpaced the regulatory frameworks designed to manage them. Current regulations often lack specific provisions for fully autonomous drone operations, which involve decision-making processes without human intervention. Issues such as accountability, decision transparency, and the ethical implications of AI-driven actions are not comprehensively addressed. This gap can lead to uncertainties and risks in the deployment of autonomous drones, requiring regulatory bodies to develop targeted regulations that cover these aspects.

2. Beyond Visual Line of Sight (BVLOS) Operations

BVLOS operations allow drones to operate beyond the direct visual range of the operator, enabling applications such as long-distance deliveries and infrastructure inspections. However, many existing regulations are based on Visual Line of Sight (VLOS) requirements, limiting the potential of BVLOS operations. Most regulatory frameworks need to evolve to establish clear guidelines, safety protocols, and certification processes for BVLOS operations, ensuring they can be conducted safely and efficiently.

3. Data Privacy and Protection

While the GDPR provides a robust framework for data protection, specific guidelines on how these rules apply to drone operations, especially those involving AI-driven data collection and processing, are often lacking. This gap creates uncertainties for drone operators and manufacturers about compliance with data privacy laws. Clear, drone-specific privacy guidelines that address issues such as data minimisation, anonymization, and secure data handling are needed to bridge this regulatory gap.

4. Integration with Manned Aviation

The safe integration of drones into airspace shared with manned aircraft presents significant challenges. Current regulations may not fully address the complexities of air traffic management systems that incorporate both manned and unmanned aircraft. Gaps exist in the coordination mechanisms, communication protocols, and conflict resolution procedures necessary to ensure safe cohabitation of airspace. Developing integrated air traffic management solutions that account for the unique characteristics of drones is essential.

5. Environmental Impact

Environmental regulations specific to drone operations are still underdeveloped. Drones can impact wildlife, noise pollution, and carbon emissions, yet there are limited regulatory guidelines addressing these environmental concerns. Comprehensive environmental impact assessments and the development of regulations that mitigate the ecological footprint of drone operations are necessary to address this gap.

Regulatory Overlaps

1. Jurisdictional Overlaps

Multiple regulatory bodies often have overlapping jurisdiction over different aspects of drone operations. For example, aviation authorities regulate airspace usage, telecommunications agencies manage communication frequencies, and data protection bodies oversee privacy issues. These overlapping jurisdictions can lead to conflicting requirements and regulatory inefficiencies. Coordination mechanisms and inter-agency agreements are essential to harmonise these overlapping regulations and provide a clear, unified framework for drone operators.

2. Conflicting Standards

Different countries and regions may have varying standards and requirements for drone operations, leading to inconsistencies and barriers to international operations. For instance, certification standards, safety protocols, and operational restrictions may differ between the EU, the US, and other regions. These discrepancies can create compliance challenges for manufacturers and operators seeking to operate across borders. Harmonising international standards through organizations like ICAO and JARUS can help mitigate these conflicts.

3. Redundant Regulations
In some cases, multiple regulatory bodies may issue similar but slightly different regulations covering the same aspects of drone operations. This redundancy can lead to confusion and increased compliance costs for operators who must navigate and comply with overlapping requirements. Streamlining regulations and eliminating redundancies can simplify compliance and enhance regulatory efficiency.

4. Divergent Enforcement Practices
Even when regulations are harmonized, differences in enforcement practices across jurisdictions can create challenges. For example, the enforcement of privacy regulations may vary significantly between countries, affecting how drone data collection is monitored and penalised. Establishing consistent enforcement practices and cooperative oversight mechanisms can ensure more uniform application of regulations.

Addressing Regulatory Gaps and Overlaps

1. Developing Comprehensive Frameworks for Autonomous Operations
To address the gap in autonomous drone operations, regulatory bodies should develop comprehensive frameworks that include specific provisions for AI and machine learning technologies. These frameworks should cover areas such as algorithmic accountability, transparency in decision-making processes, and ethical considerations. Establishing clear standards for the certification and operation of autonomous drones can provide legal certainty and enhance safety.

2. Establishing Clear Guidelines for BVLOS Operations
Regulations for BVLOS operations should include detailed safety protocols, risk assessment methodologies, and certification processes. Pilot programs and test corridors can be established to gather data and refine these regulations. International cooperation in developing BVLOS standards can facilitate global harmonization and support cross-border drone operations.

3. Creating Drone-Specific Privacy Guidelines

To address data privacy concerns, regulators should develop drone-specific guidelines that detail how GDPR and other data protection laws apply to drone operations. These guidelines should include best practices for data collection, processing, storage, and anonymization, ensuring that drone operators can comply with privacy requirements effectively.

4. Integrating Drones into Air Traffic Management Systems

Developing integrated air traffic management systems that accommodate both manned and unmanned aircraft is crucial. These systems should include advanced communication protocols, automated conflict resolution tools, and real-time tracking capabilities. Collaboration between aviation authorities, technology developers, and industry stakeholders is essential to create a seamless and safe airspace management system.

5. Implementing Environmental Regulations for Drones

Environmental impact assessments should be incorporated into the regulatory approval process for drone operations. Regulations should address issues such as noise pollution, wildlife disturbances, and carbon emissions. Encouraging the development and adoption of environmentally friendly drone technologies can also mitigate the ecological impact of drone operations.

6. Enhancing Coordination and Harmonization

To address jurisdictional overlaps and conflicting standards, regulatory bodies should enhance coordination and cooperation. Inter-agency agreements, joint task forces, and international collaboration can help harmonise regulations and enforcement practices. Establishing clear channels of communication and coordination can reduce regulatory conflicts and improve overall efficiency.

7. Simplifying Compliance through Streamlined Regulations

Streamlining regulations by eliminating redundancies and consolidating overlapping requirements can simplify compliance for drone operators. Regulatory bodies should review existing regulations to identify areas of overlap and work towards creating a unified, coherent framework. This approach can reduce compliance costs and enhance regulatory clarity.

Identifying regulatory gaps and overlaps is essential for developing an effective and coherent regulatory framework for commercial drones. By addressing these issues, regulators can enhance the clarity, efficiency, and adaptability of drone regulations. Integrating non-normative approaches, fostering international cooperation, and promoting industry collaboration are key strategies for bridging gaps and harmonising regulations. Through these efforts, the regulatory landscape can better support the safe and innovative use of drone technology in the age of artificial intelligence.

Challenges in Enforcing Drone Regulations

Enforcing drone regulations poses unique challenges due to the dynamic nature of drone operations and the complexities involved in monitoring and compliance. This subsection explores the practical challenges faced by regulatory authorities and law enforcement agencies in effectively enforcing drone regulations. It examines issues such as remote identification, airspace management, privacy concerns, and enforcement mechanisms. By analysing the challenges in enforcing drone regulations, this section aims to highlight areas that require attention and innovative solutions to ensure compliance and promote safe and responsible drone operations.

1. Remote Identification

Technological Limitations
Remote identification (Remote ID) is crucial for tracking and managing drone operations. It allows authorities to identify drones and their operators in real time. However, technological limitations such as signal interference, battery life constraints, and limited range of transmission can hinder the effectiveness of Remote ID systems. These limitations can make it difficult for enforcement agencies to consistently monitor drone activities, particularly in densely populated or rural areas.

Standardisation Issues
The lack of standardised Remote ID protocols across different jurisdictions complicates enforcement efforts. While some regions may have well-defined standards, others may not, leading to inconsistencies

in how Remote ID is implemented and enforced. This lack of uniformity can create challenges for international drone operations and cross-border enforcement.

Privacy Concerns
Implementing Remote ID systems raises significant privacy concerns. Operators may be reluctant to comply with Remote ID requirements if they perceive that their privacy is being compromised. Balancing the need for effective identification with the protection of personal data is a critical challenge for regulators.

2. Airspace Management

Integration with Manned Aviation
One of the biggest challenges in enforcing drone regulations is integrating drones safely into airspace shared with manned aircraft. Traditional air traffic control systems are not designed to handle the high volume and diverse nature of drone traffic. Developing advanced Unmanned Traffic Management (UTM) systems that can seamlessly integrate with existing air traffic control is essential but complex.

Dynamic Airspace Restrictions
Airspace restrictions can change rapidly due to various factors such as emergency response activities, temporary no-fly zones, or large public events. Enforcing these dynamic restrictions requires real-time communication and coordination between regulatory authorities and drone operators. This need for real-time adaptability poses significant enforcement challenges.

Geofencing Effectiveness
While geofencing technology can help prevent drones from entering restricted areas, its effectiveness is not absolute. Geofencing systems rely on GPS signals, which can be spoofed or jammed. Additionally, not all drones are equipped with geofencing capabilities, and those that are may not always have the most up-to-date restriction data.

3. Privacy Concerns

Surveillance and Data Collection
Drones equipped with cameras and sensors have the potential to collect vast amounts of data, raising concerns about surveillance and data privacy. Enforcing regulations that protect individuals' privacy while allowing legitimate data collection for commercial and recreational purposes is a delicate balance.

Compliance with Data Protection Laws
Ensuring that drone operators comply with data protection laws such as GDPR involves monitoring how data is collected, stored, and used. This compliance monitoring can be resource-intensive and technically challenging, especially given the high volume of data that drones can generate.

Public Awareness and Trust
Public mistrust of drones due to privacy concerns can hinder compliance and enforcement efforts. Building public awareness about the benefits of drones and the measures in place to protect privacy is crucial for gaining public support and ensuring that privacy regulations are respected.

4. Enforcement Mechanisms

Resource Constraints
Regulatory authorities and law enforcement agencies often face resource constraints that limit their ability to monitor and enforce drone regulations effectively. These constraints include limited personnel, technology, and funding. Enhancing enforcement capabilities requires significant investment in training, technology, and infrastructure.

Legal Framework and Penalties
The legal framework for enforcing drone regulations must be clear and robust, with well-defined penalties for non-compliance. However, legal processes can be slow, and the penalties may not always be sufficient deterrents. Streamlining legal processes and ensuring that penalties are proportionate and effective is necessary for strong enforcement.

326 / Drone Law 3.0

International Coordination

Drone operations are not confined by national borders, necessitating international coordination for effective enforcement. Differences in regulatory frameworks, enforcement practices, and legal standards across countries can create challenges for cross-border enforcement. International agreements and cooperation are essential to address these challenges.

Technological Solutions

Leveraging technological solutions such as automated monitoring systems, AI-driven analytics, and blockchain for tracking compliance can enhance enforcement mechanisms. However, implementing these technologies comes with its own set of challenges, including integration with existing systems, data security, and ensuring accuracy.

Innovative Solutions and Future Directions

1. Advanced Remote ID Systems

Developing more robust and reliable Remote ID systems that can operate under various conditions and across different jurisdictions is essential. These systems should ensure data security and privacy while providing accurate real-time identification of drones and their operators.

2. Integrated UTM Systems

Creating integrated UTM systems that can handle both manned and unmanned aircraft traffic will improve airspace management and enforcement. These systems should incorporate real-time data sharing, automated conflict resolution, and dynamic airspace management capabilities.

3. Public Engagement and Education

Engaging with the public to raise awareness about drone regulations and the benefits of compliant drone operations can build trust and encourage voluntary compliance. Educational campaigns and transparency about data protection measures can address privacy concerns and foster a positive perception of drone technology.

4. Cross-Border Collaboration

Strengthening international collaboration through treaties, joint task forces, and shared standards can enhance cross-border enforcement. Harmonising regulations and enforcement practices will facilitate smoother international drone operations and improve overall compliance.

5. Leveraging AI and Machine Learning

AI and machine learning can play a significant role in enhancing enforcement mechanisms. AI-driven analytics can monitor drone traffic, detect anomalies, and predict potential non-compliance, allowing for proactive enforcement. Machine learning algorithms can also improve the accuracy and efficiency of UTM systems.

6. Legal and Regulatory Reforms

Updating legal frameworks to streamline enforcement processes and introduce proportionate penalties for non-compliance is crucial. Legal reforms should also consider the evolving nature of drone technology and provide the flexibility needed to adapt to future developments.

Enforcing drone regulations involves navigating a complex landscape of technological, legal, and social challenges. Remote identification, airspace management, privacy concerns, and resource constraints are among the key issues that regulatory authorities and law enforcement agencies must address. By identifying these challenges and exploring innovative solutions, stakeholders can enhance the effectiveness of enforcement mechanisms, ensuring that drone operations are safe, compliant, and beneficial to society. Effective enforcement not only promotes safe and responsible drone operations but also supports the continued growth and innovation of the drone industry in the age of artificial intelligence.

Implications for Value Chain Actors

The regulatory and liability regime for commercial drones has significant implications for the various actors involved in the value chain. Understanding the responsibilities and liabilities of drone operators, manufacturers, suppliers, and other stakeholders is crucial to navigating the complex legal landscape. This section delves into the legal obligations, risk allocation, and potential liabilities that arise at each stage of the drone value chain, as well as the broader impacts on the insurance industry, market players, and innovation in the drone sector.

Drone operators are at the forefront of regulatory compliance and liability. They must ensure that their operations adhere to a myriad of regulations concerning safety, privacy, airspace usage, and data protection. Operators are required to obtain necessary certifications, conduct risk assessments, and maintain operational logs. Failure to comply with these regulations can result in significant penalties, including fines and operational bans. Moreover, in the event of an accident or regulatory breach, operators may be held liable for damages, which can encompass physical injuries, property damage, and breaches of privacy. The introduction of autonomous drones adds another layer of complexity, as operators must ensure that the AI systems used in these drones are reliable and comply with relevant standards.

Manufacturers play a critical role in ensuring the safety and reliability of drones. They are responsible for designing and producing drones that meet regulatory standards and can withstand operational risks. This includes integrating fail-safes, geofencing capabilities, and reliable communication systems. Manufacturers must also stay abreast of evolving regulations and update their products accordingly. Liability for manufacturers can arise from product defects, failures in design, or inadequate safety features. In such cases, manufacturers could face significant legal and financial repercussions, including recalls, lawsuits, and reputational damage. As drone technology evolves, manufacturers must also address the challenges associated with AI integration, ensuring that their AI systems are transparent, accountable, and free from bias.

Suppliers, who provide components and systems to drone manufacturers, also bear significant responsibilities. They must ensure that their products meet industry standards and contribute to the overall safety and functionality of drones. Suppliers can be held liable for defects in the components they provide, which can cascade through the value chain and impact both manufacturers and operators. Effective quality control, rigorous testing, and compliance with regulatory standards are essential to mitigate these risks. Suppliers must also collaborate closely with manufacturers to ensure seamless integration of their components into the final drone systems.

The regulatory framework for drones has profound implications for the insurance industry. Insurers must develop policies that accurately reflect the risks associated with drone operations, including potential accidents, regulatory breaches, and liability claims. As drone technology and its applications continue to evolve, insurers must adapt their products to cover new risks, such as those associated with autonomous operations and AI-driven decision-making. This requires a deep understanding of both the technological landscape and the regulatory environment. Insurers play a crucial role in risk mitigation by incentivising compliance with safety standards and best practices through premium discounts and other incentives.

Market players, including commercial enterprises that use drones for various applications, must navigate a complex regulatory landscape while pursuing innovation. Companies in sectors such as agriculture, logistics, real estate, and media are increasingly leveraging drones to enhance their operations. However, they must ensure that their use of drones complies with all relevant regulations to avoid legal and financial penalties. The regulatory framework can either facilitate or hinder innovation, depending on how well it balances safety with flexibility. Supportive regulations that encourage pilot programs, provide clear guidelines, and foster industry collaboration can stimulate innovation and growth in the drone sector.

Innovation in the drone industry is closely tied to the regulatory environment. A supportive regulatory framework that encourages experimentation and adaptation can drive technological advancements and the development of new applications. Conversely, overly restrictive regulations can stifle innovation and limit the potential of drone technology. Non-normative rules, such as industry standards and best practices, play a vital role in fostering innovation by providing a flexible regulatory environment that adapts to emerging technologies and operational models. Collaboration between regulators, industry stakeholders, and technology developers is essential to create a dynamic regulatory framework that supports both safety and innovation.

The legal landscape for commercial drones shapes the roles, responsibilities, and potential liabilities of all value chain actors. By understanding and addressing the implications of the regulatory and liability regime, these actors can navigate the complexities of the drone industry more effectively. This comprehensive approach not only ensures compliance and risk management but also promotes the sustainable growth and innovation of the commercial drone sector.

Responsibilities of Drone Operators, Manufacturers, and Suppliers

Within the complex ecosystem of commercial drone operations, this subsection delves into the specific responsibilities and liabilities borne by drone operators, manufacturers, and suppliers. It examines the legal obligations that these actors must fulfil in terms of safety, compliance, and accountability. The section analyses the potential liabilities they may face in the event of accidents, breaches of regulations, or defects in drone design or manufacturing. By clarifying the responsibilities and liabilities of value chain actors, this subsection aims to provide guidance for navigating the legal landscape and fostering responsible practices in the commercial drone industry.

Responsibilities and Liabilities of Drone Operators

1. Legal Compliance

Drone operators are required to comply with a broad spectrum of regulations that govern the operation of unmanned aircraft systems (UAS). This includes obtaining necessary certifications, adhering to operational limits such as altitude and flight zones, and ensuring that their drones are registered with relevant authorities. Operators must also comply with specific rules regarding the operation of drones in various environments, including urban areas, near airports, and over private property. Failure to comply with these regulations can result in severe penalties, including fines, suspension of operating licenses, and legal action.

2. Safety and Risk Management

Operators are responsible for ensuring the safety of their drone operations. This involves conducting pre-flight checks, maintaining the drone in good working condition, and implementing risk management strategies to prevent accidents. Operators must be proficient in the use of their drones and aware of the risks associated with their specific applications, such as aerial photography, surveying, or delivery services. In the event of an accident, operators can be held liable for damages, including physical injuries, property damage, and disruptions caused by their drone operations. Liability may be compounded if the operator is found to have been negligent or to have violated safety protocols.

3. Data Protection and Privacy

Drone operators must ensure that their activities comply with data protection laws, such as the General Data Protection Regulation (GDPR) in the European Union. This includes obtaining necessary permissions for data collection, implementing measures to protect the data collected by drones, and ensuring transparency in data usage. Operators must also be mindful of privacy concerns, avoiding unnecessary intrusions into private spaces and ensuring that data is anonymised or minimised where possible. Non-compliance with data protection regulations can result in substantial fines and legal actions, as well as damage to the operator's reputation.

Responsibilities and Liabilities of Drone Manufacturers

1. Design and Production Standards

Manufacturers are responsible for ensuring that their drones meet rigorous design and production standards. This includes compliance with safety regulations, such as those set by aviation authorities like EASA or the FAA. Drones must be designed to be safe and reliable, incorporating features such as fail-safes, geofencing, and secure communication systems. Manufacturers must also conduct thorough testing and quality control to ensure that their products meet these standards. In the event of a design or manufacturing defect, manufacturers can be held liable for damages resulting from the defect, including recalls, legal claims, and reputational harm.

2. Compliance with Certification Requirements

Manufacturers must ensure that their drones are certified by relevant regulatory bodies before they are marketed and sold. This certification process typically involves demonstrating that the drone meets specific safety, performance, and operational criteria. Manufacturers must provide detailed documentation and evidence to support their certification applications. Failure to obtain the necessary certifications can result in legal sanctions, market restrictions, and loss of consumer trust.

3. Product Liability

Manufacturers face significant liability risks related to product defects. If a drone is found to be defective in design, manufacturing, or labelling, the manufacturer can be held liable for any resulting injuries or damages. This liability can extend to compensating affected parties, covering the costs of recalls, and addressing any regulatory penalties. To mitigate these risks, manufacturers must implement robust quality assurance processes and maintain comprehensive records of their design and production activities.

Responsibilities and Liabilities of Drone Suppliers

1. Component Quality and Compliance

Suppliers who provide components and systems to drone manufacturers bear the responsibility of ensuring that their products meet industry standards and regulatory requirements. This includes providing components that are safe, reliable, and compatible with the final drone systems. Suppliers must conduct thorough testing and quality control to prevent defects and ensure compliance with relevant standards. Liability for suppliers arises if a defect in their component contributes to a failure or accident involving the drone. In such cases, suppliers can be held jointly liable with the manufacturer for damages and regulatory breaches.

2. Collaboration with Manufacturers

Effective collaboration between suppliers and manufacturers is crucial to ensuring the overall safety and performance of drones. Suppliers must work closely with manufacturers to understand the specific requirements of the drones they are supplying components for. This collaboration includes sharing technical information, participating in joint testing and validation processes, and addressing any compatibility issues. By fostering strong partnerships with manufacturers, suppliers can help ensure that their components contribute to the safe and reliable operation of the final drone products.

3. Supply Chain Accountability

Suppliers must also manage their supply chains to ensure that all materials and components sourced from third parties meet the necessary quality and regulatory standards. This involves conducting due diligence on suppliers, implementing strict procurement processes, and maintaining transparency throughout the supply chain. Supply chain disruptions or non-compliance can lead to delays, increased costs, and liability issues for suppliers. Effective supply chain management is essential to mitigating these risks and ensuring the integrity of the final drone products.

Market Players and Innovation

Market players, including commercial enterprises that use drones for various applications, must navigate a complex regulatory landscape while pursuing innovation. Companies in sectors such as agriculture, logistics, real estate, and media are increasingly leveraging drones to enhance their operations. However, they must ensure that their use of drones complies with all relevant regulations to avoid legal and financial penalties. The regulatory framework can either facilitate or hinder innovation, depending on how well it balances safety with flexibility.

Supportive regulations that encourage pilot programs, provide clear guidelines, and foster industry collaboration can stimulate innovation and growth in the drone sector. Non-normative rules, such as industry standards and best practices, play a vital role in fostering innovation by providing a flexible regulatory environment that adapts to emerging technologies and operational models. Collaboration between regulators, industry stakeholders, and technology developers is essential to create a dynamic regulatory framework that supports both safety and innovation.

Understanding the responsibilities and liabilities of drone operators, manufacturers, and suppliers is essential for navigating the complex legal landscape of the commercial drone industry. By clarifying the legal obligations and potential liabilities at each stage of the drone value chain, stakeholders can foster responsible practices and ensure compliance with regulatory requirements. The impact of the regulatory framework on the insurance industry, market players, and innovation highlights the need for a balanced approach that promotes safety, accountability, and growth in the commercial drone sector. Through collaboration and adherence to both normative and non-normative rules, the industry can continue to advance while maintaining the highest standards of safety and responsibility.

Role of the Insurance Industry

The insurance industry plays a critical role in the commercial drone sector by providing coverage for various risks associated with drone operations. Insurers must develop specialised policies that address the unique risks of drone operations, including accidents, regulatory breaches, and liability claims. As drone technology and applications evolve, insurers must continuously adapt their products to cover new risks, such as those associated with autonomous operations and AI-driven decision-making. This requires a deep understanding of both the technological landscape and the regulatory environment.

Insurance policies for drone operations typically cover physical damage to the drone, third-party liability for injuries or property damage, and coverage for regulatory fines and legal expenses. Insurers also play a role in risk mitigation by incentivising compliance with safety standards and best practices through premium discounts and other incentives. By providing comprehensive coverage and promoting responsible practices, the insurance industry helps to mitigate the financial impact of risks and support the sustainable growth of the commercial drone sector.

The regulatory and liability framework for commercial drones significantly impacts the insurance industry. As drone technology continues to evolve and its applications expand, insurance companies must adapt to the unique risks and regulatory requirements associated with drone operations. This subsection explores the implications of drone regulations on insurance coverage, risk assessment, and underwriting practices. It examines the challenges faced by insurance companies in assessing and pricing drone-related risks and the development of specialised drone insurance products. By analysing the impact on the insurance industry, this subsection aims to shed light on the evolving landscape of drone insurance and the measures taken to manage risks and promote insurability in the commercial drone sector.

Drone regulations, particularly those related to safety, privacy, and operational limits, directly influence the type and scope of insurance coverage available for drone operators. Regulatory compliance is often a prerequisite for obtaining insurance coverage, as insurers need to ensure that the drones they underwrite are operated within the bounds of the law and adhere to established safety standards. This linkage between regulatory compliance and insurance coverage ensures that only responsible and compliant operators can access insurance products, thereby promoting safer operations industry-wide.

Risk Assessment and Underwriting

1. Complexity of Risk Assessment
Assessing the risks associated with drone operations is inherently complex due to the diverse applications of drones and the varying levels

of risk they present. For instance, a drone used for agricultural monitoring may have different risk factors compared to one used for urban deliveries or industrial inspections. Insurers must consider multiple variables, including the type of drone, its operational environment, the experience and track record of the operator, and the specific regulatory requirements applicable to the operation.

2. Integration of Technological Factors

The integration of AI and autonomous technologies in drones adds another layer of complexity to risk assessment. Insurers need to evaluate the reliability and safety of AI systems, considering factors such as algorithm transparency, decision-making accuracy, and potential biases. This requires a deep understanding of both the technology and the regulatory standards governing its use. Insurers may need to collaborate with technology experts to develop robust assessment criteria for AI-driven drones.

3. Data-Driven Underwriting

Advancements in data analytics and machine learning enable insurers to adopt data-driven underwriting practices. By analysing historical data on drone operations, incidents, and regulatory compliance, insurers can identify patterns and trends that inform risk assessment and pricing. This data-driven approach allows for more accurate and personalised insurance products that reflect the specific risk profile of each drone operator.

Challenges in Assessing and Pricing Drone-Related Risks

1. Lack of Historical Data

One of the significant challenges in the drone insurance market is the relative lack of historical data on drone-related incidents and claims. Unlike traditional aviation or automotive sectors, where extensive historical data exists, the commercial drone industry is relatively new. This scarcity of data makes it difficult for insurers to develop reliable actuarial models and accurately price drone insurance products.

2. Rapid Technological Advancements

The rapid pace of technological advancements in the drone industry presents a moving target for insurers. New technologies and applications emerge regularly, each with its own set of risks and regulatory considerations. Insurers must continuously update their risk models and underwriting criteria to keep pace with these developments. This dynamic environment requires agility and a proactive approach to risk management.

3. Regulatory Variability

Variability in drone regulations across different jurisdictions adds complexity to the insurance landscape. Insurers must navigate a patchwork of regulations that may differ in terms of safety requirements, operational limits, and liability frameworks. This variability can complicate the process of underwriting international drone operations and require insurers to tailor their products to meet local regulatory standards.

Development of Specialised Drone Insurance Products

1. Comprehensive Coverage Options

To address the unique risks associated with drone operations, insurers have developed specialised drone insurance products that provide comprehensive coverage options. These products typically include liability coverage for third-party injuries and property damage, hull coverage for physical damage to the drone, and coverage for regulatory fines and legal expenses. By offering tailored insurance solutions, insurers can meet the specific needs of different drone operators and promote insurability in the sector.

2. On-Demand and Usage-Based Insurance

Innovative insurance models, such as on-demand and usage-based insurance, have emerged in response to the flexibility required by drone operators. On-demand insurance allows operators to purchase coverage for specific periods or flights, providing cost-effective solutions for occasional or project-based operations. Usage-based insurance, on the other hand, adjusts premiums based on actual flight hours and operational data, offering a more personalised and equitable pricing structure.

3. Coverage for Autonomous Operations

As autonomous drone operations become more prevalent, insurers are developing coverage options that address the specific risks associated with AI-driven decision-making. These policies may include provisions for algorithmic failures, cyber threats, and compliance with AI safety standards. By addressing the unique risks of autonomous operations, insurers can support the safe integration of AI technologies into the drone industry.

Measures to Manage Risks and Promote Insurability

1. Incentivising Compliance

Insurers play a crucial role in promoting regulatory compliance and safety standards within the drone industry. By offering premium discounts and other incentives for operators who adhere to best practices and regulatory requirements, insurers can encourage responsible behaviour and reduce overall risk. This incentivization aligns the interests of insurers, operators, and regulators, fostering a safer operational environment.

2. Collaboration with Regulators and Industry Stakeholders

Effective risk management in the drone insurance sector requires collaboration with regulators, industry associations, and technology developers. Insurers can participate in industry forums, contribute to the development of standards, and engage in dialogue with regulatory bodies to stay informed about emerging risks and regulatory changes. This collaboration helps insurers align their products with the evolving legal landscape and better serve their clients.

3. Investment in Research and Development

Investing in research and development is essential for insurers to stay ahead of the curve in the rapidly evolving drone industry. By funding studies on drone safety, operational risks, and technological advancements, insurers can enhance their understanding of the sector and develop more accurate risk models. R&D efforts also support the continuous improvement of insurance products and risk management practices.

4. Education and Training Programs

Providing education and training programs for drone operators is another effective measure to manage risks and promote insurability. Insurers can offer workshops, certification courses, and online resources that cover topics such as safety protocols, regulatory compliance, and best practices for drone operations. These programs help operators enhance their skills and knowledge, reducing the likelihood of accidents and regulatory breaches.

The regulatory and liability framework for commercial drones has a profound impact on the insurance industry. Insurers must navigate the complexities of regulatory compliance, risk assessment, and underwriting in a dynamic and rapidly evolving sector. By developing specialised insurance products, adopting data-driven underwriting practices, and collaborating with industry stakeholders, insurers can effectively manage risks and promote insurability in the commercial drone sector. The evolving landscape of drone insurance reflects the broader trends in the industry, highlighting the need for agility, innovation, and continuous adaptation to ensure the safe and responsible integration of drone technology in the age of artificial intelligence.

Implications of Drone Regulations on Insurance Coverage

Drone regulations, particularly those related to safety, privacy, and operational limits, directly influence the type and scope of insurance coverage available for drone operators. Regulatory compliance is often a prerequisite for obtaining insurance coverage, as insurers need to ensure that the drones they underwrite are operated within the bounds of the law and adhere to established safety standards. This linkage between regulatory compliance and insurance coverage ensures that only responsible and compliant operators can access insurance products, thereby promoting safer operations industry-wide.

Evolving the Legal Typology in Context

Now, reverting to our initial question, how effective are current regulatory frameworks at the national, regional, and international levels in addressing the unique challenges posed by commercial drones?

Regulatory frameworks for commercial drones are still in a state of flux, reflecting both the rapid development of drone technology and the regulatory uncertainty that comes with it. On the national level, many countries like the U.S., through the FAA's Part 107 rules, and the EU's European Union Aviation Safety Agency (EASA) regulations, offer comprehensive guidelines for drone operations. However, these frameworks tend to lag behind technological advancements, particularly with the growing sophistication of AI-powered drones and Beyond Visual Line of Sight (BVLOS) operations.

One key challenge in these frameworks is their inflexibility in accommodating the growing diversity of drone applications. Most regulations were initially designed with simple, remotely-piloted drones in mind, which do not adequately address the complexities introduced by AI-powered autonomous drones capable of independent decision-making. As such, there is a pressing need for regulatory reform that accounts for the dynamic and evolving capabilities of drones, especially in high-risk operations like package deliveries, AirTaxi's or urban air mobility.

At the international level, regulatory frameworks struggle with harmonisation across borders. The International Civil Aviation Organization (ICAO) has yet to develop comprehensive standards for drones, which leads to fragmented rules across jurisdictions. This fragmentation complicates cross-border drone operations, posing challenges for operators who wish to engage in international drone commerce or surveillance. More work is needed to align regional rules with international standards to facilitate the global expansion of drone technology.

To address whether these frameworks are adequately aligned to ensure safety, security, and accountability while encouraging innovation? It can be viewed that while current frameworks do incorporate measures to enhance safety, security, and accountability, they often do so at the expense of innovation. Many drone regulations, especially those involving certification processes or risk assessments (e.g., the EU's Specific Operations Risk Assessment - SORA), place heavy administrative burdens on operators. These rigid rules can stifle innovation by making it difficult for companies to test and deploy new drone technologies without significant regulatory delays.

To strike a balance, regulatory frameworks must evolve to adopt a more flexible, sophisticated, risk-based approach. For example, regulatory sandboxes—where companies can experiment with new technologies in a controlled environment—could be a key solution. These frameworks can facilitate rapid innovation while maintaining a strong emphasis on safety and accountability. However, balancing these interests requires a careful calibration of regulations that are sufficiently stringent to ensure safety but flexible enough to promote technological growth.

On whether there are inherent limitations of these regulatory structures in terms of scalability, flexibility, and adaptability, particularly when dealing with AI-driven drones?

Addressing how different jurisdictions handle liability in the context of commercial drones and what legal principles should guide the harmonisation of regulations to ensure both operational freedom and strict compliance with safety standards? Liability in drone operations is currently handled through a mix of traditional tort law principles—such as negligence and strict liability—and specific drone regulations. In the EU, strict liability applies to manufacturers of defective drones under the Product Liability Directive. This means that drone manufacturers can be held liable for damage caused by their products even without proof of fault, provided the product is found to be defective.

In the U.S., liability tends to follow a negligence-based model, where drone operators must prove that the damage caused was due to a breach of duty, such as flying outside permitted areas or violating safety protocols. Both models have their strengths, but the strict liability approach is often more suitable for handling incidents involving autonomous drones, where assigning fault to a human operator may be difficult. Harmonising these principles across jurisdictions will require the development of international standards that balance operational freedom with safety. One way to achieve this is through the establishment of international liability conventions similar to the Montreal Convention for manned aircraft. Such a framework could provide consistent rules on liability and compensation for cross-border drone operations.

Should we evolve toward a hybrid regulatory model, where non-normative guidelines complement formal legal frameworks? Yes, a hybrid regulatory model that combines both normative (legal) and non-normative (voluntary or industry-driven) guidelines is not only desirable but likely essential for the future of drone regulation. Non-normative guidelines, such as those developed by industry groups (e.g., ISO standards or ASTM International standards), offer the flexibility and responsiveness that formal legal frameworks often lack. They allow for quicker adaptation to technological changes and emerging risks, such as those introduced by AI and machine learning.

However, non-normative rules alone may not be enough to ensure compliance and accountability. Integrating them into the formal regulatory structure, possibly through voluntary compliance schemes that incentivise adherence to industry standards, can create a more adaptive and comprehensive system. For example, companies that adhere to non-normative standards could be granted expedited certification processes or be allowed to operate in less restricted airspaces.

Finally, in what direction should legal typologies take as drone technology continues to evolve? As drone technology evolves, legal typologies must evolve toward a more dynamic, risk-based, and sector-specific approach. The traditional "one-size-fits-all" approach to aviation regulation is no longer suitable for the wide variety of drone operations now emerging—from recreational use to AI-driven, autonomous commercial operations. Legal frameworks should be based on the specific risks posed by different drone activities, with more stringent rules for high-risk operations like BVLOS flights over urban areas, and more lenient rules for low-risk, rural operations.

Regulations will also need to incorporate advanced technological measures such as geo-fencing, UTM systems, and blockchain for tracking drone flights and ensuring compliance. The future legal typology must also address issues related to data ownership, AI decision-making, and privacy concerns, particularly as drones become integral to surveillance, delivery, and transportation systems. The legal typology for commercial drones must be an adaptable, risk-based framework that balances the need for innovation with stringent safety and accountability measures. A hybrid regulatory model that incorporates both normative and non-normative rules, as well as international harmonisation, will be key to ensuring the continued growth and safe operation of drones worldwide.

XII

12. Towards a Sophisticated Regulatory Approach

As the commercial drone industry continues to expand, the importance of a robust regulatory framework cannot be overstated. Evaluating the effectiveness of current regulations governing commercial drones is essential for understanding how well existing rules have addressed the inherent challenges and risks. This analysis aims to dissect the regulatory landscape, focusing on the progress and limitations of EU harmonized rules, and their impact on safety, privacy, and innovation.

The European Union has made significant strides in creating a harmonized regulatory framework for drones, particularly through the adoption of the Implementing Regulation (EU) 2019/947 and the Delegated Regulation (EU) 2019/945. These regulations categorise drone operations into Open, Specific, and Certified categories, each with tailored requirements designed to mitigate risk. This tiered approach allows for flexibility while maintaining safety standards across various levels of drone operations.

EU Harmonised Regulatory Framework

One of the most commendable aspects of the EU's regulatory framework is its focus on safety. By requiring drones to meet rigorous design and operational standards, the regulations aim to minimise the risk of accidents and incidents. The certification processes for different risk categories ensure that higher-risk operations undergo more stringent scrutiny, thereby protecting both the operators and the public. Moreover, the mandatory inclusion of features such as geofencing and remote identification further enhances safety by preventing unauthorized access to restricted areas and enabling real-time tracking of drones.

However, despite these advances, the current regulatory framework in the EU is not without its limitations. One of the critical areas where the regulations fall short is in the management of autonomous drone operations. The rapid advancement of AI technology has outpaced the existing regulatory provisions, leaving a gap in the effective governance of fully autonomous drones. The regulations currently lack specific guidelines on the deployment and oversight of AI-driven systems, which can lead to uncertainties and potential risks. There is a pressing need for more detailed and dynamic regulatory measures that address the unique challenges posed by autonomous drones, including algorithmic accountability and transparency in decision-making processes.

Privacy concerns also present a significant challenge within the current regulatory landscape. While the GDPR provides a robust framework for data protection, the application of these rules to drone operations is often ambiguous. The regulations need to offer clearer guidance on how drones should handle data collection, processing, and storage to ensure compliance with privacy laws. This is particularly crucial as drones increasingly incorporate advanced sensors and cameras capable of collecting vast amounts of data. Ensuring that this data is managed in a way that respects individual privacy rights is essential for maintaining public trust and fostering a responsible drone industry.

Furthermore, the current regulatory framework must better address the integration of drones into the broader air traffic management system. The coexistence of manned and unmanned aircraft in shared airspace presents complex challenges that require sophisticated solutions. The development and implementation of Unmanned Traffic Management (UTM) systems are still in their nascent stages, and there is a need for more comprehensive regulations that facilitate seamless integration. This includes establishing clear communication protocols, real-time data sharing, and automated conflict resolution mechanisms to ensure safe and efficient airspace management.

Innovation is another critical area where the current regulations exhibit both strengths and weaknesses. On one hand, the flexible categorisation of drone operations encourages innovation by allowing operators to undertake a wide range of activities within a regulated framework. On the other hand, overly stringent regulations can stifle innovation by imposing burdensome requirements that may not be proportionate to the risks involved. Achieving a balance between safety and innovation is crucial for the continued growth of the drone industry. Regulations must be adaptable and forward-looking, providing a conducive environment for experimentation and the adoption of new technologies.

The evaluation of current regulations also highlights the importance of international harmonization. The global nature of the drone industry necessitates a coordinated approach to regulation to avoid fragmentation and ensure consistent standards. Differences in regulatory frameworks across countries can create barriers to international operations and complicate compliance efforts. Strengthening international cooperation through organizations like the International Civil Aviation Organization (ICAO) and the Joint Authorities for Rulemaking on Unmanned Systems (JARUS) can help harmonise regulations and promote global interoperability.

In essence, while the current regulatory framework for commercial drones in the EU has made significant progress in enhancing safety and providing a structured approach to risk management, there are still areas that require improvement. The rapid pace of technological advancement, particularly in AI and autonomous systems, necessitates more dynamic and detailed regulations. Privacy concerns must be addressed with clearer guidelines to ensure data protection and maintain public trust. The integration of drones into the broader air traffic management system requires more sophisticated solutions to ensure seamless coexistence with manned aircraft. Additionally, fostering an environment that balances safety with innovation is essential for the sustainable growth of the drone industry. By identifying these areas for improvement and recommending targeted strategies, this chapter aims to contribute to the development of a more sophisticated regulatory approach that meets the evolving needs of the commercial drone sector in the age of artificial intelligence.

Assessing the Progress and Limitations of EU Harmonized Rules

Assessing the progress and limitations of EU harmonized rules is essential to understanding the regulatory framework governing commercial drones in the European Union. The harmonized regulations, primarily encapsulated in Implementing Regulation (EU) 2019/947 and Delegated Regulation (EU) 2019/945, have aimed to create a cohesive and comprehensive system for drone operations across EU member states. This subsection addresses critical questions about the specific successes and challenges of the current EU drone regulations in ensuring safe and responsible drone operations? How effective are the harmonized rules in standardising regulations across member states and facilitating seamless cross-border drone activities? To what extent do these regulations address key concerns such as privacy, security, and airspace safety, and what gaps or limitations still exist within the framework? How can the progress and shortcomings of EU harmonized rules provide insight into the overall strengths and weaknesses of the current regulatory approach for drones?

The specific successes of current EU regulations in ensuring safe and responsible drone operations revolve around the harmonization of standards across member states, which is a crucial development for creating consistency and predictability in the use of commercial drones. The introduction of the EU-wide framework by the European Union Aviation Safety Agency (EASA) in 2020 has been instrumental in standardising drone regulations across member states, replacing fragmented national rules that previously hindered cross-border drone activities. This harmonization allows drone operators and manufacturers to operate under a unified set of regulations, thereby facilitating commercial drone operations across the EU and ensuring that all stakeholders adhere to a common standard of safety and compliance.

One key success of these regulations is the establishment of distinct categories of drone operations—Open, Specific, and Certified—which take into account the level of risk associated with drone activities. This tiered approach provides flexibility, enabling operators to pursue a wide range of activities while aligning operational risks with the corresponding regulatory requirements. For example, low-risk drone operations can proceed with minimal regulatory oversight, while more complex or high-risk operations must undergo specific authorisations and safety assessments. This adaptability ensures that regulations remain relevant for various operational contexts, encouraging innovation while maintaining safety standards.

Despite these successes, the EU regulatory framework faces several challenges. One significant issue is the practical enforcement of these rules across different member states. Although EASA provides a harmonized legal framework, the actual implementation and enforcement are delegated to the national aviation authorities (NAAs) of each country. This can result in inconsistencies in how regulations are interpreted and applied, leading to potential discrepancies in compliance requirements, safety protocols, and legal liability across borders. Operators working in multiple EU countries may therefore encounter divergent enforcement practices, despite the EU-wide regulations. This undermines the regulatory harmonization effort and creates potential barriers to seamless cross-border drone activities.

Another challenge lies in addressing critical concerns such as privacy and security. While the current regulations provide some guidance on data protection—particularly under the EU General Data Protection Regulation (GDPR)—they do not offer a comprehensive framework specifically tailored to drone-related privacy concerns. Drones, especially those equipped with advanced surveillance technologies, pose a unique threat to privacy. The existing framework does not sufficiently account for how drone operators collect, store, and process data, which creates gaps in privacy protection. Furthermore, the increasing use of AI-powered drones exacerbates these concerns, as the capability to collect and analyze vast amounts of data in real-time becomes more sophisticated.

Security is another issue that presents challenges under the current framework. While the EU regulations include provisions for the safe operation of drones within designated airspace, there are still questions about how effectively these regulations can mitigate potential security threats, such as cyberattacks or unauthorized drone use in restricted areas. The rapid development of autonomous drones, which can operate without real-time human control, further complicates the regulatory landscape. These developments increase the risk of drones being used for unlawful purposes, such as espionage, smuggling, or even terrorism, which underscores the need for more robust security protocols.

In terms of airspace management, the EU's U-space initiative aims to provide a framework for managing drone traffic in low-altitude airspace, thereby integrating drones more seamlessly into the overall aviation system. However, the full implementation of U-space is still a work in progress, and its effectiveness in ensuring safe airspace integration for drones remains to be fully realised. There is ongoing debate on whether the existing air traffic control (ATC) systems are sufficiently equipped to manage the increasing number of drones expected in commercial operations, particularly as drone technology evolves toward more complex autonomous systems.

The EU's harmonized regulatory framework for drones represents a significant step forward in promoting safe, responsible, and innovative drone operations. However, the challenges related to enforcement, privacy, security, and airspace management highlight areas where the current framework could be improved. Future developments in drone regulation must strike a balance between fostering innovation and addressing these emerging risks. A more comprehensive, adaptive, and coordinated approach at both the EU and national levels is necessary to ensure that drone regulations can meet the evolving demands of the industry while protecting the public interest.

Identifying Areas for Improvement in Regulatory Approach

Identifying areas for improvement in the regulatory approach is crucial to enhance the effectiveness and relevance of drone regulations. As the commercial drone industry continues to evolve, regulatory frameworks must adapt to address new challenges and opportunities. This subsection identifies specific aspects of the regulatory framework that require attention and refinement. It examines potential gaps in coverage, areas of ambiguity, and challenges faced in implementing and enforcing the regulations. By pinpointing areas for improvement, this subsection aims to contribute to the development of a more robust and comprehensive regulatory approach that addresses the evolving needs and challenges of the commercial drone industry.

One of the primary areas requiring improvement is the regulation of autonomous drone operations. Current regulations lack detailed provisions for the oversight and management of fully autonomous drones, which operate with minimal human intervention. Autonomous drones, often powered by advanced AI systems, pose unique challenges related to decision-making accountability, safety, and reliability. Developing specific regulatory guidelines that address the operation, certification, and monitoring of autonomous drones is essential. This includes establishing standards for AI algorithm transparency, ensuring that autonomous systems can be audited and understood by regulatory bodies, and implementing robust safety protocols to manage the risks associated with autonomous flight.

Another critical area for improvement is the integration of drones into the existing air traffic management (ATM) systems. The coexistence of manned and unmanned aircraft in shared airspace requires sophisticated coordination to prevent conflicts and ensure safety. Current regulations provide a foundation, but there is a need for more comprehensive and technologically advanced Unmanned Traffic Management (UTM) systems. These systems should incorporate real-time data sharing, automated conflict detection and resolution, and seamless communication protocols between drones and traditional aircraft. Enhancing UTM capabilities will facilitate safer and more efficient airspace management, enabling the widespread adoption of commercial drone operations.

Privacy and data protection continue to be significant concerns in the drone regulatory landscape. While the General Data Protection Regulation (GDPR) provides a robust framework, its application to drone operations often lacks clarity. Drones equipped with high-resolution cameras and advanced sensors can collect extensive amounts of personal data, raising potential privacy issues. Clearer guidelines on how GDPR principles should be applied to drone data collection, processing, and storage are necessary. This includes specific rules on data minimisation, anonymization, and obtaining consent from individuals whose data may be captured by drones. Strengthening privacy regulations will help build public trust and ensure that drone operations do not infringe on individual privacy rights.

The issue of security also demands further attention within the regulatory framework. Drones can be vulnerable to hacking, signal interference, and other cyber threats, which can compromise their safety and operational integrity. Current regulations mandate certain security measures, but there is a need for more stringent and comprehensive security protocols. These protocols should address both physical security measures, such as tamper-resistant hardware, and cybersecurity measures, including encrypted communications and robust authentication systems. Ensuring that drones are resilient against cyber threats is essential for protecting critical infrastructure and maintaining public safety.

Flexibility and adaptability of the regulatory framework are additional areas where improvements are needed. The rapid pace of technological innovation in the drone industry means that regulations can quickly become outdated. A more flexible regulatory approach that allows for regular updates and adaptations is necessary to keep pace with technological advancements. This could involve creating regulatory sandboxes that enable operators to test new technologies and operational models in a controlled environment without being constrained by existing regulations. Such an approach would promote innovation while ensuring that new technologies are rigorously evaluated for safety and compliance before being widely adopted.

Another area for improvement is the enforcement of drone regulations. While the EU harmonized rules aim for consistency, enforcement practices vary across member states due to differences in resources, priorities, and enforcement capacities. Strengthening enforcement mechanisms and ensuring uniform application of regulations across all member states is crucial. This could involve increasing funding for regulatory bodies, enhancing training for enforcement personnel, and developing standardised enforcement protocols. Improving enforcement consistency will help ensure that all operators adhere to the same high standards of safety and compliance.

International harmonization of drone regulations is also essential for facilitating cross-border operations and ensuring global interoperability. Currently, variations in regulatory frameworks across different countries can create barriers to international drone operations and complicate compliance efforts for operators and manufacturers. Enhancing international cooperation through organizations like the International Civil Aviation Organization (ICAO) and the Joint Authorities for Rulemaking on Unmanned Systems (JARUS) can help harmonise regulations and promote a more seamless global regulatory environment. Establishing common standards and mutual recognition agreements will support the growth of the global drone industry and enable operators to expand their operations across borders with greater ease.

Stakeholder engagement and collaboration are vital for developing a more effective regulatory approach. Involving a broad range of stakeholders, including industry experts, technology developers, regulatory bodies, and end-users, in the regulatory process ensures that diverse perspectives and expertise are considered. Regular consultations, public forums, and collaborative research initiatives can help identify emerging issues, share best practices, and develop consensus-based solutions. By fostering a collaborative regulatory environment, regulators can create more informed and balanced regulations that address the needs and concerns of all stakeholders.

While the current regulatory framework for commercial drones in the EU has made significant progress, several areas require further attention and improvement. Addressing the challenges related to autonomous operations, airspace integration, privacy, security, flexibility, enforcement, international harmonization, and stakeholder engagement is crucial for developing a more sophisticated regulatory approach. By focusing on these areas, regulators can enhance the effectiveness and relevance of drone regulations, ensuring the safe, responsible, and innovative use of drone technology in the age of artificial intelligence.

Lessons from International Best Practices

Learning from international best practices can provide valuable insights into effective regulatory approaches for commercial drones. Examining how different jurisdictions have tackled the complexities of drone regulation offers valuable lessons that can enhance the EU's regulatory framework. How have international regulatory frameworks for drones facilitated collaboration between government authorities, industry stakeholders, and drone operators to promote safer and more efficient drone operations? What adaptive strategies have been used by different countries to address the challenges posed by emerging drone technologies, and how can these strategies be integrated into a more sophisticated and responsive regulatory approach for commercial drones?

To what extent can lessons from international best practices inform the development of a harmonized global regulatory framework for commercial drones, and what legal considerations need to be addressed when incorporating these practices into national or regional regulations? How do these international case studies illustrate the balance between fostering innovation in the drone industry and ensuring compliance with safety, privacy, and security standards in diverse regulatory environments?

One notable example of effective drone regulation comes from the United States, where the Federal Aviation Administration (FAA) has developed a comprehensive regulatory framework through Part 107 and the Low Altitude Authorisation and Notification Capability (LAANC) system. Part 107 provides clear guidelines for commercial drone operations, including requirements for pilot certification, operational limits, and safety protocols. The LAANC system enhances this framework by allowing drone operators to obtain real-time airspace authorisations, significantly streamlining the process and reducing administrative burdens. The integration of automated systems for airspace management demonstrates a successful adaptation of technology to regulatory needs, ensuring that safety and efficiency are maintained while facilitating the growth of the commercial drone industry.

Australia offers another compelling case study, with its Civil Aviation Safety Authority (CASA) implementing a risk-based approach to drone regulation. CASA's framework categorises drone operations based on risk, with specific requirements tailored to each category. This flexible and adaptive strategy allows for a more nuanced approach to regulation, ensuring that lower-risk operations are not overburdened with excessive requirements while higher-risk activities are subject to stricter oversight. CASA's approach also includes mandatory training and certification for drone operators, ensuring that they possess the necessary knowledge and skills to conduct safe operations. This emphasis on education and competency underscores the importance of equipping operators with the tools they need to comply with regulations and mitigate risks effectively.

In Japan, the regulatory framework for drones is characterised by a strong emphasis on innovation and industry collaboration. The Ministry of Land, Infrastructure, Transport and Tourism (MLIT) has established a regulatory environment that encourages the development and testing of new drone technologies. Japan's "Deregulated Zones" initiative allows companies to conduct experimental drone operations in designated areas with relaxed regulatory requirements. This approach fosters innovation by providing a controlled environment where new technologies can be tested and refined before broader deployment. The success of this initiative highlights the value of regulatory sandboxes in promoting technological advancement while ensuring safety and compliance.

Singapore provides a model of comprehensive airspace management and integration of drones into urban environments. The Civil Aviation Authority of Singapore (CAAS) has implemented stringent regulations for urban drone operations, focusing on safety, privacy, and public acceptance. Singapore's U-Space framework, similar to the EU's initiative, integrates drones into the national airspace through advanced traffic management systems, real-time monitoring, and automated conflict resolution. The country's proactive approach to stakeholder engagement, involving public consultations and industry partnerships, has been instrumental in addressing community concerns and building public trust. This collaborative strategy underscores the importance of involving all relevant stakeholders in the regulatory process to create a balanced and effective framework.

In the United Kingdom, the Civil Aviation Authority (CAA) has adopted a pragmatic approach to drone regulation that balances safety with innovation. The CAA's framework includes comprehensive guidelines for drone operations, mandatory registration for drones and operators, and a clear delineation of responsibilities. The introduction of the Drone and Model Aircraft Registration and Education Service (DMARES) ensures that all operators are informed about their legal obligations and best practices for safe operation. The CAA's emphasis on education and awareness campaigns has been crucial in fostering a culture of compliance and safety among drone users. Additionally, the UK's participation in international regulatory bodies and standards organizations, such as the Joint Authorities for Rulemaking on Unmanned Systems (JARUS), reflects its commitment to harmonising regulations and promoting global interoperability.

Drawing lessons from these international examples, several key themes emerge that can inform the development of a more sophisticated regulatory approach for the EU. First, the integration of automated systems for airspace management, as demonstrated by the FAA and Singapore, enhances regulatory efficiency and safety. Implementing similar technologies within the EU framework can streamline authorisation processes and improve real-time monitoring and conflict resolution.

Second, adopting a risk-based approach to regulation, as seen in Australia, allows for more tailored and proportional oversight. This strategy ensures that regulations are appropriately scaled to the level of risk, avoiding unnecessary burdens on low-risk operations while maintaining stringent controls over high-risk activities. Incorporating this approach within the EU's regulatory framework can enhance flexibility and adaptability, supporting a wider range of drone applications.

Third, fostering innovation through regulatory sandboxes and experimental zones, as practiced in Japan, can accelerate the development and deployment of new drone technologies. By creating environments where new technologies can be tested with relaxed regulatory constraints, the EU can promote innovation while ensuring that safety and compliance are rigorously evaluated.

Fourth, prioritising education and competency among drone operators, as emphasised by CASA and the CAA, is essential for maintaining high standards of safety and compliance. Mandatory training programs and certification processes ensure that operators are well-informed and capable of conducting safe operations. Enhancing educational initiatives within the EU framework can strengthen operator competency and reduce the likelihood of accidents and regulatory breaches.

Finally, the importance of stakeholder engagement and public consultation, as highlighted by Singapore, cannot be overstated. Building public trust and acceptance is crucial for the successful integration of drones into society. Engaging with communities, industry stakeholders, and regulatory bodies to address concerns and incorporate feedback ensures that the regulatory framework is balanced and reflective of diverse perspectives. This collaborative approach fosters a supportive environment for the growth of the drone industry.

Essentially, learning from international best practices provides valuable insights into effective regulatory approaches for commercial drones. By examining the successes and challenges of different jurisdictions, the EU can identify key strategies for enhancing its regulatory framework. Integrating automated systems, adopting risk-based approaches, fostering innovation, prioritising education, and engaging stakeholders are essential elements for developing a sophisticated regulatory approach that meets the evolving needs of the commercial drone industry in the age of artificial intelligence.

Countries that have implemented successful regulatory frameworks often use innovative approaches to address the unique challenges posed by drone technologies. For instance, the European Union's U-Space regulatory framework introduced a new era of harmonized air traffic management for drones, enabling safer integration into airspace. The concept of U-Space facilitates automated drone services by creating a digital environment for safe, scalable drone operations. This level of innovation not only provides structure for existing operations but also enables future advancements in autonomous drone flights. Similarly, Japan's Civil Aeronautics Act sets out clear licensing and operational rules while allowing the flexibility to adapt these rules as drone technology evolves. Countries that embrace flexible yet structured frameworks that encourage innovation while setting clear safety parameters create an environment where both safety and technological growth can thrive.

The United States Federal Aviation Administration (FAA) and the private sector's collaboration under the FAA's "Drone Integration Pilot Program" is a prime example of how effective public-private collaboration can facilitate the development of robust drone regulations.

This program encourages industry participants to provide input on regulatory development and test new technologies, fostering innovation while ensuring compliance with safety and security concerns. Collaboration between regulatory bodies, industry stakeholders, and technology developers leads to regulations that reflect the realities of the technology, ensuring practical application without stifling innovation.

Adaptive Strategies: Singapore has adopted a forward-thinking approach to its drone regulations by establishing "sandbox" environments, which allow drone operators to test new technologies and operational models in controlled, low-risk settings. This adaptive strategy allows regulators to gather data on emerging technologies and refine regulations based on real-world testing before widespread implementation. Similarly, Australia has taken a flexible approach by creating multiple categories of drone operations (standard, advanced, complex) that accommodate everything from recreational use to large-scale commercial applications. Allowing for adaptive regulatory strategies, such as sandbox initiatives, enables jurisdictions to be more responsive to technological advances and emerging risks, ensuring that regulations remain relevant and up-to-date.

The harmonization of global frameworks, as seen in the European Union's EASA regulations, demonstrates the importance of creating consistent, cross-border regulatory environments that facilitate drone operations across different regions. By standardising rules across member states, the EU has made it easier for companies to operate drones seamlessly in multiple jurisdictions, reducing bureaucratic hurdles and encouraging international growth in the industry. Likewise, ICAO's (International Civil Aviation Organization) work on the global integration of unmanned aerial systems (UAS) ensures that drone operations can expand globally while adhering to unified safety standards. Global regulatory harmonization reduces barriers for international operations and enhances compliance, creating opportunities for broader economic growth and safer, more efficient drone operations.

Canada's regulatory framework balances innovation and compliance by creating a tiered system that differentiates between basic and advanced drone operations. This allows recreational users to fly drones with fewer restrictions, while stricter regulations apply to advanced operations involving higher risks. By offering different levels of compliance based on risk, Canada promotes innovation while ensuring that high-risk operations are held to stricter safety standards. Regulatory frameworks that are stratified based on the level of risk enable both low-risk recreational activities and high-risk commercial applications to coexist, allowing innovation to flourish while maintaining high standards for safety and privacy.

Countries such as the United Kingdom emphasise public safety by integrating privacy safeguards into their drone regulations. The UK Civil Aviation Authority (CAA) has implemented rules that protect against unauthorized data collection and safeguard personal privacy, building public trust in drone technology while fostering commercial growth. Additionally, the inclusion of public safety campaigns and operator education in regulatory frameworks helps ensure responsible drone use, addressing concerns about safety and privacy. Regulatory frameworks that incorporate privacy and safety safeguards help build public trust in drone operations, which is essential for the long-term viability of the industry.

Integrating Non-Normative Rules in EU Drone Regulations

Integrating non-normative rules into the EU regulatory framework for drones offers a promising pathway to enhance flexibility, responsiveness, and innovation within the regulatory landscape. Non-normative approaches, such as industry standards, self-regulation mechanisms, and collaborative initiatives, can complement and strengthen the existing normative framework. How feasible is the application of non-normative rules in the existing EU framework, and what are the key challenges and opportunities posed by such integration? In what ways can non-normative rules complement or enhance the current normative regulations governing drone operations within the EU, and what specific legal mechanisms or structures might need to be developed to facilitate their incorporation?

Industry Standards

Industry standards play a critical role in setting technical benchmarks and operational best practices that can be widely adopted by manufacturers and operators. These standards, developed by organizations like the International Organization for Standardisation (ISO) and ASTM International, provide detailed guidelines that ensure safety, reliability, and interoperability of drone systems. In the EU, integrating industry standards into the regulatory framework could be achieved by recognising and incorporating these standards into official regulations. This approach would allow the EU to leverage the technical expertise and industry insights encapsulated in these standards, ensuring that regulations remain current with technological advancements.

One practical method for integrating industry standards is through formal recognition within the regulatory text. For example, compliance with specific ISO standards could be made a requirement for drone certification and operation. This would not only streamline the regulatory process but also ensure that all drones operating within the EU meet high safety and performance standards. Additionally, adopting industry standards can facilitate international harmonization, making it easier for EU drone operators to comply with regulations in other jurisdictions that recognize the same standards.

Self-Regulation Mechanisms

Self-regulation mechanisms enable industry stakeholders to take proactive responsibility for maintaining high standards of safety, ethics, and compliance. In the context of the EU, self-regulation could be encouraged by supporting the formation of industry associations and bodies that develop and enforce codes of conduct and best practices. These self-regulatory organizations (SROs) could be given the authority to certify operators and manufacturers who adhere to these standards, providing a market-based incentive for compliance.

For self-regulation to be effective, it must be supported by a clear legal framework that outlines the responsibilities and powers of SROs. The EU could establish guidelines for the formation and operation of these bodies, ensuring that they operate transparently and accountably. This could include requirements for regular audits, public reporting, and independent oversight to maintain credibility and public trust. By integrating self-regulation into the broader regulatory framework, the EU can create a dynamic and responsive system that encourages industry-led innovation while maintaining high standards of safety and compliance.

Collaborative Initiatives

Collaborative initiatives, such as public-private partnerships and multi-stakeholder forums, offer a valuable platform for integrating non-normative rules into the regulatory framework. These initiatives bring together diverse stakeholders, including regulators, industry leaders, academics, and civil society, to develop and refine standards and best practices. The EU can play a pivotal role in facilitating these collaborations by providing a structured environment for dialogue and cooperation.

For instance, the EU could establish advisory committees or working groups focused on specific aspects of drone regulation, such as safety, privacy, and innovation. These groups could provide ongoing input into the regulatory process, ensuring that regulations are informed by the latest technological developments and practical experiences. Additionally, collaborative research initiatives, supported by EU funding, could drive the development of new standards and solutions that address emerging challenges in the drone industry.

Feasibility and Impact

The feasibility of integrating non-normative rules into the EU regulatory framework depends on several factors, including the willingness of industry stakeholders to participate, the ability to create effective oversight mechanisms, and the compatibility of these approaches with existing legal structures. By fostering a collaborative regulatory environment, the EU can encourage stakeholder buy-in and ensure that non-normative rules are both practical and enforceable.

The impact of integrating non-normative rules can be significant, enhancing the overall effectiveness and adaptability of the regulatory framework. Industry standards and self-regulation can lead to higher levels of compliance and innovation, as stakeholders take ownership of maintaining and improving standards. Collaborative initiatives can ensure that regulations are continuously updated and aligned with the latest technological advancements, promoting a culture of ongoing improvement and responsiveness.

Balancing Flexibility and Accountability

One of the key challenges in integrating non-normative rules is balancing flexibility with accountability. While non-normative approaches offer the flexibility needed to adapt to rapid technological changes, they must be supported by mechanisms that ensure accountability and compliance.

This can be achieved by embedding non-normative rules within a broader normative framework that provides legal backing and enforcement powers.

For example, industry standards recognised within EU regulations could be subject to regular review and updates, ensuring that they remain relevant and effective. Self-regulatory bodies could operate under a legal mandate that requires transparency, accountability, and independent oversight. Collaborative initiatives could be institutionalised within the regulatory process, providing a formal mechanism for ongoing stakeholder engagement and input.

Enhancing International Cooperation

The integration of non-normative rules also presents an opportunity to enhance international cooperation and harmonization of drone regulations. By adopting globally recognised industry standards and promoting collaborative initiatives with international partners, the EU can facilitate smoother cross-border operations and compliance. This approach can reduce regulatory fragmentation and create a more cohesive global regulatory environment for drones.

The potential integration of non-normative rules into the EU regulatory framework for drones offers significant benefits in terms of flexibility, innovation, and compliance. Industry standards, self-regulation mechanisms, and collaborative initiatives can complement and enhance the existing normative framework, creating a more adaptive and effective regulatory environment. By carefully balancing flexibility with accountability and fostering a collaborative regulatory culture, the EU can ensure that its regulatory framework remains robust and responsive to the evolving needs of the commercial drone industry in the age of artificial intelligence.

Recommendations for a More Sophisticated Regulatory Approach

Based on the evaluation of current regulations, lessons from international best practices, and insights from non-normative technology management, this subsection presents recommendations for a more sophisticated regulatory approach for commercial drones. It proposes strategies and measures to address the identified limitations and gaps in the current regulatory framework. By providing recommendations, this subsection aims to contribute to the ongoing efforts to develop a comprehensive and future-proof regulatory approach that promotes safety, innovation, and responsible drone operations.

Enhancing Flexibility and Responsiveness

1. **Adopt a Risk-Based Regulatory Framework:** Develop a more nuanced risk-based approach that categories drone operations based on their potential risk levels. This approach should provide greater flexibility for low-risk operations while ensuring that high-risk activities are subject to stringent oversight. This can be achieved by refining the current categories (Open, Specific, and Certified) and introducing sub-categories that allow for more tailored regulatory requirements.

2. **Implement Regulatory Sandboxes:** Establish regulatory sandboxes that allow companies to test new technologies and operational models in a controlled environment. These sandboxes can provide a safe space for innovation, enabling regulators to monitor and evaluate new developments before they are widely adopted. This approach helps balance safety and innovation, ensuring that new technologies are rigorously assessed without stifling progress.

3. **Regularly Update Regulations:** Create mechanisms for the regular review and updating of regulations to keep pace with technological advancements. This can include setting up advisory committees that include industry experts, technologists, and regulators who meet periodically to assess emerging trends and recommend updates to the regulatory framework.

Leveraging Non-Normative Approaches

1. **Incorporate Industry Standards:** Recognize and incorporate industry standards into the regulatory framework. By mandating compliance with specific ISO, ASTM, or other internationally recognised standards, the EU can ensure that drones meet high safety and performance benchmarks. This approach leverages the expertise of industry stakeholders and promotes consistency and interoperability.

2. **Support Self-Regulation Mechanisms:** Encourage the formation of self-regulatory organizations (SROs) that can develop and enforce industry codes of conduct and best practices. Provide a clear legal framework and incentives for companies to participate in self-regulatory schemes. Ensure that these bodies operate transparently and are subject to independent oversight to maintain public trust and accountability.

3. **Facilitate Collaborative Initiatives:** Foster public-private partnerships and multi-stakeholder forums that bring together regulators, industry leaders, academics, and civil society. These collaborative platforms can be instrumental in developing and refining standards, sharing best practices, and addressing emerging challenges. Institutionalise these initiatives within the regulatory process to ensure continuous dialogue and cooperation.

Strengthening Accountability and Enforcement

1. **Enhance Compliance Monitoring:** Invest in advanced monitoring and enforcement technologies, such as remote identification systems and geofencing, to ensure real-time compliance with regulations. These technologies can automate the detection of regulatory breaches and facilitate swift enforcement actions, reducing the burden on regulatory authorities.

2. **Establish Clear Accountability Frameworks:** Define clear roles and responsibilities for all stakeholders in the drone value chain, including manufacturers, operators, and regulatory bodies. Ensure that accountability mechanisms are in place to address non-compliance and hold stakeholders responsible for their actions. This can include establishing legal provisions for penalties, fines, and corrective actions.

3. **Promote Education and Training:** Develop comprehensive education and training programs for drone operators to ensure they are knowledgeable about regulatory requirements and best practices. These programs can be integrated into the certification process and made a prerequisite for obtaining operational licenses. Continuous education initiatives can also help operators stay updated with the

latest regulations and technological developments.

Addressing Privacy and Security Concerns

4. **Strengthen Data Protection Guidelines:** Provide clearer guidance on how GDPR principles should be applied to drone operations, including data collection, processing, and storage. Develop specific rules for data minimisation, anonymization, and obtaining consent, ensuring that drone operations respect individual privacy rights.

5. **Enhance Cybersecurity Measures:** Implement robust cybersecurity requirements for drones to protect against hacking, signal interference, and other cyber threats. This includes mandating encrypted communications, secure data storage, and regular security audits. Collaboration with cybersecurity experts and the integration of best practices from the tech industry can enhance the resilience of drone operations.

6. **Foster Public Engagement:** Engage with the public to build awareness and trust in drone technology. Conduct public consultations and information campaigns to address concerns about privacy and safety. Transparency about regulatory measures and the benefits of drones can foster public acceptance and support for the industry.

Promoting International Harmonization

7. **Align with International Standards:** Work towards harmonising EU regulations with international standards and best practices. Participate actively in global regulatory bodies like ICAO and JARUS to develop and adopt common standards that facilitate cross-border operations and compliance.

8. **Develop Mutual Recognition Agreements:** Establish mutual recognition agreements with other jurisdictions to simplify the regulatory process for international drone operations. These agreements can reduce redundancy and streamline compliance for operators looking to expand their activities across borders.

9. **Encourage Global Collaboration:** Promote international collaboration through joint research initiatives, shared regulatory frameworks, and collaborative enforcement actions. By working together with international partners, the EU can address global challenges and support the development of a cohesive regulatory environment for drones.

By incorporating these recommendations, the EU can develop a more sophisticated regulatory approach that effectively addresses the evolving needs and challenges of the commercial drone industry. Enhancing flexibility and responsiveness, leveraging non-normative approaches, strengthening accountability and enforcement, addressing privacy and security concerns, and promoting international harmonization are all critical components of a comprehensive and future-proof regulatory framework. These measures will not only ensure the safe and responsible use of drone technology but also support innovation and growth in the industry, positioning the EU as a leader in the global drone market in the age of artificial intelligence.

The Dutch Diamond Model as a Collaborative Blueprint.

As drone technology advances and commercial applications proliferate, crafting a regulatory framework that balances innovation with public safety, privacy, and environmental considerations becomes essential. This chapter explores how the Dutch Diamond Model—a framework emphasising collaboration among government, knowledge institutions, industry, and civil society—can shape a dynamic regulatory environment for commercial drones. Through the integration of diverse stakeholders, the Dutch Diamond Model provides a robust blueprint for achieving an adaptable regulatory approach, aligning well with the principles outlined in Drone Law 3.0. By examining each pillar of the Dutch Diamond Model, we illustrate how the model's collaborative essence can drive regulatory innovation and foster a sustainable, responsive legal framework for drones in an AI-powered age.

The Dutch Diamond Model has long been recognised for its unique ability to foster innovation and sustainable growth through coordinated efforts among four primary pillars: government, knowledge institutions, industry, and civil society. This model positions collaboration as central to creating a regulatory environment where diverse perspectives are not only acknowledged but integrated to ensure comprehensive, nuanced, and forward-looking regulations. This collaborative model is particularly suited to the challenges of regulating disruptive technologies like drones, where single-agency approaches often struggle to keep pace with rapid advancements and multifaceted societal impacts.

Government: Crafting Adaptive and Forward-Thinking Policies

Government plays a foundational role in the Dutch Diamond Model, with responsibility for establishing the regulatory frameworks that protect public interest while allowing innovation to thrive. In the drone sector, governments are tasked with creating flexible and adaptive regulations that accommodate the fast-evolving capabilities of drone technology, such as AI-driven navigation, real-time data collection, and autonomous operation. Given the pace at which the drone industry develops, traditional regulatory frameworks often fall behind, stifling innovation or creating uncertainty for businesses.

To counter this, governments can adopt regulatory sandboxes and pilot programs, allowing real-world testing under monitored conditions. This experimental regulatory approach, which has been effectively implemented in sectors like fintech, could enable governments to test drone technologies in a controlled environment before enacting formal regulations.

For instance, the Civil Aviation Authority of the Netherlands has taken proactive steps by creating limited regulatory exemptions in designated zones, enabling companies to pilot innovative drone applications without immediately navigating a complex regulatory landscape. Such measures illustrate how governments can work with industry and knowledge institutions to establish adaptable policies that are informed by empirical data and real-world outcomes, aligning with the adaptable legal principles advocated in Drone Law 3.0.

Knowledge Institutions: Driving Research and Development

Knowledge institutions, including universities, research centers, and technical institutes, contribute critical research that advances drone technology and informs the broader regulatory landscape. Within the

Dutch Diamond Model, knowledge institutions play an instrumental role by collaborating with government and industry to conduct research on areas like AI ethics, drone safety, environmental impact, and data privacy. Their work provides a foundational understanding of the complex issues posed by drone operations, enabling more nuanced regulatory approaches.

In the drone industry, partnerships between knowledge institutions and companies can generate cutting-edge insights into both technical and ethical considerations, from algorithm transparency and bias mitigation in autonomous drones to noise reduction strategies for urban air mobility. Knowledge institutions also provide essential input into standard-setting initiatives and policy development processes, ensuring that regulatory frameworks are grounded in scientific evidence and best practices. Collaborations between entities such as Delft University of Technology and drone manufacturers in the Netherlands exemplify how academic research can be directly applied to develop responsible, innovative drone solutions, with regulatory standards and technological advancements evolving in tandem.

Industry: The Engine of Innovation and Standardisation

Industry stakeholders, including drone manufacturers, operators, and service providers, are at the forefront of technological innovation in the drone sector. Within the Dutch Diamond Model, the industry is not only seen as a source of innovation but also as a key partner in the regulatory process, providing insights that inform standards and best practices. Industry-led standardisation efforts are crucial for creating harmonized guidelines that promote both safety and efficiency across borders, while self-regulation initiatives enable companies to adopt higher safety and ethical standards that exceed baseline regulatory requirements.

In recent years, industry associations such as the Global UTM Association (GUTMA) and the Commercial Drone Alliance have led the charge in developing non-normative guidelines for drone operations, addressing issues like unmanned traffic management (UTM), beyond visual line of sight (BVLOS) operations, and geofencing. Such standards not only enhance operational safety but also allow regulators to leverage industry expertise when formulating policies. The close collaboration between industry stakeholders and regulatory bodies in developing these standards exemplifies the Dutch Diamond Model in action, demonstrating how private-sector innovation can shape proactive, adaptable regulatory frameworks.

Civil Society: Building Public Trust and Acceptance

Civil society's role in the Dutch Diamond Model is essential, particularly in sectors like drone technology where public perception and trust are critical. Civil society groups, including NGOs, community organizations, and advocacy groups, provide valuable feedback on issues like privacy, safety, and environmental impact, helping regulators understand and address societal concerns. Engaging civil society in the regulatory process fosters transparency and builds trust in drone operations, which is particularly important as drones become more integrated into everyday life, from delivery services to infrastructure inspections.

Public trust is a recurring theme in Drone Law 3.0, as drones raise unique concerns about surveillance, noise pollution, and airspace safety. Addressing these issues requires transparent communication and inclusive policymaking processes that incorporate community perspectives. For example, the city of Amsterdam has engaged local residents in discussions about urban air mobility (UAM) and airspace usage, aiming to address public concerns about noise and privacy before implementing UAM services. This proactive engagement approach exemplifies how civil society's involvement can foster greater acceptance of new technologies while ensuring that regulations are responsive to public needs.

The Dutch Model in Action and Beyond

The Dutch Diamond Model's collaborative approach has inspired similar regulatory innovations worldwide. In Singapore, the government has worked with industry and knowledge institutions to develop a regulatory sandbox for drone deliveries, facilitating real-world testing and iterative regulation. Similarly, in Japan, collaboration among government agencies, industry players, and academic researchers has led to the development of a comprehensive UTM system, addressing airspace integration challenges for both drones and manned aircraft. These case studies demonstrate the effectiveness of the Dutch model's principles in fostering regulatory innovation, with stakeholders working together to develop solutions that are both technically viable and socially acceptable.

Integrating the Dutch Diamond Model

Incorporating the Dutch Diamond Model into EU drone regulation presents a powerful opportunity to build a flexible, responsive framework that can keep pace with technological advancements while addressing public safety, privacy, and environmental concerns. By formalising collaboration among government, knowledge institutions, industry, and civil society, EU regulators can create a regulatory environment that promotes responsible innovation and ensures a high standard of public accountability. This model could support the EU in developing a sophisticated, future-ready approach to drone regulation that aligns with the principles outlined in Drone Law 3.0, ensuring that legal frameworks evolve alongside technological progress.

The Dutch Diamond Model offers a compelling framework for addressing the regulatory challenges of the commercial drone industry, balancing innovation with public safety and ethical considerations. By emphasising collaboration among government, knowledge institutions, industry, and civil society, this model aligns with Drone Law 3.0's vision for an adaptive, integrated regulatory landscape. As the EU and other jurisdictions seek to develop comprehensive drone regulations, the Dutch Diamond Model provides a roadmap for fostering sustainable, responsible growth in an era of rapid technological advancement. This collaborative approach not only enhances regulatory effectiveness but also ensures that drone law evolves to meet the needs of a complex, interconnected world where technology, society, and policy must work hand in hand.

XIII

13. Drone Regulation in the Foreseeable Future

This chapter provides a comprehensive recapitulation of the key findings and arguments discussed throughout the book on the future of drone regulation in simple and actionable points. It reiterates how the current regulatory frameworks have addressed the rapid expansion of drone technology and re-examines the legal implications of various regulatory approaches and assesses their adequacy in managing the complexities of emerging drone technologies. The chapter reflects on the impact of the current liability structures, considering whether they are fit for purpose in mitigating the risks posed by drones, particularly in cases of autonomous or AI-driven technologies.

Furthermore, the chapter showcases how non-normative, voluntary rules—developed through industry standards and best practices—can complement legally binding regulations, and delves into the practicality and feasibility of merging these approaches to build a more adaptive regulatory framework. It revisits the benefits and shortcomings of risk-based regulations that allow for flexibility but pose challenges in consistency and enforcement. Finally, it re-assesses the gaps in the current regulations, such as those related to privacy, data protection, and the ethical implications of AI-driven drones, and outlines potential solutions for addressing these gaps.

Regulatory Landscape for Commercial Drones

The regulatory landscape for commercial drones in the European Union is characterised by a comprehensive and multi-layered framework designed to ensure the safe and responsible integration of drones into European airspace. This framework includes harmonized EU-wide regulations, national variations to address specific regional needs, and various areas of overlap that highlight the complexity of the regulatory environment. By summarising the regulatory landscape, this subsection provides a comprehensive understanding of the current state of drone regulations in the EU, integrating the key elements discussed in earlier chapters such as registration requirements, operational limitations, and certification standards.

The foundation of the EU's regulatory approach to commercial drones is built on the Implementing Regulation (EU) 2019/947 and the Delegated Regulation (EU) 2019/945. These regulations, developed by the European Union Aviation Safety Agency (EASA), provide a harmonized set of rules that apply across all EU member states. The primary goal of these regulations is to ensure a high level of safety while facilitating the growth and innovation of the drone industry.

Registration Requirements

A fundamental component of the regulatory framework is the requirement for the registration of drones and their operators. All drones weighing 250 grams or more, as well as drones equipped with sensors capable of capturing personal data, must be registered with the relevant national aviation authority. This registration process helps to ensure accountability and traceability, enabling authorities to track drone operations and enforce compliance with safety standards. Operators must also register, providing their personal details and obtaining an operator ID, which must be displayed on their drones.

Operational Limitations

The EU regulations categorise drone operations into three primary categories: Open, Specific, and Certified, each with its own set of operational limitations and requirements.

- **Open Category:** This category covers low-risk operations and is divided into three subcategories (A1, A2, A3) based on the level of risk associated with the operation. Drones in the Open category do not require specific authorisation if they comply with the defined limitations, such as maximum altitude (120 meters), maintaining visual line of sight (VLOS), and avoiding operations over uninvolved people or in restricted areas. This category is designed to facilitate hobbyist and small-scale commercial operations without imposing excessive regulatory burdens.

- **Specific Category:** This category covers higher-risk operations that require a risk assessment and specific authorisation from the national aviation authority. Operators must submit an operational risk assessment (ORA) outlining the risks and mitigation measures for their planned operations. The Specific category provides flexibility for more complex and innovative uses of drones, such as beyond visual line of sight (BVLOS) flights and operations in urban environments, by ensuring that appropriate safety measures are in place.

- **Certified Category:** This category is intended for the highest-risk operations, comparable to manned aviation. It includes requirements for drone certification, operator certification, and potentially airworthiness certification. Operations in this category may involve large drones carrying passengers or dangerous goods, and they are subject to the most stringent regulatory oversight to ensure safety and security.

Certification Standards

Certification standards are crucial for ensuring that drones meet rigorous safety and performance criteria. Drones in the Specific and Certified categories must undergo a certification process that includes technical evaluations, safety assessments, and compliance with established standards. EASA has developed comprehensive guidelines for the certification process, ensuring that certified drones can operate safely in diverse and potentially challenging environments.

National Variations and Overlaps

While the harmonized EU regulations provide a consistent framework, individual member states have the authority to implement additional requirements or variations to address specific national needs and contexts. For example, some countries may impose stricter regulations for drone operations in urban areas or near critical infrastructure. These national variations can create a complex regulatory environment, particularly for operators conducting cross-border operations. It is essential for operators to be aware of both the harmonized EU rules and any additional national requirements that may apply to their operations.

Areas of Overlap

The regulatory landscape for drones also includes overlaps with other regulatory domains, such as data protection, telecommunications, and environmental protection. For instance, drones equipped with cameras and sensors must comply with the General Data Protection Regulation (GDPR) to ensure that data collected during operations is handled lawfully and transparently. Similarly, the use of radio frequencies for drone communication must adhere to regulations set by telecommunications authorities to prevent interference with other critical communication systems. Environmental regulations may also apply, particularly in terms of noise pollution and wildlife protection, necessitating a holistic approach to compliance.

Integration with Non-Normative Approaches

The incorporation of non-normative approaches, such as industry standards and self-regulation mechanisms, further enhances the regulatory framework. Industry standards developed by organizations like ISO and ASTM International provide detailed technical guidelines that complement the normative regulations. Self-regulation mechanisms, including codes of conduct and best practices established by industry associations, encourage operators and manufacturers to maintain high standards voluntarily. Collaborative initiatives, such as public-private partnerships and stakeholder forums, facilitate ongoing dialogue and cooperation, ensuring that regulations evolve in line with technological advancements and practical experiences.

Challenges and Opportunities

Despite the comprehensive nature of the regulatory framework, several challenges remain. The rapid pace of technological innovation in the drone industry often outstrips the ability of regulatory processes to adapt, leading to gaps and ambiguities in the regulations. Privacy and security concerns require continuous attention, particularly as drones become more capable and widely used. Ensuring consistent enforcement across member states is another challenge, as varying resources and priorities can lead to discrepancies in how regulations are applied and monitored.

However, these challenges also present opportunities for further refinement and enhancement of the regulatory framework. By integrating insights from non-normative technology management, learning from international best practices, and fostering a collaborative regulatory environment, the EU can develop a more sophisticated and adaptable approach. This approach should balance safety and innovation, ensuring that the commercial drone industry can thrive while maintaining high standards of safety, security, and public trust.

The regulatory landscape for commercial drones in the EU is characterised by a comprehensive and multi-faceted framework designed to ensure safety and promote innovation. The harmonized regulations, combined with national variations and non-normative approaches, provide a robust foundation for managing the complexities of drone operations. By addressing the identified challenges and leveraging opportunities for improvement, the EU can continue to lead in the development of a forward-thinking regulatory environment that supports the safe and responsible use of drone technology in the age of artificial intelligence.

Principles of Liability for Drone Operators

Drone operators bear significant responsibility under the current regulatory framework. The principles of liability for operators are designed to ensure that they maintain high standards of safety and compliance. Operators must adhere to strict operational guidelines, including maintaining visual line of sight (VLOS), observing altitude restrictions, and avoiding restricted areas. Failure to comply with these regulations can result in severe penalties, including fines, suspension of operating licenses, and legal action.

Operators are also liable for any damage or injury caused by their drones. This liability extends to both property damage and personal injury. In the event of an accident, operators may be required to compensate victims for their losses. This principle of fault liability ensures that operators take necessary precautions to prevent accidents and operate their drones responsibly. Additionally, operators must ensure that their drones are properly maintained and fit for flight, as negligence in maintenance can also lead to liability issues.

Liability for Manufacturers and Suppliers

Manufacturers and suppliers of drones and their components also face significant liability risks. They are responsible for ensuring that their products meet all regulatory standards and are safe for use. This includes rigorous testing and quality control processes to identify and mitigate any potential defects. If a drone or its components are found to be defective and cause harm, manufacturers and suppliers can be held liable under product liability laws.

Product liability for manufacturers and suppliers is based on the principle that they are best positioned to prevent harm by ensuring their products are safe and reliable. This includes liability for design defects, manufacturing defects, and inadequate warnings or instructions. For example, if a drone's control system malfunctions due to a design flaw, the manufacturer could be held liable for any resulting accidents. Similarly, if a supplier provides substandard components that fail during operation, they could be liable for the consequences.

Vicarious Liability

The concept of vicarious liability extends the responsibility for drone operations to parties who may not be directly operating the drone but have control or influence over the operation. This includes employers who may be held liable for the actions of their employees operating drones within the scope of their employment. Vicarious liability ensures that entities benefiting from drone operations also bear responsibility for ensuring those operations are conducted safely and in compliance with regulations. For example, a logistics company employing drone operators to deliver packages can be held vicariously liable for any damages caused by the drones during delivery operations. This principle encourages companies to implement strict operational guidelines, training programs, and monitoring systems to ensure compliance and mitigate risks.

Risk Mitigation Strategies

Effective risk mitigation strategies are essential for managing the risks identified in drone operations. These strategies can include both technological solutions and operational practices designed to reduce the likelihood and impact of identified risks.

Technological Solutions

Technological solutions play a crucial role in risk mitigation. For instance, geofencing technology prevents drones from entering restricted or hazardous areas by creating virtual boundaries that the drone cannot cross. Remote identification systems enable real-time tracking and identification of drones, facilitating enforcement and enhancing security. Advanced sensors and AI algorithms can also enhance situational awareness and decision-making capabilities, enabling drones to detect and avoid obstacles autonomously. Redundant systems and fail-safe mechanisms ensure that drones can safely land or return to their point of origin in the event of a malfunction.

Operational Practices

Operational practices complement technological solutions by ensuring that drone operations are conducted safely and in compliance with regulations. These practices include regular maintenance and inspection of drones, thorough training programs for operators, and the implementation of standard operating procedures (SOPs) for various operational scenarios.

For higher-risk operations, such as those in the Specific and Certified categories, detailed contingency plans must be developed to address potential emergencies. These plans should include procedures for handling loss of control, communication failures, and other critical incidents. Regular drills and simulations can help operators prepare for these scenarios and ensure a swift and effective response.

The Importance of Accountability and Risk Management

Addressing accountability and managing risks are paramount in the drone industry. Ensuring that all stakeholders, including operators, manufacturers, and suppliers, are held accountable for their roles and responsibilities is essential for maintaining high safety standards. This accountability extends to ensuring that all operations are conducted in compliance with regulatory requirements and best practices.

Effective risk management involves continuous monitoring and reassessment of risks and mitigation strategies. As technology and operational environments evolve, so too must the approaches to risk management. This dynamic approach ensures that the regulatory framework remains relevant and effective in mitigating emerging risks.

The examination of liability and risk-based approaches highlights the critical importance of accountability and risk management in the drone industry. The principles of liability ensure that operators, manufacturers, and suppliers maintain high standards of safety and compliance. Risk-based regulations, through the integration of risk assessment, identification of risks, and implementation of mitigation strategies, provide a flexible and adaptive framework that enhances safety while promoting innovation. By addressing these key aspects, the EU can continue to develop a sophisticated regulatory approach that supports the safe and responsible use of drone technology.Assessment of Non-Normative Rules and Regulatory Sophistication.

The assessment of non-normative rules and regulatory sophistication is summarised in this subsection. It highlights the definition and characteristics of non-normative rules, their role in regulating disruptive technologies, and the potential benefits and challenges they present. The summary also encompasses the evaluation of current drone regulations in terms of sophistication and the exploration of incorporating insights from non-normative technology management. By providing an assessment of non-normative rules and regulatory sophistication, this subsection underscores the significance of adaptive and innovative approaches in drone regulation.

Enhancing Drone Regulations for Sustainable and Smart Mobility

This subsection outlines the path forward for enhancing drone regulations in the European Union to foster sustainable and smart mobility. It emphasises the need for continuous improvement and adaptation in the regulatory framework to accommodate the evolving landscape of drone operations. The path forward encompasses various dimensions, including recommendations for improving EU drone regulations, promoting safety, privacy, and sustainability in drone operations, and embracing innovation while ensuring responsible drone use. By charting the path forward, this section provides a roadmap for the future development and enhancement of drone regulations in the European Union.

Recommendations for Improving EU Drone Regulations

This subsection outlines the path forward for enhancing drone regulations in the European Union to foster sustainable and smart mobility. It emphasises the need for continuous improvement and adaptation in the regulatory framework to accommodate the evolving landscape of drone operations. The path forward encompasses various dimensions, including recommendations for improving EU drone regulations, promoting safety, privacy, and sustainability in drone operations, and embracing innovation while ensuring responsible drone use. By charting the path forward, this section provides a roadmap for the future development and enhancement of drone regulations in the European Union.

Continuous Improvement and Adaptation

To remain effective and relevant, EU drone regulations must undergo continuous improvement and adaptation. The rapid pace of technological advancements in the drone industry necessitates a dynamic regulatory approach that can evolve alongside these developments. Regular reviews and updates to the regulatory framework are essential to address emerging challenges and incorporate new technologies.

One approach to achieving continuous improvement is to establish a permanent advisory committee composed of industry experts, technologists, regulators, and other stakeholders. This committee could provide ongoing insights and recommendations for regulatory updates, ensuring that the framework remains aligned with industry innovations and best practices. Additionally, implementing a formal process for public consultations and stakeholder feedback can help identify areas for improvement and ensure that regulations are responsive to the needs of all stakeholders.

Recommendations for Improving EU Drone Regulations

1. **Streamline Regulatory Processes:** Simplify and streamline regulatory processes to reduce administrative burdens on operators and manufacturers. This includes expediting the approval process for new technologies and operations, reducing paperwork, and leveraging digital platforms for regulatory submissions and communications.

2. **Enhance Flexibility in Risk-Based Categories:** Refine the risk-based categories (Open, Specific, Certified) to allow for greater flexibility and adaptability. Introduce sub-categories or special provisions for emerging technologies and innovative use cases, ensuring that regulations can accommodate a wide range of operations without compromising safety.

3. **Integrate Advanced Monitoring Technologies:** Invest in advanced monitoring technologies, such as AI-driven analytics and blockchain for compliance tracking, to enhance oversight and enforcement capabilities. These technologies can provide real-time data on drone operations, enabling proactive risk management and swift enforcement actions.

4. **Develop Comprehensive Data Protection Guidelines:** Provide clear and comprehensive guidelines on data protection and privacy for drone operations. Ensure that these guidelines align with the GDPR and address specific concerns related to data collection, storage, and processing by drones.

5. **Promote International Harmonization:** Work towards greater harmonization of drone regulations with international standards and best practices. Engage in active collaboration with global regulatory bodies, such as ICAO and JARUS, to develop common standards that facilitate cross-border operations and compliance.

Promoting Safety, Privacy, and Sustainability

Promoting safety, privacy, and sustainability in drone operations is paramount for the responsible integration of drones into the EU's airspace and society. Regulations should prioritise these principles to ensure public trust and support for drone technology.

Safety: Safety remains the cornerstone of drone regulations. Ensuring that all drone operations are conducted safely requires a multifaceted approach that includes robust certification processes, stringent operational guidelines, and continuous monitoring and enforcement.

Investing in research and development to advance safety technologies, such as collision avoidance systems and fail-safe mechanisms, can further enhance the safety of drone operations.

Privacy: Privacy concerns must be addressed through clear regulations that protect individuals' rights while enabling legitimate drone operations. Operators should be required to implement privacy-by-design principles, ensuring that data protection is integrated into the design and operation of drones. Transparency measures, such as public notifications and consent mechanisms, can help build public trust and mitigate privacy risks.

Sustainability: Sustainability should be a key consideration in the development of drone regulations. Encouraging the use of environmentally friendly technologies, such as electric propulsion systems and renewable energy sources, can reduce the environmental impact of drone operations. Regulations should also promote sustainable practices, such as minimising noise pollution and protecting wildlife, to ensure that drone technology contributes positively to environmental sustainability.

Embracing Innovation and Ensuring Responsible Drone Use

To foster innovation while ensuring responsible drone use, regulations must strike a balance between encouraging technological advancements and maintaining high standards of accountability and safety.

1. **Supporting Innovation:** Regulatory sandboxes and pilot programs can provide a controlled environment for testing new technologies and operational models. These initiatives allow regulators to work closely with innovators to understand the implications of new developments and develop appropriate regulatory responses. Providing financial incentives and funding for research and development can also stimulate innovation in the drone industry.

2. **Ensuring Accountability:** Accountability mechanisms are essential for maintaining public trust and ensuring responsible drone use. This includes clear definitions of liability for operators,

manufacturers, and suppliers, as well as robust enforcement mechanisms to address non-compliance. Transparency and public reporting requirements can enhance accountability by providing visibility into drone operations and regulatory actions.

Fostering a Collaborative Regulatory Environment

Collaboration between regulators, industry stakeholders, and other relevant parties is critical for developing a regulatory framework that is both effective and adaptive. Multi-stakeholder forums and advisory committees can facilitate ongoing dialogue and cooperation, ensuring that regulations are informed by diverse perspectives and expertise.

Educating and Engaging the Public

Public education and engagement are vital for building support for drone technology and addressing concerns. Awareness campaigns, educational programs, and public consultations can help inform the public about the benefits and risks of drone operations, as well as the regulatory measures in place to ensure safety and privacy. Engaging the public in the regulatory process can also foster a sense of ownership and trust in the regulatory framework.

The path forward for enhancing drone regulations in the European Union involves a comprehensive and multi-dimensional approach that emphasises continuous improvement, safety, privacy, sustainability, and innovation. By adopting a flexible and adaptive regulatory framework, promoting international harmonization, and fostering collaboration and public engagement, the EU can create a robust and future-proof regulatory environment for commercial drones. These measures will ensure that drone technology continues to evolve and integrate seamlessly into European airspace and society, supporting sustainable and smart mobility in the age of artificial intelligence.

Promoting Safety, Privacy, and Sustainability in Drone Operations

Promoting safety, privacy, and sustainability in drone operations is crucial for the future of the drone industry. This subsection emphasises the importance of integrating measures to ensure safe and responsible drone operations while safeguarding privacy rights and promoting environmental sustainability. By highlighting the significance of these factors, this subsection underscores the importance of balancing technological advancement with societal values and concerns.

Safety in Drone Operations: Ensuring safety in drone operations is paramount to gaining public trust and facilitating the widespread adoption of drone technology. Comprehensive safety guidelines and standards must be established and rigorously enforced to minimise the risks associated with drone use.

Comprehensive Safety Guidelines: The development of comprehensive safety guidelines should involve a multi-faceted approach that includes stringent operational protocols, mandatory training and certification for operators, and robust maintenance and inspection regimes for drones. These guidelines should be regularly updated to reflect technological advancements and emerging risks.

- **Operational Protocols:** Establish clear operational protocols that dictate safe flying practices, such as maintaining visual line of sight (VLOS), adhering to altitude and speed limits, and avoiding restricted areas. These protocols should be tailored to different categories of drone operations, ensuring that higher-risk activities are subject to more stringent controls.

- **Training and Certification:** Require all drone operators to undergo formal training and obtain certification before conducting commercial operations. Training programs should cover essential topics such as airspace rules, emergency procedures, and the ethical use of drones. Continuous education and re-certification should be mandated to keep operators updated with the latest regulations and best practices.

- **Maintenance and Inspection:** Implement mandatory maintenance and inspection schedules for drones to ensure they remain in safe working condition. Operators should be required to document all maintenance activities and inspections, providing proof of compliance upon request. Regular audits by regulatory authorities can help verify adherence to these standards.

Privacy Protection in Drone Operations: Protecting privacy rights is critical to addressing public concerns and ensuring the ethical use of drones. Privacy protection measures should be integrated into the design and operation of drones, with a focus on transparency, data minimisation, and consent.

Transparency and Accountability: Enhancing transparency and accountability in drone operations is essential for protecting privacy. Operators should be required to provide clear information about their data collection practices, including the types of data collected, the purposes for which it is used, and how it is stored and protected.

- **Public Notifications:** Require operators to issue public notifications before conducting drone operations that involve data collection. These notifications should include details about the flight plan, the data being collected, and the duration of the operation. Publicly accessible registries of drone operations can enhance transparency and allow individuals to stay informed about drone activities in their vicinity.

- **Data Minimisation and Anonymization:** Implement data minimisation principles to ensure that only the necessary data is collected during drone operations. Data anonymization techniques should be employed to protect the identities of individuals captured by drone sensors. These measures help reduce the risk of privacy violations and unauthorized data use.

- **Consent Mechanisms:** Establish consent mechanisms for data collection, particularly in scenarios where drones capture identifiable information. Operators should obtain explicit consent from individuals before collecting their data, where feasible. For instance, drones conducting surveillance or monitoring in residential areas should seek consent from residents.

Sustainability in Drone Operations: Promoting environmental sustainability in drone operations is vital to minimising the ecological footprint of drone technology and supporting broader sustainability goals. Sustainable practices should be integrated into all aspects of drone design, manufacturing, and operation.

Environmentally Friendly Technologies: Adopting environmentally friendly technologies can significantly reduce the environmental impact of drone operations. This includes the use of electric propulsion systems, renewable energy sources, and recyclable materials in drone manufacturing.

- **Electric Propulsion Systems:** Encourage the development and use of electric propulsion systems for drones, which produce lower emissions and noise compared to traditional combustion engines. Electric drones are particularly well-suited for urban environments and short-range missions, contributing to cleaner and quieter skies.

- **Renewable Energy Sources:** Promote the use of renewable energy sources for charging drone batteries and powering ground control stations. Solar, wind, and other renewable energy options can help reduce the carbon footprint of drone operations.

- **Recyclable Materials:** Encourage manufacturers to use recyclable and sustainable materials in the production of drones. This can help minimise waste and promote a circular economy, where materials are reused and recycled rather than discarded.

Sustainable Operational Practices: Implementing sustainable operational practices is crucial for minimising the environmental impact of drone flights. This includes measures to reduce noise pollution, protect wildlife, and optimise energy use.

- **Noise Reduction:** Establish noise reduction standards for drones to minimise the impact of noise pollution on communities and wildlife. Operators should be required to use noise-mitigating technologies and adhere to quiet flying practices, particularly in noise-sensitive areas.

- **Wildlife Protection:** Implement guidelines to protect wildlife from the potential disturbances caused by drone operations. This includes restrictions on flying near sensitive habitats, nesting sites, and during critical periods such as breeding seasons. Drones should be equipped with sensors and software to detect and avoid wildlife.

- **Energy Optimization:** Encourage operators to optimise energy use during drone flights. This can involve planning efficient flight paths, minimising idle time, and using energy-saving modes. Operators should also monitor and report their energy consumption, promoting accountability and continuous improvement.

Balancing Technological Advancement with Societal Values

Balancing technological advancement with societal values is essential for the responsible integration of drones into society. This balance ensures that the benefits of drone technology are realised without compromising safety, privacy, and sustainability.

Ethical Considerations: Incorporating ethical considerations into the design and operation of drones is crucial for maintaining public trust and ensuring responsible use. This includes adhering to principles of fairness, transparency, and accountability in all aspects of drone operations.

- **Fairness:** Ensure that drone operations do not disproportionately impact certain communities or individuals. This includes avoiding discriminatory practices in data collection, analysis, and use. Operators should be mindful of the social and economic implications of their activities, striving to promote inclusivity and equity.

- **Transparency:** Maintain transparency in all aspects of drone operations, from data collection and processing to decision-making and accountability. Operators should be open about their practices and provide clear information to the public, fostering trust and understanding.

- **Accountability:** Establish robust accountability mechanisms to ensure that operators are held responsible for their actions. This includes clear liability frameworks, enforcement of regulations, and mechanisms for addressing grievances and complaints.

Public Engagement: Engaging with the public is vital for building support for drone technology and addressing societal concerns. Public engagement initiatives can help educate individuals about the benefits and risks of drones, as well as the regulatory measures in place to ensure safety and privacy.

- **Education Campaigns:** Launch public education campaigns to inform individuals about drone technology, its applications, and the regulatory framework. These campaigns can help dispel myths and misconceptions, fostering a more informed and supportive public.

- **Public Consultations:** Conduct public consultations to gather input and feedback on drone regulations and practices. These consultations can provide valuable insights into public concerns and preferences, helping to shape more effective and acceptable regulations.

- Community Involvement: Involve local communities in decision-making processes related to drone operations. This can include seeking input on flight paths, noise mitigation measures, and other aspects of drone use that directly impact communities.

Promoting safety, privacy, and sustainability in drone operations is crucial for the future of the drone industry. By integrating comprehensive safety guidelines, privacy protection measures, and sustainable practices, the EU can ensure that drone technology is used responsibly and ethically. Balancing technological advancement with societal values and concerns is essential for gaining public trust and support, fostering a positive and sustainable future for the drone industry. Through continuous improvement, collaboration, and public engagement, the EU can lead the way in developing a regulatory framework that promotes safe, private, and sustainable drone operations in the age of artificial intelligence.

Embracing Innovation and Ensuring Responsible Drone Use

Innovation plays a key role in the advancement of the drone industry. As drone technology continues to evolve, it brings with it significant opportunities for various sectors, including agriculture, logistics, healthcare, and environmental monitoring. However, the rapid pace of innovation also presents challenges, particularly in ensuring that drone operations are conducted responsibly and ethically. This subsection emphasises the need to embrace innovation while ensuring responsible drone use. It discusses the importance of fostering a supportive regulatory environment that encourages innovation, research, and development. It also emphasises the need for promoting responsible practices through education, awareness campaigns, and industry collaboration. By advocating for the balance between innovation and responsibility, this subsection addresses the challenges and opportunities presented by the rapidly evolving landscape of drone technology.

Fostering a Supportive Regulatory Environment

A supportive regulatory environment is essential for fostering innovation in the drone industry. Regulations must strike a balance between ensuring safety and enabling technological advancements. Overly restrictive regulations can stifle innovation, while overly lenient regulations can lead to safety and ethical concerns. Therefore, a nuanced approach that promotes both innovation and responsibility is necessary.

Regulatory Sandboxes and Pilot Programs: One effective way to foster innovation is through the use of regulatory sandboxes and pilot programs. These initiatives allow companies to test new technologies and operational models in a controlled environment with relaxed regulatory constraints. Regulatory sandboxes provide a safe space for experimentation, enabling regulators to observe and evaluate new developments in real-time.

For instance, the European Union could establish drone-specific regulatory sandboxes where companies can conduct experimental flights and operations. These sandboxes can focus on emerging technologies such as autonomous drones, AI-driven decision-making systems, and advanced delivery models. By closely monitoring these pilots, regulators can gather valuable data and insights, which can inform the development of future regulations. This approach not only supports innovation but also ensures that new technologies are rigorously tested and validated before being widely adopted.

Incentivising Research and Development: Providing financial incentives and funding for research and development (R&D) is another crucial strategy for promoting innovation. The EU can establish grants, subsidies, and tax incentives for companies engaged in drone-related R&D. This financial support can help accelerate the development of new technologies and applications, driving growth and competitiveness in the drone industry.

Collaborative research initiatives that bring together academia, industry, and government can also play a significant role in advancing drone technology. Joint research projects can leverage the expertise and resources of multiple stakeholders, fostering innovation and addressing complex challenges. By promoting a collaborative R&D ecosystem, the EU can ensure that drone technology continues to evolve in a direction that benefits society as a whole.

Promoting Responsible Practices: While fostering innovation is essential, it is equally important to ensure that drone operations are conducted responsibly. Promoting responsible practices involves educating operators, raising public awareness, and encouraging industry collaboration.

Education and Certification Programs: Education and certification programs are fundamental to promoting responsible drone use. Mandatory training and certification for drone operators ensure that they possess the necessary knowledge and skills to operate drones safely and ethically. These programs should cover topics such as airspace rules, privacy considerations, data protection, and emergency procedures.

Continuous education initiatives are also vital for keeping operators updated with the latest regulations, technologies, and best practices. By requiring periodic re-certification, regulators can ensure that operators maintain high standards of competency and professionalism.

Awareness Campaigns: Public awareness campaigns can play a significant role in promoting responsible drone use. These campaigns can educate the public about the benefits and risks of drone technology, as well as the regulatory measures in place to ensure safety and privacy. By increasing public understanding and acceptance of drones, these campaigns can foster a supportive environment for drone innovation.

Awareness campaigns can also address specific concerns, such as privacy and noise pollution, by informing the public about how these issues are being managed. Transparency about regulatory measures and industry practices can help build trust and confidence in drone technology.

Industry Collaboration: Collaboration within the drone industry is crucial for promoting responsible practices. Industry associations and organizations can develop codes of conduct, best practices, and ethical guidelines that members are encouraged to follow. These self-regulatory measures can complement formal regulations, providing additional layers of accountability and responsibility.

Industry collaboration can also facilitate the sharing of knowledge and resources. For example, companies can collaborate on developing common safety standards, interoperability protocols, and data protection frameworks. By working together, industry stakeholders can address shared challenges and promote a culture of responsibility and excellence.

Balancing Innovation and Responsibility: Balancing innovation and responsibility is a complex but essential task. It requires a regulatory framework that is both flexible and robust, capable of adapting to new technologies while maintaining high standards of safety and ethics.

Risk-Based Regulatory Approaches: Adopting a risk-based regulatory approach can help achieve this balance. By categorising drone operations based on their level of risk, regulators can tailor requirements to match the specific risks associated with different activities. Low-risk operations can benefit from streamlined processes and reduced regulatory burdens, while high-risk operations can be subject to more stringent oversight.

Risk-based regulations can also incorporate dynamic elements that allow for regular updates and adjustments. This flexibility ensures that regulations remain relevant and effective as technology evolves. By continuously assessing and managing risks, regulators can support innovation while safeguarding public safety and interests.

Ethical Considerations: Ethical considerations should be integrated into the regulatory framework to ensure that innovation does not come at the expense of societal values. This includes principles of fairness, transparency, and accountability in all aspects of drone operations.

- **Fairness:** Ensure that the benefits and burdens of drone technology are distributed equitably across society. Avoid practices that disproportionately impact certain communities or individuals, and strive to promote inclusivity and equity.

- **Transparency:** Maintain transparency in regulatory processes and industry practices. Provide clear information about data collection, use, and protection, and involve the public in decision-making processes related to drone operations.

- **Accountability:** Establish robust accountability mechanisms to hold operators, manufacturers, and suppliers responsible for their actions. Implement enforcement measures to address non-compliance and ensure that ethical standards are upheld.

-

Public Engagement: Engaging with the public is vital for balancing innovation and responsibility. Public consultations, forums, and surveys can provide valuable insights into public concerns and preferences. By involving the public in the regulatory process, regulators can ensure that drone policies reflect societal values and priorities.

Public engagement initiatives can also help build trust and support for drone technology. By demonstrating a commitment to transparency, safety, and ethics, regulators and industry stakeholders can foster a positive relationship with the public, paving the way for the responsible integration of drones into society.

Embracing innovation and ensuring responsible drone use are both essential for the sustainable growth of the drone industry. A supportive regulatory environment that encourages innovation, research, and development can drive technological advancements and economic growth. At the same time, promoting responsible practices through education, awareness campaigns, and industry collaboration is crucial for maintaining high standards of safety, privacy, and ethics. By balancing innovation and responsibility, the EU can create a regulatory framework that supports the safe, ethical, and sustainable use of drone technology. This balanced approach will enable the drone industry to thrive while addressing societal values and concerns, ensuring that the benefits of drone technology are realised in a way that is responsible and inclusive. Through continuous improvement, collaboration, and public engagement, the EU can lead the way in developing a forward-thinking regulatory environment that embraces innovation while safeguarding public interests in the age of artificial intelligence.

A Path to Drone Law 3.0

In light of the foregoing comprehensive discussions, this book concludes that the EU harmonized rules applicable to commercial drones are evolving but not yet sufficiently adaptive and sophisticated in their regulatory approach in the age of artificial intelligence. The current regulatory framework, while comprehensive and risk-based, remains predominantly within the realm of a regulatory mindset (Law 2.0). This approach, characterised by rules and compliance measures managed by regulatory bodies, does not fully address the dynamic and disruptive nature of drone technology. The evolution of drone regulations needs to advance towards a technology mindset (Law 3.0), where the emphasis is on integrating technological solutions that preclude the practical option of non-compliance.

Our earlier discussions uncovered technology management and the technology mindset termed Law 3.0. This non-normative approach leverages technological tools to manage and mitigate risks associated with disruptive technologies like drones. Rules set with a technology mindset are sophisticated because they balance liability and regulation by steering innovation towards eliminating the possibility of non-compliance. In essence, Law 3.0 represents a technology fix to a technology problem.

In recent years, drone technologies have become ubiquitous. The sophisticated and versatile nature of drones makes them effective in providing a wide range of services in a more sustainable and climate-friendly manner. However, despite their transformational potential, drones present significant risks and negative externalities. They can collide and crash, causing bodily and property harm, infringe on privacy, disturb wildlife, and threaten avian species.

Generally, the default mindset synonymous with black letter law (Law 1.0) is not flexible enough to accommodate disruptive technological innovations like drones. While strict liability, product liability, and fundamental rights principles offer some responses to the risks posed by drones, the default mindset has been disrupted and often lacks the precise tools needed to address these risks comprehensively. Therefore, the EU has adopted a regulatory mindset (Law 2.0), where the focus has shifted from courtrooms to regulatory and compliance offices.

The EU's harmonized rules for commercial drones are based on a risk-based approach. These rules categorise drone operations by risk levels and develop appropriate codes and requirements proportionate to the operational risks. While this regulatory mindset has led to significant advancements, including empowering competent authorities to conduct risk assessments and issue operational authorisations, it still retains a rule-based nature that leaves room for non-compliance.

To fully address the challenges posed by drones, the future regulatory approach must evolve to embrace a technology mindset (Drone Law 3.0). This approach focuses on removing the practical option of non-compliance through technological solutions. Legal rules in Drone Law 3.0 will continue to prescribe risk-based, precautionary, and retributive measures but will emphasise the use of technological instruments. These include designing drones with features like detect-and-avoid systems, obstacle avoidance, anonymization and data erasure software, and noise cancellation technologies.

Moreover, Drone Law 3.0 necessitates urban planning and infrastructure design that mitigates risks associated with drones. Deploying drone surveillance systems capable of detecting, assessing, and responding to risks in real-time will be crucial. Technological instruments like drone shield systems can protect geographical areas from unauthorized drone operations, effectively removing the option of operating drones unlawfully.

The integration of automated technologies in drone traffic management, as exemplified by the U-space airspace, is a model for Drone Law 3.0. U-space leverages digital automation technologies to provide a range of non-normative interventions such as geo-awareness, flight authorisation, electronic registration, electronic identification, detect-and-avoid capabilities, tactical deconfliction, dynamic geofencing, and collaborative interfaces. These technologies facilitate compliance and safety, ensuring that drones operate within regulated parameters.

The sophisticated nature of U-space should be leveraged to offer fair and flexible access to airspace for commercial drone operators, fostering a market for services like drone delivery, emergency rescue, healthcare, insurance, and data management. Collaboration between national competent authorities, EASA, and SESAR will be essential in maintaining and enhancing U-space, ensuring compliance with essential requirements and promoting innovative drone services while safeguarding safety and security.

Sustaining a harmonious internal market with drones while preventing a dystopian future cannot be achieved with regulatory rules alone. Regulatory frameworks must evolve to incorporate a technology mindset, effectively managing the risks posed by drones. Only through technological solutions can the EU ensure the responsible and innovative use of drones, aligning with societal values and mitigating potential negative impacts.

The evolution of drone regulations must transition towards a technology mindset (Drone Law 3.0) to address the sophisticated nature of drone technology and the associated risks comprehensively. By integrating technological solutions into the regulatory framework, the EU can create a robust, adaptive, and forward-thinking regulatory environment that promotes safety, privacy, sustainability, and innovation. This balanced approach will ensure the safe and responsible use of drone technology, supporting the growth of a sustainable and smart mobility ecosystem in the age of artificial intelligence.

XIV

Operational Guide.

14. Step by Step Guide to Compliant Drone Operations

Recreational Drones

Navigating the rules of recreational drone operations can be both exciting and challenging, given the myriad of regulations, guidelines and stakeholders that vary by regions and countries. This step-by-step guide aims to demystify the process, ensuring that drone enthusiasts can enjoy their flights while adhering to local laws and maintaining safety standards. Whether you are a hobbyist capturing stunning aerial photographs or a tech enthusiast experimenting with the latest drone models, understanding and complying with the rules is crucial.

This comprehensive guide covers recreational drone operations across major jurisdictions including the European Union, United Kingdom, United States, Australia, Canada, China, India, Japan, Singapore, South Africa, and Nigeria. Each section delves into the specific steps required for safe and compliant drone flights, such as registration processes, obtaining necessary certifications, checking airspace restrictions, and adhering to operational limitations. By providing detailed insights and practical advice tailored to each location, this guide ensures that drone pilots can fly with confidence and responsibility, fully aware of the legal and safety considerations pertinent to their activities. Whether flying in urban environments, scenic landscapes, or near sensitive infrastructure, this guide offers the essential knowledge and resources needed to navigate the skies safely and legally.

European Union

Step 1: Understand the Regulatory Framework

In the European Union, drone operations are governed by the European Union Aviation Safety Agency (EASA). The regulations are designed to ensure the safety of drone flights while minimising risks to people and property. The EASA's regulations for recreational drones are categorised under the Open Category, which is divided into three subcategories: A1 (Fly over people), A2 (Fly close to people), and A3 (Fly far from people).

Step 2: Register Your Drone

All drones weighing 250 grams or more, or equipped with a camera, must be registered with the National Aviation Authority (NAA) of the operator's country unless the drone is without a camera and weighs less than 250 grams, or it is a toy drone with a CE marking for toys. Registration can typically be completed online via the respective NAA websites. For example, in Germany, you can register through the Luftfahrt-Bundesamt (LBA), in France, registration is done through the Directorate General for Civil Aviation (DGAC) and in the Netherlands, registration is done through Netherlands Vehicle Authority (RDW).

Step 3: Obtain a Drone Pilot Certificate

Operators of drones in subcategories A1 and A3 need to pass an online theoretical knowledge exam provided by their NAA. The exam covers basic aviation safety, airspace restrictions, and privacy considerations.

Step 4: Obtain Drone Insurance

For recreational drones weighing more than 250 grams, obtaining insurance is highly recommended, though not strictly required by law. Insurance can cover third-party liabilities and damages caused by the drone. Several insurers offer tailored policies for recreational drone pilots.

Step 5: Check Airspace Restrictions

Before flying, drone pilots must check for any airspace restrictions. This can be done using apps like GoDrone in the Netherlands or apps provided by local NAAs, which show no-fly zones, restricted areas, and other pertinent information.

Step 6: Follow Operational Limitations

- A1 Category: Drones weighing less than 250 grams can be flown over people but not over crowds.
- A2 Category: Drones weighing up to 2 kg must maintain a safe horizontal distance of at least 30 meters from uninvolved people.
- A3 Category: Drones weighing up to 25 kg must be flown far from people and residential areas.

Step 7: Set Up Remote ID

All drones within the EU must comply with Remote ID regulation For example, in the Netherlands, drone operators must register with the Netherlands Vehicle Authority (RDW). Apply for a drone operator registration number from RDW, which must be displayed visibly on all drones. Ensure your drone has built-in Remote ID or install an external module, update the firmware, input registration details, and enable Remote ID broadcasting. Finally, conduct a test flight to verify the system's functionality, ensuring compliance with EU regulations.

Step 7: Safety Measures

- Always maintain visual line of sight (VLOS).
- Do not fly higher than 120 meters (400 feet).
- Respect privacy and avoid flying over private property without

permission.

- Ensure your drone is in good working condition before each flight.

Useful Resources

- EASA website: easa.europa.eu
- EASA Drone Portal: easa.europa.eu/domains/civil-drones
- RDW Netherlands: rdw.nl
- Drone Rule.eu: dronerules.eu/
- UAV Coach: uavcoach.com/drone-laws-in-europe
- EASA Insurance Guidelines: EASA Insurance
- Drone Insurance Providers: Consult local insurance companies.

United Kingdom

Step 1: Understand the Regulatory Framework

The Civil Aviation Authority (CAA) governs drone operations in the UK. Recreational drone operations fall under the Open Category, similar to the EASA framework.

Step 2: Register Your Drone and Obtain an Operator ID

All drones weighing more than 250 grams must be registered with the CAA. Drone operators must also obtain an Operator ID and a Flyer ID. Registration can be completed online through the CAA's registration portal.

Step 3: Pass the Online Theory Test

To obtain a Flyer ID, operators must pass an online theory test that assesses their knowledge of safe flying practices and regulations.

Step 4: Obtain Drone Insurance

While not legally required, it is advisable for recreational drone users to have insurance if their drone weighs more than 250 grams. This insurance typically covers third-party liability and accidental damage.

Step 5: Check Airspace Restrictions

Use the Drone Assist app to check for airspace restrictions and ensure it is safe to fly in your desired location.

Step 6: Follow Operational Limitations

- Fly below 120 meters (400 feet).
- Always maintain VLOS.
- Do not fly within 50 meters of people, vehicles, buildings, or vessels not under your control.
- Stay at least 150 meters away from residential, recreational, commercial, and industrial areas.

Step 7: Set Up Remote ID

To set up Remote ID for your drone in the UK, you need both a flyer ID, which requires passing an online theory test, and an operator ID. The operator ID must be displayed visibly on your drone. This process is essential for legal compliance and can be completed through the CAA's registration portal. Next, check if your drone has built-in Remote ID capability. If not, you may need to purchase and install an external Remote ID module. Update your drone's firmware via the manufacturer's app to ensure it supports Remote ID functionality. Using the app, enter your registration details, including the operator ID, and enable the Remote ID broadcasting feature. Finally, conduct a test flight to verify that the Remote ID system is functioning correctly, ensuring that your drone broadcasts its identification and location information in compliance with UK regulations.

Useful Resources

- CAA website: caa.co.uk
- Drone Assist app: droneassist.co.uk
- CAA Insurance Information: CAA Insurance
- DroneCover Club and other UK-based drone insurers.

United States

Step 1: Understand the Regulatory Framework

The Federal Aviation Administration (FAA) regulates drone operations in the United States. Recreational drone pilots must follow the FAA's rules for model aircraft.

Step 2: Register Your Drone

All drones weighing between 0.55 pounds (250 grams) and 55 pounds (25 kg) must be registered with the FAA. Registration can be completed online through the FAA's DroneZone portal.

Step 3: Follow the Recreational UAS Safety Test (TRUST)

Recreational drone pilots must pass the FAA's TRUST exam, which covers the basic safety and regulatory knowledge required to operate drones safely.

Step 4: Obtain Drone Insurance

While not mandatory, insurance for recreational drones from 250 grams and above is strongly recommended. Coverage usually includes liability for property damage and personal injury.

Step 5: Check Airspace Restrictions

Use the FAA's B4UFLY app to check for airspace restrictions and ensure it is safe to fly in your chosen location.

Step 6: Follow Operational Limitations

- Fly below 400 feet.
- Always maintain VLOS.
- Do not fly over people or moving vehicles.
- Obtain authorisation before flying in controlled airspace (Class B, C, D, and E).
- Respect privacy and avoid flying over private property without permission.

Step 7: Set Up Remote ID

To set up Remote ID for your drone in the United States, first ensure your drone is registered with the FAA through the FAADroneZone website, providing necessary details such as make and model, contact information, and the Remote ID serial number if applicable. Update the drone's firmware via the manufacturer's app to support Remote ID broadcasting, then input your FAA registration number and operator ID in the app's settings to enable broadcasting. Conduct a test flight to verify the Remote ID system is functioning correctly, ensuring your drone's identification and location information are properly broadcasted. Regularly check for firmware updates and maintain accurate registration details to ensure continuous compliance with FAA regulations.

Useful Resources
* FAA website: faa.gov/uas
* B4UFLY app: faa.gov/uas/recreational_fliers/where_can_i_fly/ b4ufly/
* FAA Insurance Recommendations: FAA
* Drone insurance providers like Verifly and SkyWatch.AI.

Canada

Step 1: Understand the Regulatory Framework
Transport Canada regulates drone operations in Canada. Recreational drone pilots must follow the rules set out in the Canadian Aviation Regulations.

Step 2: Register Your Drone
All drones weighing between 250 grams and 25 kilograms must be registered with Transport Canada. Registration can be completed online through the Transport Canada Drone Management Portal.

Step 3: Obtain a Drone Pilot Certificate
Recreational drone pilots must obtain a Basic or Advanced Drone Pilot Certificate by passing an online exam administered by Transport Canada.

Step 4: Obtain Drone Insurance
Although not required by Transport Canada, insurance for drones from 250 grams and above is advised. Policies can include liability coverage and damage to the drone itself.

Step 5: Check Airspace Restrictions
Use the NAV CANADA Drone Site Selection Tool to check for airspace restrictions and ensure it is safe to fly in your chosen location.

Step 6: Follow Operational Limitations

- Fly below 122 meters (400 feet).
- Always maintain VLOS.
- Do not fly closer than 30 meters (100 feet) to bystanders.
- Stay away from airports, heliports, and other restricted areas.

Step 7: Set Up Remote ID

To set up Remote ID for your drone in Canada, begin by ensuring your drone is registered with Transport Canada through their Drone Management Portal, especially if it weighs between 250 grams and 25 kilograms. Registration requires you to provide details like the drone's make, model, serial number, and your contact information. This process costs $5 and must be completed before flying your drone. While Canada does not yet have a mandatory Remote ID requirement similar to the United States, Transport Canada and other bodies are working towards integrating Remote ID capabilities in the future. It is advisable to stay informed on these developments through sources such as the Drone Pilot Association of Canada (DPAC), which provides updates and guidance on the potential implementation of Remote ID regulations.

Once Remote ID requirements are in place, you'll likely need to either ensure your drone has built-in Remote ID capabilities or install an external Remote ID broadcast module. This module must be configured to broadcast your drone's identification and location information. Keep an eye on Transport Canada's website and related regulatory announcements for detailed compliance instructions as these regulations evolve.

For now, make sure your drone is marked with its registration number and that you follow all current operational regulations, such as maintaining visual line of sight and respecting airspace restrictions.

Useful Resources
- Transport Canada website: tc.canada.ca
- Drone Site Selection Tool: nrc-cnrc.github.io
- NAV Canada: navcanada.ca/en/drone-operations.aspx
- Drone Pilot Canada: drone-pilot.ca
- Canadian drone insurance providers like CoverDrone, Zensurance. SkyWatch AI and CapriCMW Insurance.

China

In China, the Civil Aviation Administration of China (CAAC) oversees drone regulations, ensuring that both recreational and commercial drone pilots operate safely and in compliance with national standards.

Step 1: Understanding the Regulatory Framework
The CAAC has established a comprehensive set of regulations that all drone operators must adhere to. These regulations are designed to ensure the safe integration of drones into the airspace and to protect public safety and privacy.

Step 2: Registering Your Drone
All drones weighing more than 250 grams must be registered with the CAAC. This process involves providing detailed information about the drone and its operator, which can typically be done online through the CAAC's registration portal. This requirement helps maintain accountability and ensures that all drones are traceable.

Step 3: Flight Certification and Permissions
For recreational use, no formal pilot certification is generally required, but understanding and following safety guidelines is mandatory. However, certain activities, such as flying in controlled airspace or beyond visual line of sight (BVLOS), may require special permissions.

Step 4: Obtain Drone Insurance
In China, insurance for recreational drones from 250 grams and above is not required but is recommended to cover potential third-party liabilities and damages.

Step 5: Checking Airspace Restrictions
Before taking off, drone pilots must verify airspace restrictions. This can be done using various apps and online platforms provided by the CAAC or third-party services. These tools highlight no-fly zones, temporary flight restrictions, and other critical airspace information to help pilots avoid unauthorized areas.

Step 6: Operational Limitations

Recreational drone pilots in China must follow specific operational rules to ensure safety:

- Maximum Altitude: Drones must not be flown above 120 meters (400 feet) to avoid interfering with manned aircraft.
- Visual Line of Sight (VLOS): Operators must maintain a direct line of sight with their drones at all times.
- Proximity to Airports: Drones must be kept at least 5 kilometres away from airports and heliports.
- Crowds and Sensitive Areas: Flying over people, sensitive infrastructure, and urban areas without proper authorisation is prohibited.

Step 7: Set Up Remote ID

To set up Remote ID for your drone in China, you must first ensure your drone is registered with the Civil Aviation Administration of China (CAAC). Registration is required for all drones weighing more than 250 grams. The registration process must be completed on the CAAC's website, which requires a Chinese cell phone number and a WeChat account. This process involves providing your personal identification number, contact details, and drone information, including make, model, and serial number.

China's new regulations, effective January 1, 2024, require drones used by individuals or companies to be equipped with Remote ID technology. This technology enables the drone to broadcast its identification and location information during flight, ensuring compliance with safety and security measures set by Chinese authorities. Once registered, you must ensure your drone's firmware is updated to support Remote ID functionality, if built-in, or install an external Remote ID module if necessary. Follow the manufacturer's instructions for configuring the module to broadcast the required information. Conduct a test flight to verify that the Remote ID system is functioning correctly, ensuring your drone's identification and location information are properly broadcasted. Regularly check for updates and maintain compliance with Chinese regulations. For more detailed guidelines, refer to the CAAC website and official regulatory updates.

Step 8: Safety Measures
Safety is a top priority in drone operations. Pilots must ensure their drones are in good working condition before each flight, including checking the battery, motors, and control systems. You must also be aware of weather conditions, as adverse weather can significantly impact flight safety.

Useful Resources
- CAAC website: caac.gov.cn
- CAAC Drone Registration: UAV Portal
- DJI FlySafe: dji.com/flysafe
- UAV Coach China: uavcoach.com/drone-laws-in-china
- Chinese drone insurance providers

India

Step 1: Understanding the Regulatory Framework
Recreational drone operations in India are regulated by the Directorate General of Civil Aviation (DGCA). The regulations are designed to ensure the safe operation of drones while promoting their recreational use.

Step 2: Drone Classification and Registration
India classifies drones into five categories based on their maximum takeoff weight: Nano (less than 250 grams), Micro (250 grams to 2 kg), Small (2 kg to 25 kg), Medium (25 kg to 150 kg), and Large (over 150 kg). Recreational drones typically fall into the Nano and Micro categories. All drones must be registered on the Digital Sky platform.

Step 3: Obtain an Unmanned Aircraft Operator Permit (UAOP)
For Nano drones used recreationally, no permit is required. For Micro drones, users must obtain a UAOP from the DGCA. This involves submitting an application through the Digital Sky platform and providing details about the drone and its intended use.

Step 4: Operational Guidelines
- Maximum Altitude: Nano drones can fly up to 50 feet, while Micro

drones can fly up to 200 feet.

- Visual Line of Sight (VLOS): Operators must maintain a direct line of sight with their drones.
- No-Fly Zones: Avoid flying near airports, military bases, and other sensitive areas. Specific no-fly zones are listed on the Digital Sky platform.

Step 5: Drone Insurance

Although not legally required, it is advisable to obtain insurance for drones over 250 grams to cover potential third-party liabilities and damages.

Step 5: Pre-Flight Checks and Safety Measures

Before each flight, conduct a thorough pre-flight check, including verifying the drone's battery levels, motor functionality, and sensor calibration. Ensure that the operating environment is safe and free from obstacles.

Step 6: Set Up Remote ID

To set up Remote ID for your drone in India, you must register your drone with the Digital Sky platform managed by the Directorate General of Civil Aviation (DGCA). Registration is mandatory for all drones except those classified as Nano drones (weighing less than 250 grams and operating below 15 meters). The registration process requires providing details such as the drone's make, model, serial number, and your personal identification information. India's drone regulations also mandate that drones must have a Unique Identification Number (UIN) and must be equipped with a Digital Sky-enabled Remote ID module. This module broadcasts the drone's identification and location information, ensuring compliance with Indian airspace regulations. Update your drone's firmware through the manufacturer's app to ensure it supports Remote ID broadcasting. Enter your registration details, including the UIN, into the app and enable the Remote ID broadcasting feature. If using an external module, follow the manufacturer's instructions to attach and configure it. Conduct a test flight to verify the Remote ID system's functionality, ensuring your drone's information is correctly broadcasted. Stay informed on regulatory updates from the DGCA to ensure ongoing compliance. For more information, visit the Digital Sky platform and the DGCA's drone regulations page.

By following these steps, you will ensure that your drone operations comply with the respective regulations in China and India, promoting safety and accountability in their airspaces.

Useful Resources
*   Digital Sky platform: Digital Sky
*   DGCA drone guidelines: DGCA

Australia

Step 1: Understanding the Regulatory Framework
Recreational drone operations in Australia are regulated by the Civil Aviation Safety Authority (CASA). The regulations are designed to ensure the safe use of drones while allowing for recreational enjoyment.

Step 2: Drone Classification and Registration
Australia classifies drones based on their weight. Drones weighing over 250 grams must be registered with CASA. Registration can be completed online through the myCASA portal.

Step 3: Obtain an Operator Accreditation
For recreational drone users, obtaining an operator accreditation from CASA is required. This involves completing an online training course and passing a knowledge test.

Step 4: Operational Guidelines
*   Maximum Altitude: Drones must not be flown above 120 meters (400 feet).
*   Visual Line of Sight (VLOS): Operators must maintain a direct line of sight with their drones.
*   Proximity to People and Property: Drones must not be flown within 30 meters of people.
*   No-Fly Zones: Adhere to restrictions around airports, heliports, and other sensitive areas.

Step 5: Drone Insurance

While not mandatory, it is recommended to obtain insurance for drones over 250 grams to cover potential liabilities and damages.

Step 6: Pre-Flight Checks and Safety Measures

Before each flight, conduct a thorough pre-flight check, including verifying the drone's battery levels, motor functionality, and sensor calibration. Ensure that the operating environment is safe and free from obstacles.

Step 7: Set Up Remote ID

To set up Remote ID for your drone in Australia, you must first register your drone with the Civil Aviation Safety Authority (CASA) through the myCASA portal, especially if you are using the drone for business purposes or as part of your job. This includes drones weighing more than 250 grams. Registration requires details such as the drone's make, model, serial number, and your personal identification information. The registration process is straightforward and can be completed online, with a fee of AU$40 for drones weighing over 500 grams.

Although Australia is still developing its Remote ID regulations, you should stay informed about upcoming mandates. Currently, Remote ID technology is expected to become a requirement to help monitor drone activities and ensure accountability. Once mandated, you will need to ensure your drone has built-in Remote ID capabilities or install an approved external Remote ID module.

After registration and ensuring your drone's firmware supports Remote ID, configure the system by entering your registration details and enabling the Remote ID broadcasting feature. Conduct a test flight to verify that the system functions correctly. For ongoing compliance, regularly check for updates on CASA's website and ensure your drone's information is up to date.

Useful Resources

- CASA website: casa.gov.au
- myCASA portal: myCASA
- CASA Drone Safety Apps: Can I Fly There?

Singapore

In Singapore, the Civil Aviation Authority of Singapore (CAAS) regulates drone operations, ensuring that both recreational and commercial activities are conducted safely and responsibly.

Step 1: Understanding the Regulatory Framework
The CAAS has established a set of regulations that all drone operators must adhere to, which are designed to ensure safety and minimise risks to the public.

Step 2: Registering Your Drone
All drones weighing more than 250 grams must be registered with the CAAS. This process can be completed online, where operators must provide details about themselves and their drones. Registration helps maintain accountability and ensures that drones are operated by responsible individuals.

Step 3: Flight Certification and Permissions
For recreational use, operators do not need formal certification, but they must follow CAAS guidelines. However, for certain activities, such as flying in controlled airspace or at night, special permits may be required.

Step 4: Obtain Drone Insurance
While insurance is not mandatory for recreational drones from 250 grams and above, it is recommended to safeguard against potential liabilities and damages.

Step 5: Checking Airspace Restrictions
Before flying, drone operators must check for airspace restrictions using the OneMap.sg portal or other CAAS-provided tools. These resources indicate no-fly zones, temporary restrictions, and other critical information to help pilots avoid unauthorized areas.

Step 6: Operational Limitations
Recreational drone pilots in Singapore must adhere to several operational rules:
- Maximum Altitude: Drones must not be flown above 60 meters

(200 feet) without special permission.

- Visual Line of Sight (VLOS): Operators must maintain a direct line of sight with their drones.
- Proximity to Airports: Drones must be kept at least 5 kilometres away from airports and airbases.
- Sensitive Areas: Flying over populated areas, emergency response zones, and sensitive infrastructure without authorisation is prohibited.

Step 7: Set Up Remote ID

In Singapore, to set up Remote ID for your drone, you must first register your drone with the Civil Aviation Authority of Singapore (CAAS) through their online portal. This registration is mandatory for all drones weighing more than 250 grams. The registration process involves providing details such as the drone's make, model, serial number, and your personal identification information. Singapore is in the process of implementing Remote ID regulations, which will require drones to have Remote ID capabilities to broadcast identification and location information. Ensure your drone has built-in Remote ID capabilities or install an approved external module. Update your drone's firmware through the manufacturer's app to support Remote ID, enter your registration details, and enable the Remote ID broadcasting feature. Conduct a test flight to ensure the Remote ID system is functioning correctly. Staying informed about the latest regulations and compliance requirements is crucial. Regularly check the CAAS website for updates on Remote ID implementation and ensure your drone's information is current.

Step 8: Safety Measures

Safety is paramount in drone operations. Pilots should ensure their drones are in good working condition and be mindful of weather conditions. Pre-flight checks should include verifying the battery, motors, and control systems to ensure a safe flight.

Useful Resources
- CAAS website: caas.gov.sg
- OneMap.sg: OneMap
- SkySafe Singapore: skysafe.io/dronelaws/singapore

South Africa

In South Africa, the South African Civil Aviation Authority (SACAA) regulates drone operations to ensure safety and compliance with national standards.

Step 1: Understanding the Regulatory Framework
The SACAA has established comprehensive regulations that all drone operators must follow to ensure safe and responsible flying.

Step 2: Registering Your Drone
All drones must be registered with the SACAA. The registration process involves providing detailed information about the drone and its operator, which can be done online through the SACAA's portal. Registration helps maintain accountability and ensures traceability of drones.

Step 3: Flight Certification and Permissions
Recreational drone pilots do not need formal certification but must adhere to SACAA guidelines. Special permissions may be required for specific activities such as flying in controlled airspace or at night.

Step 4: Obtaining Drone Insurance

Insurance for recreational drones from 250 grams and above is not required by law but is advisable to cover third-party liabilities and damages.

Step 5: Checking Airspace Restrictions
Drone operators must verify airspace restrictions using SACAA-provided resources or third-party tools. These tools indicate no-fly zones, temporary flight restrictions, and other critical airspace information to help pilots avoid unauthorized areas.

Step 6: Operational Limitations
Recreational drone pilots in South Africa must follow several operational rules:
* Maximum Altitude: Drones must not be flown above 120 meters (400 feet).
* Visual Line of Sight (VLOS): Operators must maintain a direct

line of sight with their drones.
- Proximity to Airports: Drones must be kept at least 10 kilometres away from airports.
- Sensitive Areas: Flying over people, national key points, and other sensitive areas without authorisation is prohibited.

Step 7: Set Up Remote ID

To set up Remote ID for your drone in South Africa, start by registering your drone with the South African Civil Aviation Authority (SACAA). Registration is mandatory for all drones used for commercial purposes or weighing more than 250 grams. You need to provide details such as the drone's make, model, serial number, and your personal information. This process can be completed online through the SACAA's portal. Although Remote ID is not yet mandatory in South Africa, it is expected that regulations will be updated to include this requirement. You should ensure that your drone is equipped with built-in Remote ID technology or an approved external module. After registration and configuration, conduct a test flight to ensure that the Remote ID system is functioning correctly. For more details and updates, visit the SACAA website.

Step 8: Safety Measures

Ensuring the safety of drone operations is crucial. Pilots must check their drones before each flight, including the battery, motors, and control systems, and be aware of weather conditions.

Useful Resources
- SACAA website: caa.co.za
- Drone Safety South Africa: dronesafety.co.za
- UAV Coach South Africa: uavcoach.com/drone-laws-in-south-africa

Nigeria

In Nigeria, the Nigerian Civil Aviation Authority (NCAA) regulates drone operations, ensuring that both recreational and commercial activities comply with national standards.

Step 1: Understanding the Regulatory Framework
The NCAA has established a comprehensive set of regulations that all drone operators must adhere to, ensuring safety and minimising risks to the public.

Step 2: Registering Your Drone
All drones must be registered with the NCAA. The registration process involves providing detailed information about the drone and its operator, which can typically be done online through the NCAA's registration portal. Registration helps maintain accountability and ensures that drones are operated by responsible individuals.

Step 3: Flight Certification and Permissions
Recreational drone pilots do not need formal certification but must follow NCAA guidelines. Special permissions may be required for specific activities such as flying in controlled airspace or at night.

Step 4: Obtaining Drone Insurance

Insurance for recreational drones from 250 grams and above is not required by law but is advisable to cover third-party liabilities and damages.

Step 5: Checking Airspace Restrictions
Drone operators must verify airspace restrictions using NCAA-provided resources or third-party tools. These tools indicate no-fly zones, temporary flight restrictions, and other critical airspace information to help pilots avoid unauthorized areas.

Step 6: Operational Limitations
Recreational drone pilots in Nigeria must adhere to several operational rules:
- Maximum Altitude: Drones must not be flown above 120 meters.

- Visual Line of Sight (VLOS): Operators must maintain a direct line of sight with their drones.
- Proximity to Airports: Drones must be kept at least 5 kilometres away from airports.
- Sensitive Areas: Flying over populated areas, national key points, and other sensitive areas without authorisation is prohibited.

Step 7: Set Up Remote ID

To set up Remote ID for your drone in Nigeria, you must first ensure your drone is registered with the Nigerian Civil Aviation Authority (NCAA) if it weighs more than 250 grams or is equipped with a camera. Currently, there is no online registration system, so you need to contact the NCAA directly for manual registration. You can reach them via email at RPAregistration@rpas-wg.org.ng or call +234 9091 390626 for support.

While Remote ID is not mandatory in Nigeria, it is recommended for both hobbyist and commercial drone operators to enhance safety and accountability. Ensuring your drone has built-in Remote ID capabilities or installing an external module can be beneficial. Regularly check for updates on NCAA regulations to stay compliant with any future mandates. For more detailed information, visit the NCAA's RPAS portal.

Step 8: Safety Measures
Safety is paramount in drone operations. Pilots should ensure their drones are in good working condition, including checking the battery, motors, and control systems, and be mindful of weather conditions.

Useful Resources
- NCAA website: ncaa.gov.ng
- Drone Laws Nigeria: dronelaws.ng
- FlySafe Nigeria: flysafe.ng/

Commercial Drones

Commercial drone operations represent a significant leap forward in how businesses can harness the power of aerial technology. Unlike recreational drones, commercial drones are used for a wide array of purposes, including surveying, agriculture, infrastructure inspection, delivery services, and media production. This diversity necessitates a robust and comprehensive regulatory framework to ensure safety, compliance, and the successful integration of drones into various sectors. This guide provides an exhaustive overview of the steps required for commercial drone operations across several key jurisdictions: the European Union, the United Kingdom, the United States, Canada, China, Singapore, South Africa, and Nigeria. Each section will delve deeply into the specific regulatory requirements, certifications, operational guidelines, and resources available for commercial drone operators in these locations. Whether you are a drone service provider, a business looking to integrate drones into your operations, or a consultant guiding clients through the regulatory landscape, this guide will serve as an indispensable resource for navigating the complexities of commercial drone operations.

European Union

Step 1: Understanding the Regulatory Framework
Commercial drone operations in the European Union are governed by the European Union Aviation Safety Agency (EASA). The regulations are designed to ensure safety while promoting innovation and efficiency. These regulations fall under three categories: Open, Specific, and Certified, with commercial operations typically falling under the Specific and Certified categories.

Step 2: Registering Your Drone
All commercial drones must be registered with the National Aviation Authority (NAA) of the operator's country and Remote Identification number must be obtained. This process requires providing detailed information about the drone, its intended use, and the operator's credentials. Registration can usually be completed online through the respective NAA websites.

Step 3: Obtaining an Operational Authorisation

For operations falling under the Specific category, operators must obtain an operational authorisation from their NAA. This involves submitting a risk assessment known as the Specific Operations Risk Assessment (SORA). The SORA evaluates the risk of the operation and outlines the necessary mitigations. The process includes:

- Describing the operation in detail.
- Conducting a risk assessment.
- Implementing risk mitigations and safety measures.

SORA evaluates risks at all phases of the flight and determines necessary mitigations to ensure safety. Based on the SORA, a Specific Assurance and Integrity Level (SAIL) is assigned, categorising the operation as low, medium, or high risk. Operations that fall into the low-risk category often fit into Standard Scenarios (STS) predefined by EASA. These scenarios include specific operational limits, training requirements, and equipment standards. Operators simply need to follow the STS Operations Manuals and make an operational declaration without needing additional approvals.

Medium risk operations require obtaining an Operational Authorisation from the NAA. This involves a comprehensive application process where operators must demonstrate that all necessary risk mitigations are in place. Additionally, a Design Verification Process may be required to certify that the drone meets safety specifications. High risk operations, categorised as SAIL V and VI, necessitate a rigorous aircraft certification process according to airworthiness standards (PART21). These operations also require pre-coordination and final coordination with relevant authorities to ensure safety.

Step 4: Obtaining Drone Insurance

Commercial drone operators in the EU must have insurance that complies with Regulation (EC) No 785/2004, which sets out minimum insurance requirements for air carriers and aircraft operators. This insurance must cover third-party liability and, in some cases, cargo insurance. It is essential for operators to ensure their insurance policy covers all potential risks associated with their commercial activities.

Step 5: Pilot Certification
Commercial drone pilots must obtain a Remote Pilot Certificate from their NAA. This typically involves passing a theoretical knowledge exam and a practical flight test. The training covers various aspects, including air law, flight performance, operational procedures, and human factors.

Step 6: Operational Limitations and Safety Measures
- Maximum Altitude: Drones must not be flown above 120 meters (400 feet).
- Visual Line of Sight (VLOS): Operators must maintain a direct line of sight with their drones unless specific permissions are granted for Beyond Visual Line of Sight (BVLOS) operations.
- Proximity to People and Property: Maintain safe distances from people, vehicles, and buildings. Specific distances vary based on the risk assessment.
- No-Fly Zones: Adhere to no-fly zones, including areas around airports, military installations, and other sensitive locations.

Useful Resources
- EASA website: easa.europa.eu
- EU Drone Port: eudroneport.com
- UAS operations: EASA UAS Operations
- DroneRules.eu: DroneRules.eu
- EASA Insurance Guidelines: EASA Insurance

United Kingdom

Step 1: Understanding the Regulatory Framework
The Civil Aviation Authority (CAA) regulates commercial drone operations in the UK. The framework is similar to the EU's EASA regulations but includes UK-specific requirements and procedures.

Step 2: Registering Your Drone and Obtaining an Operator ID
Commercial drones must be registered with the CAA, and operators need to obtain an Operator ID. This process involves providing detailed information about the drone and its intended commercial use.

Registration is done online through the CAA's portal.

Step 3: Obtaining a Permission for Commercial Operations (PfCO)
Commercial drone operators must obtain a Permission for Commercial Operations (PfCO) or, more recently, an Operational Authorisation under the Specific category. This involves:

- Completing a Theory Course and Passing the Test: To operate drones commercially, you must complete a recognised drone training course provided by a CAA-approved National Qualified Entity (NQE). These courses typically last 2-3 days and cover the rules and regulations governing commercial drone operations, including air law, flight performance, and operational procedures. At the end of the course, you must pass a multiple-choice theory test. This test assesses your knowledge of the covered material. Some NQEs also offer practical flight training, which is beneficial if you have no prior drone flying experience.
- Preparing an operations manual detailing the intended: An Operations Manual is a legal requirement by the CAA. This document details every aspect of your commercial drone operations, including the instructions and actions needed for safe flight. Writing this manual can be a daunting task, but there are templates and professional services available to assist you. Additionally, you will need Flight Reference Cards (FRCs) for each drone you intend to fly commercially. These cards provide quick reference information for safe operation and emergency procedures. The Operations Manual and FRCs must be thorough and tailored to your specific operations.

Step 4: Obtaining Drone Insurance
Before flying commercially, you must have public liability insurance that complies with Regulation (EC) 785/2004. This insurance should cover damages caused to third parties during drone operations. It is essential to ensure that your insurance policy covers all potential risks associated with your commercial activities. Various insurance providers offer policies tailored to drone operations, and it's crucial to select one that meets the CAA's requirements.

Step 5: Pilot Certification and Operational Evaluation
Commercial drone pilots must hold a General VLOS Certificate (GVC) or an A2 Certificate of Competency (A2 CofC) for specific operations. After completing your theory course and passing the test, you need to undergo a practical flight assessment, known as the Operational Evaluation. This evaluation tests your ability to safely operate your drone under various conditions and scenarios. You should be comfortable flying your aircraft and have all the necessary safety equipment to carry out the flight to a professional standard.

Step 6: Submitting Final Documentation and Paying the CAA's Fee
Finally, you need to submit several documents to the CAA to obtain your Permission for Commercial Operations (PfCO). These documents include a completed Operations Manual with an original signature, a completed SRG1320 Form, a copy of your annual Commercial Drone Insurance Policy, and a stock image of your drone. Some NQEs will submit these documents on your behalf. Additionally, you must pay a submission fee, which is £173 for the first time and £130 for renewals. Payment can be made via bank details provided in the SRG123 Form or over the phone. After the CAA processes your payment, you should receive your PfCO within approximately 28 days, allowing you to legally carry out commercial drone operations.

Useful Resources
- CAA website: caa.co.uk
- DroneSafe UK: dronesafe.uk
- NATS Drone Assist app: NATS Drone Assist

United States

Step 1: Understanding the Regulatory Framework
The Federal Aviation Administration (FAA) regulates commercial drone operations in the United States under Part 107 of the Federal Aviation Regulations. These regulations ensure the safety of airspace and people on the ground.

Step 2: Registering Your Drone
All drones used for commercial purposes must be registered with the

FAA. Registration involves providing details about the drone and its owner, and it must be renewed every three years. Registration is done through the FAA's DroneZone portal.

Step 3: Obtaining a Remote Pilot Certificate
Commercial drone pilots must obtain a Remote Pilot Certificate by passing the FAA's Part 107 exam. The exam covers topics such as airspace regulations, weather, drone performance, and emergency procedures. Pilots must also undergo a TSA security background check.

Step 4: Obtaining Drone Insurance
Commercial operators must have liability insurance that covers third-party damages. The Federal Aviation Administration (FAA) strongly recommends, though does not mandate, comprehensive drone insurance for commercial operations.

Step 5: Waivers and Authorisations
For operations that exceed the limitations of Part 107, such as flying at night or beyond visual line of sight (BVLOS), operators must apply for waivers through the FAA. The application process involves demonstrating how the operation will be conducted safely.

Step 6: Operational Limitations and Safety Measures
- Maximum Altitude: Drones must not fly above 400 feet unless within 400 feet of a structure.
- Visual Line of Sight (VLOS): Maintain direct visual contact with the drone.
- Proximity to People and Property: Avoid flying over people and vehicles.
- No-Fly Zones: Adhere to restrictions around airports, military bases, and other sensitive areas.

Useful Resources
- FAA website: faa.gov/uas
- FAA Part 107 Exam Guide: faa.gov/uas/commercial_operators/part_107
- B4UFLY app: B4UFLY
- American drone insurance providers like Verifly and SkyWatch.AI.

Canada

Step 1: Understanding the Regulatory Framework
Transport Canada regulates commercial drone operations under the Canadian Aviation Regulations (CARs). The regulations categorise operations into basic and advanced, each with specific requirements.

Step 2: Registering Your Drone
All drones weighing between 250 grams and 25 kilograms must be registered with Transport Canada. The registration process involves providing information about the drone and its operator and can be completed online.

Step 3: Obtaining a Pilot Certificate
Commercial drone pilots must obtain either a Basic or Advanced Operations Certificate. The Basic Operations Certificate is for lower-risk operations, while the Advanced Operations Certificate is for higher-risk operations such as flying in controlled airspace or near people. Certification involves passing an online exam and, for advanced operations, a flight review.

Step 4: Obtaining Drone Insurance
Commercial drone operators must have insurance that includes liability coverage as per Transport Canada guidelines. This insurance is essential for obtaining Special Flight Operations Certificates (SFOC) for higher-risk operations.

Step 5: Flight Certification and Permissions
For advanced operations, operators may need additional permissions, including Special Flight Operations Certificates (SFOC) for activities that fall outside standard regulations.

Step 6: Operational Limitations and Safety Measures
- Maximum Altitude: Drones must not be flown above 122 meters (400 feet).
- Visual Line of Sight (VLOS): Maintain a direct line of sight with the drone.
- Proximity to People and Property: Maintain safe distances, with

specifics depending on the category of operation.
• No-Fly Zones: Comply with restrictions around airports, heliports, and other sensitive areas.

Useful Resources
• Transport Canada website: tc.canada.ca
• Drone Management Portal: Drone Management Portal
• NAV Canada Drone Flight Planning: NAV Canada
• Transport Canada Insurance Guidelines: Transport Canada
• Canadian drone insurance providers like Drone Insurance Canadian drone insurance providers like CoverDrone, Zensurance. SkyWatch AI and CapriCMW Insurance.

China

Step 1: Understanding the Regulatory Framework
The Civil Aviation Administration of China (CAAC) oversees commercial drone operations. Regulations are stringent, reflecting China's emphasis on safety and security.

Step 2: Registering Your Drone
All commercial drones must be registered with the CAAC. The process involves submitting detailed information about the drone, its intended use, and the operator. Registration is conducted online via the CAAC's UAV portal.

Step 3: Obtaining Operational Permits
Commercial drone operators must obtain a UAV operating permit from the CAAC. This requires submitting an application that includes a detailed operational plan, risk assessment, and evidence of pilot qualifications.

Step 4: Obtaining Drone Insurance
Commercial drone operators must have liability insurance. The Civil Aviation Administration of China (CAAC) requires operators to provide proof of insurance when applying for operational permits.

Step 5: Pilot Certification

Commercial drone pilots in China must pass a theoretical exam and a practical flight test to obtain certification. Training covers air law, operational procedures, safety, and emergency protocols.

Step 6: Operational Limitations and Safety Measures

- Maximum Altitude: Drones must not be flown above 120 meters (400 feet).
- Visual Line of Sight (VLOS): Maintain direct visual contact with the drone.
- Proximity to People and Property: Maintain safe distances from people, vehicles, and buildings. Specific distances vary based on the operational risk assessment.
- No-Fly Zones: Adhere to no-fly zones, including areas around airports, military installations, and other sensitive locations.

Step 7: Safety Measures

Safety is a paramount concern in commercial drone operations. Operators must conduct pre-flight checks, including verifying battery levels, motor functionality, and sensor calibration. They must also monitor weather conditions to avoid adverse weather that could impact flight safety.

Useful Resources

- CAAC website: caac.gov.cn
- DJI FlySafe: DJI FlySafe
- UAV Coach - China: UAV Coach China Guide.
- Chinese drone insurance providers.

India

Step 1: Understanding the Regulatory Framework
Commercial drone operations in India are regulated by the DGCA. The regulations ensure the safe and compliant use of drones for commercial purposes.

Step 2: Drone Classification and Registration
Similar to recreational drones, commercial drones must be registered on the Digital Sky platform. This applies to all categories of drones.

Step 3: Obtain an Unmanned Aircraft Operator Permit (UAOP)
Commercial operators must obtain a UAOP from the DGCA. This involves submitting a detailed application through the Digital Sky platform, including an operations manual, risk assessment, and proof of pilot qualifications.

Step 4: Pilot Certification
Commercial drone pilots must obtain a Remote Pilot License (RPL) by completing a DGCA-approved training program and passing both a theoretical exam and practical flight test.

Step 5: Drone Insurance
Commercial drone operators must have liability insurance that covers potential third-party damages. This is mandatory for obtaining a UAOP.

Step 5: Operational Guidelines
*   Maximum Altitude: Follow the altitude limits specified in the UAOP.
*   Visual Line of Sight (VLOS): Maintain direct visual contact unless BVLOS (Beyond Visual Line of Sight) operations are specifically permitted.
*   Proximity to People and Property: Maintain safe distances as specified in the UAOP.
*   No-Fly Zones: Adhere to all no-fly zones and obtain necessary permissions for restricted areas.

Step 6: Pre-Flight Checks and Safety Measures
Conduct thorough pre-flight checks, including verifying the drone's components, assessing the environment, and ensuring compliance with all safety protocols.

Useful Resources
* Digital Sky platform: Digital Sky
* DGCA drone guidelines: DGCA

Australia

Step 1: Understanding the Regulatory Framework
Commercial drone operations in Australia are regulated by CASA. The regulations ensure that commercial drone activities are conducted safely and compliantly.

Step 2: Drone Classification and Registration
All commercial drones must be registered with CASA. Registration is done through the myCASA portal and requires providing detailed information about the drone and its intended use.

Step 3: Obtain a ReOC (Remote Operator Certificate)
Commercial operators must obtain a Remote Operator Certificate (ReOC) from CASA. This involves submitting an application that includes an operations manual, risk assessment, and evidence of pilot competence.

Step 4: Pilot Certification
Commercial drone pilots must hold a Remote Pilot License (RePL). This requires completing an approved training course and passing both a theoretical exam and practical flight test.

Step 5: Drone Insurance
Commercial drone operators must have liability insurance that covers potential third-party damages. This is a mandatory requirement for obtaining an ReOC.

Step 6: Operational Guidelines
*   Maximum Altitude: Drones must not be flown above 120 meters (400 feet) unless specifically authorised.
*   Visual Line of Sight (VLOS): Maintain direct visual contact unless BVLOS operations are specifically permitted.
*   Proximity to People and Property: Maintain safe distances as specified in the ReOC.
*   No-Fly Zones: Adhere to all no-fly zones and obtain necessary permissions for restricted areas.

Step 7: Pre-Flight Checks and Safety Measures
Conduct thorough pre-flight checks, including verifying the drone's components, assessing the environment, and ensuring compliance with all safety protocols.

Useful Resources
*   CASA website: casa.gov.au
*   myCASA portal: myCASA
*   CASA Drone Safety Apps: Can I Fly There?

Singapore

Step 1: Understanding the Regulatory Framework
The Civil Aviation Authority of Singapore (CAAS) regulates commercial drone operations, ensuring safety and compliance through a comprehensive set of guidelines and requirements.

Step 2: Registering Your Drone
All drones weighing more than 250 grams must be registered with the CAAS. The registration process involves submitting detailed information about the drone and its operator and can be completed online.

Step 3: Obtaining Operator Permits
Commercial drone operators must obtain an Operator Permit (OP) from CAAS. This involves submitting an application detailing the intended operations, including risk assessments and safety measures. Additionally, for specific activities such as flying near people or in

controlled airspace, an Activity Permit (AP) may be required.

Step 4: Obtaining Drone Insurance
Commercial drone operators are required to have liability insurance. The Civil Aviation Authority of Singapore (CAAS) mandates this as part of the process for obtaining Operator Permits and Activity Permits.

Step 4: Pilot Certification
Commercial drone pilots must obtain a Remote Pilot License (RPL) by completing a training course approved by CAAS and passing a theoretical exam and practical assessment. The training covers air law, operational procedures, and emergency protocols.

Step 5: Operational Limitations and Safety Measures
- Maximum Altitude: Drones must not be flown above 60 meters (200 feet) without special permission.
- Visual Line of Sight (VLOS): Maintain direct visual contact with the drone.
- Proximity to People and Property: Maintain safe distances, with specifics depending on the risk assessment.
- No-Fly Zones: Adhere to restrictions around airports, military bases, and other sensitive areas.

Step 6: Safety Measures
Safety measures include conducting pre-flight checks, verifying the integrity of the drone's components, and ensuring the operating environment is safe. Operators must also consider weather conditions and avoid flying in adverse weather.

Useful Resources
- CAAS website: caas.gov.sg
- OneMap.sg: OneMap
- SkySafe - Singapore Drone Laws: SkySafe Singapore Drone Laws

South Africa

Step 1: Understanding the Regulatory Framework
The South African Civil Aviation Authority (SACAA) oversees commercial drone operations, providing detailed guidelines to ensure safety and compliance.

Step 2: Registering Your Drone
All commercial drones must be registered with the SACAA. The registration process requires submitting detailed information about the drone and its operator. Registration is typically completed online.

Step 3: Obtaining an ROC (Remote Operator Certificate)
Commercial drone operators must obtain a Remote Operator Certificate (ROC) from the SACAA. This involves submitting an application that includes a comprehensive operations manual, safety management system, and evidence of pilot competence.

Step 4: Obtaining Drone Insurance
Commercial operators must have public liability insurance. The South African Civil Aviation Authority (SACAA) requires proof of insurance for obtaining a Remote Operator Certificate (ROC).

Step 5: Pilot Certification
Commercial drone pilots must obtain a Remote Pilot License (RPL) by completing an approved training course and passing both a theoretical exam and a practical flight test. The training covers areas such as air law, navigation, flight performance, and operational procedures.

Step 6: Operational Limitations and Safety Measures
- Maximum Altitude: Drones must not be flown above 120 meters (400 feet).
- Visual Line of Sight (VLOS): Maintain direct visual contact with the drone.
- Proximity to People and Property: Maintain safe distances, with specifics depending on the operational risk assessment.
- No-Fly Zones: Comply with restrictions around airports, military

installations, and other sensitive areas.

Step 7: Safety Measures
Ensuring safety involves conducting thorough pre-flight checks, verifying the condition of the drone, and assessing the operating environment. Operators must also consider weather conditions and avoid flying in adverse weather.

Useful Resources
* SACAA website: caa.co.za
* Drone Safety - South Africa: Drone Safety South Africa
* UAV Coach - South Africa: UAV Coach South Africa Guide
* South African drone insurance providers.

Nigeria

Step 1: Understanding the Regulatory Framework
The Nigerian Civil Aviation Authority (NCAA) regulates commercial drone operations, ensuring that operators adhere to national safety standards.

Step 2: Registering Your Drone
All commercial drones must be registered with the NCAA. The registration process involves providing detailed information about the drone and its intended use, which can typically be done online.

Step 3: Obtaining Operational Permits
Commercial drone operators must obtain an operational permit from the NCAA. This requires submitting an application that includes a detailed operational plan, risk assessment, and proof of pilot qualifications.

Step 4: Obtaining Drone Insurance
Commercial drone operators must have liability insurance. The Nigerian Civil Aviation Authority (NCAA) requires proof of insurance when applying for operational permits.

Step 5: Pilot Certification

Commercial drone pilots must obtain a Remote Pilot License (RPL) by completing an approved training course and passing both a theoretical exam and a practical flight test. The training covers essential areas such as air law, flight performance, and operational procedures.

Step 6: Operational Limitations and Safety Measures
- Maximum Altitude: Drones must not be flown above 120 meters (400 feet).
- Visual Line of Sight (VLOS): Maintain direct visual contact with the drone.
- Proximity to People and Property: Maintain safe distances, with specifics based on the operational risk assessment.
- No-Fly Zones: Adhere to restrictions around airports, national key points, and other sensitive areas.

Step 7: Safety Measures
Safety measures include conducting thorough pre-flight checks, ensuring the drone is in good working condition, and assessing the operating environment. Operators must also consider weather conditions and avoid flying in adverse weather.

Useful Resources
- NCAA website: ncaa.gov.ng
- Drone Laws Nigeria: Drone Laws Nigeria
- FlySafe Nigeria: FlySafe Nigeria

Notes

Notes

Legislation

Commission Delegated Regulation (EU) 2019/945 of 12 March 2019 on unmanned aircraft systems and on third-country operators of unmanned aircraft systems <https://eur-lex.europa.eu/legal-content/EN/TXT/?uri=CELEX:32019R0945>

Commission Implementing Regulation (EU) 2019/947 of 24 May 2019 on the rules and procedures for the operation of unmanned aircraft<https://eur-lex.europa.eu/legal-content/EN/TXT/?uri=CELEX:32019R0947>

Convention on Damage Caused by Foreign Aircraft to Third Parties on the Surface 1952

Council Directive 85/374/EEC of 25 July 1985 on the approximation of the laws, regulations and administrative provisions of the Member States concerning liability for defective products 1985 [31985L0374]

Directive 1999/34/EC of the European Parliament and of the Council of 10 May 1999 amending Council Directive 85/374/EEC on the approximation of the laws, regulations and administrative provisions of the Member States concerning liability for defective products 1999 [31985L0374]

European Commission, Commission Implementing Regulation (EU) 2021/664 of on a regulatory framework for the U-space. <https://eur-lex.europa.eu/legal-content/EN/TXT/?uri=CELEX%3A32021R0664> 2021 [C/2021/2671] International Civil Aviation Organization (ICAO), Convention on Civil Aviation ('Chicago Convention') 1944

Regulation (EU) 2018/1139 of the European Parliament and of the Council of 4 July 2018 on common rules in the field of civil aviation and establishing a European Union Aviation Safety Agency <https://eur-lex.europa.eu/legal-content/EN/TXT/?uri=celex%3A32018R1139> 2018

Regulation (EU) 2021/1119 of the European Parliament and of the Council of

30 June 2021 establishing the framework for achieving climate neutrality and amending Regulations (EC) No 401/2009 and (EU) 2018/1999 ('European Climate Law') PE/27/2021/REV/1<https://eur-lex.europa.eu/legal-content/EN/TXT/?uri=CELEX:32021R1119>

Regulation (EU) of the European Parliament and of the Council on on jurisdiction and the recognition and enforcement of judgments in civil and commercial matters 2012 [No 1215/2012]

European Commission, Commission Implementing Regulation (EU) 2021/664 of on a regulatory framework for the U-space. <https://eur-lex.europa.eu/legal-content/EN/TXT/?uri=CELEX%3A32021R0664> 2021 [C/2021/2671]

Books

Brownsword R, Law, Technology and Society: Re-Imagining the Regulatory Environment (Routledge 2019)

Brownsword R, Law 3.0: Rules, Regulation, and Technology (1st Edition, Routledge 2020) <https://doi.org/10.4324/9781003053835>

Francis Fukuyama, Our Posthuman Future (2002) London: Profile

Communications

European Commission, 'COM/2014/0207, Communication from the Commission to the European Parliament and the Council on Civil Use of Remotely Piloted Aircraft Systems in a Safe and Sustainable Manner.' <https://eur-lex.europa.eu/legal-content/EN/TXT/?uri=CELEX%3A52014DC0207> accessed 2 October 2021

European Commission, 'COM/2015/0598, Communication from the Commission and the European Parliament, the European Council, the Council, the European Economic and Social Committee and the Committee of Regions On Aviation Strategy for Europe' <https://eur-lex.europa.eu/legal-content/EN/ALL/?uri=CELEX%3A52015DC0598> accessed 30 October 2021

European Commission, 'COM/2019/640, Communication from the Commission and the European Parliament, the European Council, the Council, the European Economic and Social Committee and the Committee of Regions on the European Green Deal.' <https://eur-lex.europa.eu/legal-content/EN/TXT/?qid=1588580774040&uri=CELEX:52019DC0640> accessed 29 October 2021

European Commission, 'COM/2020/789, Communication from the Commission and the European Parliament, the European Council, the Council, the European Economic and Social Committee and the Committee of Regions on Sustainable and Smart Mobility Strategy – Putting European Transport on Track for the Future.' <https://eur-lex.europa.eu/legal-content/EN/TXT/?uri=CELEX%3A52020DC0789> accessed 29 October 2021

European Commission, 'COM/2014/0207, Communication from the Commission to the European Parliament and the Council, A New Era for Aviation Opening the Aviation Market to the Civil Use of Remotely Piloted Aircraft Systems in a Safe and Sustainable Manner.' <https://eur-lex.europa.eu/legal-content/EN/TXT/?uri=CELEX%3A52014DC0207> accessed 30 October 2021

European Commission, 'A Drone Strategy 2.0 for Europe to Foster Sustainable and Smart Mobility' <https://ec.europa.eu/info/law/better-regulation/have-your-say/initiatives/13046-A-Drone-strategy-20-for-Europe-to-foster-sustainable-and-smart-mobility_en> accessed 29 October 2021

European Commission, 'A Drone Strategy 2.0 for Europe to Foster Sustainable and Smart Mobility' <https://ec.europa.eu/info/law/better-regulation/have-your-say/initiatives/13046-A-Drone-strategy-20-for-Europe-to-foster-sustainable-and-smart-mobility_en> accessed 29 October 2021

European Commission, 'Advanced Technologies for Industry – Product Watch and Drones for Less Intensive Farming and Arable' (European Commission 2021) Product Watch Report EA-03-20-851-EN-N <https://ati.ec.europa.eu/sites/default/files/2021-02/ProductWatch_SatelliteDrone.pdf>

European Commission, 'Drone Infographics: A Look into the Aviation of the Future' (Mobility and Transport - European Commission, 22 September 2016) <https://ec.europa.eu/transport/modes/air/drones-infographics_en> accessed 9 October 2021

Executive Agency for Small and Medium-sized Enterprises (European Commission) Now known as and others, Advanced Technologies for Industry: Product Watch: Satellites and Drones for Less Intensive Farming and Arable Crops (Publications Office of the European Union 2021) <https://data.europa.eu/doi/10.2826/152441> accessed 17 October 2021

Directorate-General for Internal Policies of the Union (European Parliament) and Marzocchi O, Privacy and Data Protection Implications of the Civil Use of Drones: In Depth Analysis (Publications Office of the European Union 2015) <https://data.europa.eu/doi/10.2861/28162> accessed 2 October 2021

Data Protection Working Party., 'Article 29, Opinion 01/2015 on Privacy and Data Protection Issues Relating to the Utilisation of Drones. Adopted on WP 231.' (16 June 2015) <https://ec.europa.eu/justice/article-29/documentation/opinion-recommendation/index_en.htm> accessed 2 October 2021

'EU Regulations Stakeholders - Drone Rules' <https://dronerules.eu/lt/professional/eu_regulations_stakeholders> accessed 3 January 2022

'Safe Operations of Drones in Europe' (EASA) <https://www.easa.europa.eu/newsroom-and-events/news/safe-operations-drones-europe> accessed 3 January 2022

SESAR, U-Space: Aiming to Enable Complex Drone Operations with a High Degree of Automation (2018) <https://youtu.be/>

Academic Journals

Alamouri A, Lampert A and Gerke M, 'An Exploratory Investigation of UAS Regulations in Europe and the Impact on Effective Use and Economic Potential' (2021) 5 Drones 63 <https://www.mdpi.com/2504-446X/5/3/63>

accessed 17 October 2021

Allouch A and others, 'UTM-Chain: Blockchain-Based Secure Unmanned Traffic Management for Internet of Drones' (2021) 21 Sensors (Basel, Switzerland) 3049

Antolini A, 'Aviation – Policies on Drones in the Transport Sector: EU Publishes Delegated Regulation and Implementing Regulation on the Rules and Procedures for the Operation of Unmanned Aircraft (Drones)' <https://www.jttri.or.jp/document/2019/andrea28.pdf>

Balasingam M, 'Drones in Medicine—The Rise of the Machines' (2017) 71 International journal of clinical practice (Esher) e12989

Bassi E, 'Urban Unmanned Aerial Systems Operations: On Privacy, Data Protection, and Surveillance' (2019) 36 Law in Context: A Socio-Legal Journal 61 <https://heinonline.org/HOL/P?h=hein.journals/lwincntx36&i=177> accessed 2 October 2021

Bassi E, 'From Here to 2023: Civil Drones Operations and the Setting of New Legal Rules for the European Single Sky' [2020] Journal of Intelligent & Robotic Systems 493 <https://doi-org.proxy.uba.uva.nl/10.1007/s10846-020-01185-1> accessed 28 October 2021

Brownsword R, 'In the Year 2061: From Law to Technical Management' (2015) 7 Law Innovation & Technology

Chamola V and others, 'A Comprehensive Review of the COVID-19 Pandemic and the Role of IoT, Drones, AI, Blockchain, and 5G in Managing Its Impact' (2020) 8 IEEE access 90225

Clavell GG, 'Drones, Commercial Applications Of', The SAGE Encyclopedia of Surveillance, Security, and Privacy (SAGE Publications, Inc 2018) <https://sk.sagepub.com/reference/the-sage-encyclopedia-of-surveillance-security-privacy/i4528.xml> accessed 1 November 2021

De Miguel Molina B and Oña M, 'The Drone Sector in Europe', Ethics and Civil Drones: European Policies and Proposals for the Industry (2018) <https:/

/library.oapen.org/bitstream/handle/20.500.12657/27815/1002190.pdf?
sequence=1#page=16> accessed 28 October 2021

De Schrijver S, 'Commercial Use of Drones: Commercial Drones Facing Legal
Turbulence: Towards a New Legal Framework in the EU' (2019) 16 US-China
Law Review 338 <https://heinonline.org/HOL/P?h=hein.journals/
uschinalrw16&i=338> accessed 8 October 2021

Deagon A, 'The Tools That B(l)Ind: Technology as a New Theology' (2021) 3
Law, Technology and Humans 82 <https://heinonline.org/HOL/P?h=hein.
journals/lwtchmn3&i=88> accessed 9 October 2021

Dorsey S, 'They Are Watching You: Drones, Data & the Unregulated
Commercial Market Notes' (2018) 70 Federal Communications Law Journal
351 <https://heinonline.org/HOL/P?h=hein.journals/fedcom70&i=383>
accessed 17 October 2021

Downey C, 'Roger Brownsword (2020) Law 3.0: Rules, Regulation and
Technology. Abingdon: Routledge.' (2021) 3 Law, Technology and Humans
151 <https://doi.org/10.5204/lthj.1838> accessed 28 October 2021

Du H and Heldeweg MA, 'Responsible Design of Drones and Drone Services:
Legal Perspective Synthetic Report' <https://research.utwente.nl/en/
publications/responsible-design-of-drones-and-drone-services-legal-
perspective> accessed 2 October 2021

Du H and Heldeweg MA, 'Responsible Design of Drones and Drone Services
- A Synthetic Report' (Social Science Research Network 2017) SSRN
Scholarly Paper ID 3096573 <https://papers.ssrn.com/abstract=3096573>
accessed 1 October 2021

Dupont QFM and others, 'Potential Applications of UAV along the
Construction's Value Chain' (2017) 182 Procedia Engineering 165 <https://
www.sciencedirect.com/science/article/pii/S1877705817312912> accessed 4
November 2021

Eller KH, 'Is "Global Value Chain" a Legal Concept?: Situating Contract Law
in Discourses Around Global Production' (2020) 16 European Review of

Contract Law 3 <https://www.degruyter.com/document/doi/10.1515/ercl-2020-0002/html> accessed 3 January 2022

Filcak R, Považzan R and Viaud V, 'Delivery Drones and The' (European Environment Agency 2019) <https://www.eea.europa.eu/publications/delivery-drones-and-the-environment/file> accessed 3 January 2022

Gehra B, Leiendecker J and Lienke G, 'Compliance by Design: Banking's Unmissable Opportunity' (Boston Consulting Group 2017) Whitepaper <https://image-src.bcg.com/Images/Compliance-by-Design-Dec2017_tcm9-198779.pdf> accessed 7 July 2021

Giliker P, 'What Do We Mean By EU Tort Law?' (2018) 9 Journal of European Tort Law 1 <https://www.degruyter.com/document/doi/10.1515/jetl-2018-0104/html> accessed 3 November 2021

Gogarty B and Hagger M, 'The Laws of Man over Vehicles Unmanned: The Legal Response to Robotic Revolution on Sea, Land and Air Refereed Article' (2008) 19 Journal of Law, Information and Science 73 <https://heinonline.org/HOL/P?h=hein.journals/jlinfos19&i=125> accessed 7 October 2021

Hoffmann T and Prause G, 'On the Regulatory Framework for Last-Mile Delivery Robots' (2018) 6 Machines 33 <https://www.mdpi.com/2075-1702/6/3/33> accessed 7 October 2021

Kellington WL, 'Drones' (2017) 49 667

Kirrane L-MC, 'Civil Liability Arising Out of the Commercial Ownership and Operation of Drones' (Lexology, 15 April 2016) <https://www.lexology.com/library/detail.aspx?g=064b7d54-fd69-4312-bc0b-73ca521de1b5> accessed 3 November 2021

Kiršienė J and others, 'Rethinking the Implications of Transformative Economic Innovations: Mapping Challenges of Private Law' (2019) 12 Baltic journal of law & politics 47

Koch BA, 'Liability for Emerging Digital Technologies: An Overview' (2020) 11 Journal of European Tort Law 115 <https://www.degruyter.com/document/

doi/10.1515/jetl-2020-0137/html> accessed 3 November 2021

Konert A and Dunin T, 'A Harmonized European Drone Market? – New EU Rules on Unmanned Aircraft Systems' (2020) 5 Adv. Sci. Technol. Eng. Syst. J. 93 <https://astesj.com/v05/i03/p12/> accessed 28 October 2021

McEvoy S, 'The Future Is Now: Embracing Technology in the Construction Sector' (www.ashurst.com, 2017) <https://www.ashurst.com/en/news-and-insights/insights/the-future-is-now-embracing-technology-in-the-construction-sector/> accessed 1 October 2021

Moguel E and others, 'Towards the Use of Unmanned Aerial Systems for Providing Sustainable Services in Smart Cities' (2018) 18 Sensors 64 <https://www.mdpi.com/1424-8220/18/1/64> accessed 3 October 2021

Monterossi MW, 'Liability for the Fact of Autonomous Artificial Intelligence Agents. Things, Agencies and Legal Actors' (2020) 20 Global Jurist <https://www.degruyter.com/document/doi/10.1515/gj-2019-0054/html> accessed 3 November 2021

Prof. dr. W. Dewulf, 'Best Practices for Drone Operations in Europe: How to Optimise The Drone Operations for Urban Areas and Remote Islands in Indonesia' (Universiteit Antwerpen, 17 December 2020) <https://balitbanghub.dephub.go.id/file/472> accessed 28 October 2021

Sarrión J, 'Actual Challenges for Fundamental Rights Protection in the Use of Drone Technology' [2018] SSRN Electronic Journal

Simo FZ, 'Drones and Privacy-Related Issues/Concerns: Alice in Techno-Land' (2019) 8 Acta Universitatis Sapientiae: Legal Studies 89 <https://heinonline.org/HOL/P?h=hein.journals/ausapls8&i=91> accessed 2 October 2021

Stoica A-A, 'Emerging Legal Issues Regarding Civilian Drone Usage' (2018) 12 Challenges of the Knowledge Society 692

Stolaroff J and others, 'Energy Use and Life Cycle Greenhouse Gas Emissions of Drones for Commercial Package Delivery' 9 Nature Communications 409

<https://www.nature.com/articles/s41467-017-02411-5> accessed 29 October 2021

Wang BT, 'The Machine Metropolis: Introduction to the Automated City' in Brydon T Wang and CM Wang (eds), Automating Cities: Design, Construction, Operation and Future Impact (Springer 2021) <https://doi.org/10.1007/978-981-15-8670-5_1> accessed 1 October 2021

Wendehorst C, 'Strict Liability for AI and Other Emerging Technologies' (2020) 11 Journal of European Tort Law 150 <https://www.degruyter.com/document/doi/10.1515/jetl-2020-0140/html> accessed 3 November 2021

Wyber S, 'A Balancing Act: The Right to Be Forgotten and Libraries 1.'" (2018) 27 Journal of Information Ethics 81 <https://www.proquest.com/docview/2161594688?pq-origsite=gscholar&fromopenview=true> accessed 30 October 2021

Zhang L and others, 'Study on Pavement Defect Detection Based on Image Processing Utilizing UAV' (2019) 1168 Journal of Physics: Conference Series 042011 <https://doi.org/10.1088/1742-6596/1168/4/042011> accessed 1 October 2021

Reports

Albawaba Ltd, 'Belgium: The EU Drone Policy' (2016) MENA Report

Amsterdam Drone Week, 'Unmanned Air Mobility and the Role of Cities' (2021) <https://www.amsterdamdroneweek.com/news/uam-and--the-role-of-cities/> accessed 3 November 2021

Behr VS, 'Frankfurt: Air fight at the Commerzbank tower - falcon is subject to drone' FrankfurterRundschau (Frankfurt, 24 November 2021) <https://www.fr.de/frankfurt/frankfurt-skyline-luftkampf-commerzbank-turm-drohne-falke-tod-amtsgericht-91134990.html> accessed 28 December 2021

KIOS Research And Innovation Center Of Excellence, Drones' Footage of a Search-and-Rescue Exercise Scenario (Zenodo 2021) <https://search.datacite.org/works/10.5281/zenodo.5500655> accessed 2 October 2021

Drone Industry Insights UG, 'DIAS - The Swiss Drone Industry 2021' (2021) Industry Report <https://www.s-ge.com/sites/default/files/article/downloads/dias_-_the_swiss_drone_industry_report_2021.pdf>

DroneRules, 'Privacy-by-Design Guide: A DroneRules.Eu Professional Resource for Drone Manufacturers' (2019) <https://dronerules.eu/assets/files/DRPRO_Privacy_by_Design_Guide_EN.pdf>
Dronerules.eu, 'Drone Rules for Recreational Users: Privacy Handbook' <https://dronerules.eu/assets/handbooks/PrivacyHandbook_EN.pdf> accessed 2 October 2021

Hit Radio FFH, 'Peregrine falcon killed by drone' (Rhine Main, 23 November 2021) <https://www.ffh.de/nachrichten/hessen/rhein-main/285916-prozess-um-drohnen-unfall-bruetender-wanderfalke-auf-commerzbank-tower-getoetet.html> accessed 23 November 2021

Index

A

Adaptive guidelines 278–279, 296

Adaptive regulation 47, 162, 173, 363

AeroVironment 57, 70–72

Agriculture 7–10, 20–21, 59–60, 65, 67, 69–71, 73–81, 83, 85, 90, 92–94, 96, 107, 117, 143, 151, 170, 174–176, 179, 208, 236, 307, 329, 333, 335, 384, 413

Air Taxi 6, 33–34, 72–73

Air Traffic Management 40–41, 98, 113, 146, 179, 245, 290, 305, 320, 322, 343–344, 348, 353

Airspace 16, 19, 34, 39, 42–43, 48–50, 53–54, 56, 75–76, 85, 94, 96–106, 108, 110, 112, 113–120, 123, 135, 136–137, 140–142, 145–147, 149, 155, 157–158, 162, 173–174, 182, 186, 190–191, 198–203, 205, 221–222, 228, 232, 238, 239, 241–242, 244–245, 253, 261–263, 266, 282, 289–290, 297, 301, 305–309, 311, 313, 315, 318, 320, 322–324, 326–328, 341, 343–348, 350–353, 365, 368, 376, 378–379, 385, 389, 394–396, 398–401, 404–405, 407, 409, 411, 417–419, 425

Airspace authorisation 305, 315, 350

Airspace management 98, 116, 145–146, 158, 305, 307, 318, 322–324, 326–327, 344, 346–348, 350–352

Alphabet 16, 81

Amazon Prime Air 14, 17–19, 57, 70, 72, 310–312

Amsterdam 36, 39, 110, 365, 438, 449

Artificial Intelligence 4, 19, 29, 66–67, 77, 113, 134–135, 144, 146, 153, 164–167, 169, 174, 184, 271, 279, 296, 302–303, 308, 318, 323, 327, 338, 344, 350, 353, 358, 362, 371, 378, 383, 388, 390, 437

AI integration 173–174, 293, 328

AI-driven surveillance 183

Algorithmic bias 134, 171–172, 174, 179–180, 183–184, 284

Autonomous navigation 65, 69, 134, 152, 169, 174–176, 188, 234, 270

Decision-making 35, 66, 69, 77, 113, 125, 128, 132–134, 159, 162, 167, 170–172, 174, 178–181, 183–184, 229–230, 237, 272, 286, 299, 312, 319, 321, 329, 334–335, 337, 339, 341, 343, 347, 373, 382, 384

Decision-making transparency 134, 171, 179–181, 184, 312

Ethical considerations 115, 125, 134, 254, 256, 284–285, 321, 364, 366, 382, 387

Image recognition 177–178

Machine learning 19, 29, 66, 77–80, 134, 150, 155, 169, 174–175, 199, 211, 234, 299, 311–312, 321, 327, 335, 340

Obstacle avoidance 16, 70, 77, 169, 389

Obstacle detection 66, 175, 304, 310–312

Unpredictable AI behaviours 179, 181–182

Assembly 45, 66, 68, 195, 200

ASTM International 280, 288, 301, 316, 340, 355, 370

Augmented reality 21, 35

Australia 5, 31, 83, 85, 102–103, 109, 126, 128, 351–352, 354, 393, 405–406, 423

B

Battery 5, 14–16, 57, 69–70, 74–76, 151, 178, 212, 216, 323, 381, 403–404, 406, 408, 410, 412, 421

Beyond visual line of sight (BVLOS) 16, 24, 109, 117–118, 120, 197, 232, 319, 339, 364, 369, 401, 415, 418

Big Data 80

Biometric 40, 44, 258

About the Author

Saheed Babajide Okuboyejo is an accomplished international technology lawyer, author, and entrepreneur. A Solicitor and Advocate of the Supreme Court of Nigeria and a member of the Nigerian Bar Association in Lagos, he holds an LL.B. in Common Law from the University of Lagos, an MBA from the Gordon Institute of Business Science (GIBS) at the University of Pretoria, South Africa and an LL.M. in European Private Law from the University of Amsterdam, the Netherlands.

With a career grounded in sustainability, tech start-ups, renewable energy, and smart mobility, Okuboyejo brings a depth of legal expertise to emerging and transformative technologies. His practice spans disruptive sectors, including commercial drones, clean tech, robotics, generative AI, blockchain, self-driving vehicles, bioengineering, and space exploration.

In his debut book, Drone Law 3.0, Okuboyejo provides a thorough critique of the legal landscape for commercial drones, examining diverse regulatory challenges, addressing complex legal questions, and proposing institutional frameworks for effective oversight. This essential work delivers clarity on nuanced legal concepts, making it a must-read for policymakers, legal professionals, technology leaders, manufacturers, service providers and drone enthusiasts. With a clear and engaging style, Drone Law 3.0 is poised to become a vital resource in understanding the evolving legal frameworks that shape drone technology.

www.ingramcontent.com/pod-product-compliance
Lightning Source LLC
Chambersburg PA
CBHW021025210326
41598CB00016B/907